GeoComputation

GeoComputation

Edited by
Stan Openshaw and
Robert J. Abrahart

London and New York

First published 2000 by Taylor & Francis
11 New Fetter Lane, London EC4P 4EE

Simultaneously published in the USA and Canada
by Routledge
29 West 35th Street, New York, NY 10001

Taylor & Francis is an imprint of the Taylor & Francis Group

Typeset in Times by Graphicraft Limited, Hong Kong
Printed and bound in Great Britain by T.J. International Ltd, Padstow, Cornwall

Every effort has been made to ensure that the advice and
information in this book is true and accurate at the time of
going to press. However, neither the publisher nor the authors
can accept any legal responsibility or liability for any errors or
omissions that may be made. In the case of drug administration,
any medical procedure or the use of technical equipment
mentioned within this book, you are strongly advised to
consult the manufacturer's guidelines.

British Library Cataloguing in Publication Data
A catalogue record for this book is available from the
British Library

Library of Congress Cataloging in Publication Data
Openshaw, Stan.
    GeoComputation / Stan Openshaw & Robert J. Abrahart.
        p.   cm.
    Includes bibliographical references.
    1. Information storage and retrieval systems – Geography.
    2. Geographic information systems. I. Abrahart, Robert J.,
    1956– . II. Title.
G70.2.0645 1999
910'.285–dc21   99-29605
                                                      CIP

ISBN 0-7484-0900-9

Contents

Contributors

Robert J. Abrahart, School of Earth and Environmental Sciences, University of Greenwich, Medway Campus, Central Avenue, Chatham Maritime, Kent ME4 4TB, United Kingdom. Email: bob@ashville.demon.co.uk

Michael Batty, Centre for Advanced Spatial Analysis, University College London, 1–19 Torrington Place, London WC1E 6BT, United Kingdom. Email: m.batty@ucl.ac.uk

George Benwell, Spatial Information Research Centre, Department of Computer and Information Science, University of Otago, P.O. Box 56, Dunedin, New Zealand. Email: gbenwell@gandalf.otago.ac.nz

Roger Bivand, Department of Geography, Norwegian School of Economics and Business Administration, Breiviksveien 40, N-5045 Bergen, Norway. Email: roger.bivand@nhh.no

Antonio Camara, Environmental Systems Analysis Group, New University of Lisbon, P-2825, Monte de Caparica, Portugal. Email: asc@mail.fct.unl.pt

Gary Diplock, GMAP Ltd., 15 Blenheim Terrace, Leeds LS2 9HN, United Kingdom. Email: gary.diplock@gmap.co.uk

Manfred M. Fischer, Department of Economic Geography and Geoinformatics, Vienna University of Economics & Business Administration, Rossauer Lände 23/1, A-1090 Vienna, Austria. Email: manfred.fischer@wu-wien.ac.at

Peter Fisher, Department of Geography, University of Leicester, Leicester LE1 7RH, United Kingdom. Email: pff1@le.ac.uk

A. Stewart Fotheringham, Department of Geography, Daysh Building, University of Newcastle, Newcastle-Upon-Tyne, NE1 7RU, United Kingdom. Email: stewart.fotheringham@newcastle.ac.uk

Mark Gahegan, Department of Geography, The Pennsylvania State University, 302 Walker Building, University Park, PA 16802, USA. Email: mark@geog.psu.edu

Mike Kirkby, School of Geography, University of Leeds, Leeds LS2 9JT, United Kingdom. Email: mike@geog.leeds.ac.uk

Paul Longley, School of Geographical Sciences, University of Bristol, University Road, Bristol BS8 1SS, United Kingdom.
Email: paul.longley@bristol.ac.uk

Anne Lucas, Department of Geography, University of Bergen, Breiviksveien 40, N-5045 Bergen, Norway. Email: anne.lucas@geog.uib.no

Bill MacMillan, School of Geography, University of Oxford, Mansfield Road, Oxford OX1 3TB, United Kingdom. Email: bill.macmillan@geog.ox.ac.uk

Tony Moore, Plymouth Marine Laboratory, Prospect Place, West Hoe, Plymouth, Devon PL1 3DH, United Kingdom.
Email: abm@wpo.nerc.ac.uk

Shane Murnion, Department of Geography, University of Portsmouth, Buckingham Building, Lion Terrace, Portsmouth PO1 3HE, United Kingdom. Email: murnions@geog.port.ac.uk

Stan Openshaw, Centre for Computational Geography, School of Geography, University of Leeds, Leeds LS2 9JT, United Kingdom. Email: stan@geog.leeds.ac.uk

J. Miguel Remidio, Environmental Systems Analysis Group, New University of Lisbon, P-2825, Monte de Caparica, Portugal.
Email: jmr@virtual.dcea.fct.unl.pt

Joanna Schmidt, Information Systems Services, University of Leeds, Leeds LS2 9JT, United Kingdom. Email: j.g.schmidt@leeds.ac.uk

Henk J. Scholten, Department of Regional Economics, Faculty of Economics and Econometrics, Vrije Universiteit/Free University, Amsterdam, The Netherlands. Email: henk@geodan.nl

Ian Turton, Centre for Computational Geography, School of Geography, University of Leeds, Leeds, United Kingdom. Email: ian@geog.leeds.ac.uk

Preface

GeoComputation (GC) is a follow-on revolution that is occurring after the introduction of geographical information systems (GIS). It is expected to gather speed and momentum in the first decade of the 21st century. The rationale behind this latest revolution is quite simple. When we have finished creating our GIS databases, set up our digital spatial data libraries, and expanded them to include just about everything that can be linked into a two- or three-dimensional geographical co-ordinate system then we are all set for GC. We are 'under starter's orders' because the next logical step is to do something useful or meaningful with the enormous volumes of accumulated geo-data. The new challenges inherent in GC are:

1. to start to make better and fuller use of the vast data riches that have been generated through GIS, remote sensing, and associated developments in IT, which have acted to complement each other;
2. to exploit the value of this information resource beyond the original data gathering objectives that resulted in its creation;
3. to expand the notion of geographic information science beyond its role of data gathering, towards analysis, modelling and real-world problem solving.

Geographical science is more than the 'surveyor's art', but try telling that to a modern Department of Surveying now renamed as the Department of Geoinformatics! Data are a means to an end and not an end in themselves. Drawing maps and recording the position and nature of spatial objects is important. This is the first step. Subsequent modelling and analysis are able to progress into areas of research that are far more interesting, and where there exists the potential to do much more useful or more rewarding things with that data! GC is neither an equation-fest nor a computer gigaflop speed contest. GeoComputation is about using the various different types of geo-data and about developing relevant geo-tools within the overall context of a 'scientific' approach. It is about solving academic, theoretical, and applied problems. It is about converting interesting computer science 'playthings' and artificial intelligence 'toys' into useful operational tools that can be usefully applied to geo-information. It is about using the available statistical, mathematical, artificial intelligence and computational intelligence tool-kits, to do some practical work that is beneficial or useful, and not about partaking in esoteric undertakings that are often judged or labelled as being something akin to a 'minority sport'. It is also about finding new uses for the emerging new generations of super-supercomputers. There are signs that our existing high performance computing sciences could soon run out of appropriate problems that are able to keep multi-teraflop hardware fully occupied. Yet as we know there are untouched areas in GC, related work for instance in 'people modelling', that present such large and complex computational challenges that terraflop computers are barely the entry level of computing that is required.

GeoComputation is also about exploiting the data gathering capabilities and continuing related developments in earth observation technologies. The main thread throughout this discussion should now be quite clear. GeoComputation is about doing useful things. It is not about the adoption of new gadgets and gizmos, or about the provision of clever technical demonstrations, following the spirit of a *Tomorrow's World* type presentation – the 'cor blimey' factor to the nth degree!

GeoComputation is seen to contain three inter-related components:

1. *geographical or environmental data*, although there is no constraint on what form it takes and it does not have to come from a GIS;
2. *modern computational technologies* which attempt to compute a solution to some problem that involves some form of geographical or environmental data, and could include statistical, mathematical, numerical and/ or smart technologies, or a hybrid combination of these items;
3. *high performance computing hardware* which spans the hardware spectrum from workstations, to workstations with multiple processors, to SMP and vector computers and to massively parallel supercomputers.

GeoComputation research will often embrace aspects of all three. The concept of evolution is also important. There are ample examples, from other computational sciences, where an improvement in the quality of the science is observed to have a close relationship in some form or other with the speed of *available* hardware. There is a common historical evolution or trajectory. The world's most computation-intensive simulations, in physics or meteorology or chemistry, started on PC-sized hardware. These then expanded and grew, as ambitions developed, as computer hardware speeds increased, and as access was gained to sufficient resources to support the ever-increasing demand for large-scale computation. Likewise with current experimental GC developments. Research might well start on a 450 MHz PC because there is nothing else available. You start small, think big, and build in scalability. You can then grow and expand in scale and resolution, perhaps sufficiently to keep the world's fastest and biggest supercomputers going for six months at a time! It is interesting that in the UK the entire supercomputing allocation for social science is only 5% of the total for all science! However, it takes a while to create new research agendas and to acquire the resources that are needed for large-scale GC. But it is clear that time is on our side. The rapid speed of current microprocessors is forever increasing and expanding that which is possible or what can be done with ease – thus postponing the need for a large chunk of high performance computing. However, a time will soon come, when microprocessor speeds will have reached their asymptotic maximum. To gain additional speed and power will then require access to massively parallel hardware. By then the case will have been made and the resources will exist. Because, if nothing else, the range of topics covered within GC will be of sufficient importance to require high priorities in future research funding.

GeoComputation is a wonderful new word! Somehow it seems to be a term that once discovered has instant application to a vast range of different topics and subject areas. Perhaps this breadth of content and instant recognition of application relevance will be its eventual undoing. It could be doomed to failure through the absence of a single owner discipline. This might be so. But right now it is great fun. It also seems to be of almost unlimited scope and purpose. There is a feeling that the early years of the 21st century will see an explosion of interest in the subject. The main driving force is the need to solve important problems. Hard problems exist in numerous different areas of geographical and environmental science and the objective is to compute solutions in data-rich situations using both dumb and smart computational tools. There is a direct link between the rise of GC and the vast explosion in data on the one hand and an equivalent but somewhat unnoticed eruption in the speed of high performance hardware. There is a real challenge to discover how best to use the increasing and available data riches via supercomputer hardware with terraflop speeds and terrabyte memories. Moreover, in a practical sense, what is possible in terms of present or near-future computation is fast becoming less dependent on computer speeds or computer memories or data deficiencies. The vital missing ingredient is the computational tools that are needed to exploit the new opportunities that now exist for tackling new and old problems in the context of multiple different subjects and different problem areas.

This book seeks to establish general principles and to build solid foundations for further research and subsequent growth in GC. There is a companion volume that contains extra material and additional insights. There is also an associated international conference series that is now moving towards its fifth annual meeting and a dedicated web site. The web page service provides up-to-date details on the latest news about unfolding developments and related matters. The main page is a pointer to international mirror sites that contain the conference proceedings. *http://www.ashville.demon.co.uk/geocomp/*

Stan Openshaw and Bob Abrahart
Leeds, August 1999

Acknowledgements

The idea for this book originated from the Organising Committee for the First International Conference on GeoComputation; as if initiating a new conference series and attempting to endorse or promote a new scientific paradigm was not enough to keep us occupied!

Each chapter is an invited contribution from one or more specialists in that field. Support and guidance from these cutting-edge researchers, with their individual strengths and detailed knowledge, has been instrumental in helping to formulate the roots of this new scientific focus.

Editorial assistance from various colleagues has also been important. Particular recognition must go to Maureen Rosendale who performed the collation and correction operations, and to Linda See who assisted with the onerous task of proof-reading final scripts.

We would also like to thank our wives and families, who have once more had to endure the mutual burden of a substantial project, which has taken its toll in terms of time and attention. The completion of this edited volume would not have been possible without their wholehearted support and on-going encouragement.

Chapter 1

GeoComputation

Stan Openshaw

1.1 Introduction

GeoComputation (GC) is new, exciting, and here, but what is it? Some writers seem to think it has been around as long as there have been computers being used in geography. Others that GC is more or less a 'brand new' invention. There is seemingly an understandable confusion so the purpose of this chapter is to examine some of the alternative definitions, identify the more appropriate ones, and then outline some examples of what it may mean in practice.

1.2 Origins

GeoComputation is linked by name to what is broadly termed computational science with which it is clearly related and shares many of its aims. Computational science is a relatively new multidisciplinary paradigm for doing science in the late 20th century. As yet there is no general consensus as to a precise definition of what computational science actually is. In broad terms, computational science involves using computers to study scientific problems and it seeks to complement the use of theory and experimentation in scientific investigation. It seeks to gain understanding principally through the use and analysis of mathematical models and computer simulation of processes performed using, and often totally dependent upon, the availability of high performance computers. It is a largely or wholly computational approach to scientific investigation in which computer power is used to supplement and perhaps in some areas supplant more traditional scientific tools. Indeed once computer hardware became fast enough and big enough and numerical methodologies clever or flexible enough, then a computational paradigm provided a substitute for physical experimentation. It allows the visualization of hitherto unseen scientific processes, and it offers a basis for the simulation of complex systems which are too difficult for economical study by any other route. Computation permits the investigator to test theory by simulation, to create new theory by experimentation, to obtain a view of the previously invisible, to explore the previously unexplorable, and to model the previously unmodellable. There is clearly considerable potential here that will be released in the new millennium as computer speeds increase and a computational paradigm becomes a more common paradigm for doing science in many more areas of scientific interest. It is probably as unavoidable as it is inevitable but with the greatest developments having to wait for both faster computers and new generations of computationally minded scientists.

So in science there is now a strong and growing trend favouring a computational paradigm. Indeed, many scientific experiments and investigations that were once performed in a laboratory, a wind tunnel, or in the 'field' are now being increasingly augmented or replaced by purely computational alternatives. A common feature of computational science is that there appears to be an underlying implicit belief that the quality of the science depends in some way on the speed of the fastest available computers. As computers have become faster, so computational science has emerged as a powerful and increasingly indispensable method of analysing a variety of problems in research, process development, and manufacturing. It is now being widely advocated and increasingly accepted as a third methodology in engineering and scientific research that fills a gap between physical experiments and analytical approaches. Computer simulations now provide both qualitative and quantitative insights into many phenomena that are too complex to be dealt with by analytical methods and which are too expensive or dangerous to study by physical experiments. For example, the prohibition of atmospheric and underground nuclear weapons testing has stimulated the need to be able to simulate nuclear explosions by numerical means. In the US this military need has resulted in the Accelerated Strategic Computing Initiative (ASCI) which will eventually serve many more civilian applications than purely military ones. Indeed it has already spawned the first teraflop computers. In 1998 these were about 30 times faster than previous machines, e.g. the Cray T3E 1200 at Manchester, UK. So it is likely that the early years of the 21st century will see increasing availability of terraflop supercomputers, hardware that will be useful in many other areas including GC. As high performance computing (HPC) becomes faster so it stimulates entirely new areas of application which were previously computationally infeasible and generally unthinkable.

The emergence of computational science is not a particularly new phenomenon although it is one which has gathered speed throughout the 1990s. The availability of high performance computers, high performance graphic workstations, and high speed networks, coupled with major advances in algorithms and software, has brought about a silent revolution in the way many scientific and engineering investigations are performed. In the UK most of the Research Councils now (late 1998) have for the first time a six-year programme of committed baseline investment in HPC following the inauguration of the SCAR service late in 1998. It remains to be seen how these developments will diffuse into the social sciences in general and geography in particular.

Nevertheless, it should be readily apparent that there are similar attractions for a computational style of approach in geography and the social sciences. GeoComputation can be regarded, therefore, as the application of a computational science paradigm to study a wide range of problems in geographical and earth systems (the 'geo') contexts. Note that the 'geo' includes human as well as physical systems. This extension of the computational paradigm is such an obvious development that it may be a surprise to discover that the word 'GeoComputation' which seems so generically applicable was only recently invented. Such is the power of language that a word is almost instantly able to describe whole areas of research that have existed for two decades or more before the term was invented. A similar claim could be made for GIS which when it entered common usage aptly and almost instantly described many pre-existing areas of research and provided a central focus for their subsequent development and dissemination.

The word GeoComputation first 'appeared' in the author's spell checker dictionary after coffee time discussions relating to a computational geography conference being planned for 1996 in Leeds. Two years earlier the School of Geography in Leeds had created a Centre for Computational Geography. The original intention was to develop a new computational paradigm for doing human geography but subsequently the human focus was de-emphasized as it became readily apparent that it was equally applicable to many areas of physical geography. Openshaw (1994a, b, 1995a) describes various attempts at devising a computational human geography (CHG) research agenda. However, a major 'difficulty' was that the ideas and methodologies being advocated in CHG seemed to be far more attractive to physical geographers than to human geographers! This 'difficulty' is neatly and instantly resolved by using the term GeoComputation. The new word is also a useful device for further broadening the attractions of a computational paradigm. However, the study of earth systems is now of interest to a whole host of disciplines, many of which share common interests and common methodologies. Would they now feel left out in the development of computational geography? The words computational geography were just too parochial and restricting. It also limited the scope of the computational to geographers, and would have excluded other disciplines that may have wanted to be involved because they too study 'geo' contexts. Maybe only a non-geographer would have dared say as much. So it was that an 'ice geographer' (Tavi Murray), who (either deliberately or accidentally is now lost in the coffee flavoured mists of time) invented the term GeoComputation as a more meaningful alternative to 'computational geography'. At a stroke, she changed the original words 'computational human geography' into something far more general (GeoComputation) that was instantly understood and which applied equally to many physical and human phenomena and was also inherently multidisciplinary. The subsequent use of a capital C in the middle of the word GeoComputation can be attributed to Bob Abrahart. It is designed to emphasize the importance of the computation component and hence emphasize this very distinctive characteristic. It is a pity that the first book on GeoComputation dropped the upper-case middle C; it is said because the publishers did not like it! Here we think that it is of sufficient importance as a distinctive logo and trade-mark and have put it back!

1.3 So what is distinctive about GeoComputation?

GeoComputation is not just the application of computers in geography. Nor is it just about computation for its own sake. It is meant to imply the adoption of a large-scale computationally intensive scientific paradigm as a tool for doing all manner of geographical research. Some will now claim they have been doing GC for 10 or 30 years or more. This is certainly possible, but if they were then, until 1996, it was certainly called something else; terms such as mathematical modelling, simulation, statistical modelling all spring to mind. There are three aspects which makes GC special.

Firstly, there is an emphasis on the 'geo' subjects. This is partly a disciplinary focus to the areas of interest but it is more than geography. GeoComputation is concerned with geographical or spatial information of all types but until recently the distinctiveness of geographical data had been lost. In much of the quantitative work in geography, the geo-aspects were either completely missing or underdeveloped and underemphasized. It may now appear really weird that so many of the quantitative methods used in geography were (and still are) geo-poor! Somehow the 'geo' was left out, except as a description of the data source being used. Geographical data were, it seems, the same as any other data, and methods used in the more advanced physical sciences could be imported unchanged into geography. Indeed they were, and this provided the basis for the quantitative revolution in geography, a process that lasted about 30 years (from the early 1960s onwards) and saw the introduction of a whole host of statistical and mathematical analysis and modelling methods. Unfortunately, many geographers were slow to realize the unique and special features of geographical data that limited or rendered invalid many of these early quantitative and science-based tools. Mention can be made here that spatial data constitute populations (rather than samples) of spatially dependent (rather than independent) data. Likewise, the lack of applicability of stationarity assumptions and the substitution of the local for the global all massively complicate geographical study. Gradually as data environments became richer and computers faster, then more spatially appropriate methods have started to emerge but much of the damage still needs to be undone. The 'geo' is extremely important and distinctive and, sadly, it has historically been de-emphasized far too much because of the additional complexity that a more explicit recognition would have caused. Today this is no longer an acceptable excuse.

Secondly, the computation subphrase in GC is also special. It is the intensity of the computation that is especially distinctive. As computers become faster the very concept of a solution changes. GC is about finding new or better solutions to existing problems via a computational route. It also involves thinking about tackling new classes of problems that previously were unthinkable or insoluble. The hope is that it is now becoming possible to compute solutions to previously insoluble problems, to compute improved solutions to other problems, and to begin to identify new ways of working and new subjects for study that are computational in their methodology.

Thirdly, just as important and maybe even more significant is the underlying mindset. Computers can be used for many different purposes and to argue that number crunching is a distinctive feature ignores the fact that a spreadsheet is also a number crunching device, whilst processing 100 million billion characters would not seemingly involve many megaflops! Computation implies a very particular paradigm based on numerical approximation rather than analytical precision. It can be based on data-driven high-performance computer-powered inductive tools rather than data free, analytically based, deductive methods. It involves trying to compute solutions to problems that could not previously be solved at all. It is based on substituting vast amounts of computation as a substitute for missing knowledge or theory and even to augment intelligence. It could be data driven in a data mining sense, or it could be entirely data free with large-scale computer experimentation being used as a purely theoretical tool for understanding how complex systems work via modelling and simulation of their dynamics and behaviours.

1.4 How does GeoComputation relate to quantitative geography?

It is quite simple. GC includes all the tools of quantitative geography but it also encompasses far more than quantitative geography ever did as it has a very different emphasis. It also has the potential to negate most of the weaknesses and to re-express quantitative geography in a form more suitable for the 21st century; see Openshaw (1998a). Also, GC offers a new perspective and a paradigm for applying science in a geographical context. It is true that superficially it does not seem to be much different from quantitative geography. It is merely (some critics will argue) the application of computationally intensive approaches to the problems of doing physical and human geography in particular and the geosciences in general. Indeed it is, but it is important to appreciate what this means and to try and understand some of the potential power of adopting a computational paradigm that views statistical methods as only one small part of a far more flexible and expansive tool-box.

GeoComputation is concerned with new computational techniques, algorithms, and paradigms that are dependent upon and can take advantage of HPC. In fact it involves four leading edge technologies:

1. GIS, which creates the data;
2. artificial intelligence (AI) and computational intelligence (CI), which provide smart tools;
3. HPC, which provides the power;
4. science, which provides the philosophy.

So GC is much more than just using computers in geography and neither is it an attempt to rejuvenate a moribund quantitative geography in the style of 30 years ago! It is simultaneously a tool, a paradigm, and a way of thinking. It is not just about using computers to do arithmetic; it is the *scale* or *amount* of arithmetic computation that makes it distinctive! As high performance computers become faster and larger so the attractions of developing computational approaches increase its appeal and feasibility. It can be argued that GC is *new* because until recently computers were neither fast enough nor possessed of sufficient memory to make a GC paradigm into a practical proposition capable of general application. This is no longer the case as the 1990s have witnessed the development of highly parallel supercomputers; for instance, the Cray T3D (around 1994) has 512 processors, 32 gigabytes of memory, and is rated at about 40 gigaflops, whilst the Cray T3E (around 1998) has 576 processors, 148 gigabytes of RAM and is rated at about 122 gigaflops of sustained speed. It is likely that very soon HPC will be offering over 1 million gigaflops of sustained computation as the new teraflop computing era dawns.

Do we actually need extra computational power? Most geographers probably think that their PC with all the power of a mid-1980s mainframe is more than they need to run their 1970s and early 1980s vintage modelling and statistical technologies! In some ways they are correct but this is the wrong perspective to apply in an era where HPC offers or promises 10 000 or more times that level of performance. Nearly all the mathematical models and statistical analysis tools used today in geography come from an era of either manual calculation or slow-small computers. They use short-cuts, numerous simplifications, etc., to minimize the amount of computation that is performed. Indeed most of our computational technology is old fashioned and outmoded, and likely to yield far poorer results than more leading edge tools. However it is important to be able to demonstrate that if we perform 10 000 or several million times more computation that the benefits are worthwhile. If GC is to survive then it is this challenge that needs to be convincingly addressed.

Macmillan (1998) observes that there are strong elements of continuity between 'geocomputation' and the established traditions of quantitative geography because they share the same scientific philosophy. The view here is that GC can easily encompass quantitative geography if it so wished and if there was some virtue in so doing. However, GC is more than quantitative geography ever aspired to. Today's quantitative geography can be regarded as a repository for various legacy statistical and mathematical technologies that reflect an era of slow computers. Quantitative geography was a computationally minimizing technology, reflecting its origins in a hand-computer era. Analytical approximation and clever mathematical manipulation had to substitute for the lack of computing power. The absence of data fostered a theoretical perspective because there was seldom any other possibility. It is this technology and outlook that still survives in modern quantitative geography. The idea of running a computer program for a month or a year is still something quite alien. Indeed it is only in the last five years that computing environments have changed by such a degree that large-scale computation is now routinely feasible. In 1998 it can be calculated that 12 hours on a 512 processor Cray T3E parallel supercomputer is broadly equivalent to somewhere between 4 and 8 years of non-stop computing on a top-end workstation or PC. This is the emerging world within which GC is located; it was never the world of quantitative geography where such vast amounts of computation were seldom envisaged. The challenge for GC is to develop the ideas, the methods, the models and the paradigms able to use the increasing computer speeds to do 'useful', 'worthwhile', 'innovative' and 'new' science in a variety of geo-contexts.

1.5 What do others say about GeoComputation?

The definitions of GC described here were those which were expressed by the author during the time of the First International Conference in Geo-Computation held in Leeds in September 1996. Since then, thinking about the subject has intensified following two other conferences. The question now is to what extent do subsequent writers agree or disagree with these suggestions? Although, to be fair, this work has not previously been published and is thus largely unknown to them.

Rees and Turton (1998 p. 1835) define 'geocomputation' as '. . . the process of applying computing technology to geographical problems'. At first sight this definition would appear to suggest that GC is equivalent to doing geography with a computer. However, it has to be understood that there is an important distinction between 'doing geography with a computer' (which could be using a computer to map data) to 'solving geographical problems with new computing power' (which is what Rees and Turton wrote about).

Couclelis notes that 'geocomputation just means the universe of computational techniques applicable to spatial problems' (Couclelis, 1998a, p. 18).

Indeed if you accept her definitions then we have been doing 'geocomputation' for years without realizing it (see also Couclelis, 1998a, p. 19). However, it is important not to confuse using computers in geography with what GC is really about. Whilst understandable, such confusion greatly underplays the novelty of GC. GeoComputation certainly involves the use of computers in geography but its subject domain is not limited to geography nor is it merely another term for quantitative geography or geo-computer applications. There is a major paradigm shift occurring behind the scenes that is affecting why the computing is being applied. The style of GC envisaged today would have been both impossible and also unnecessary in the 1960's mainframe era when computers first started to be applied (as a replacement for hand calculation and electric calculators) to problems in geography. The really new idea behind GC is the use of computation as a front-line problem solving paradigm which as such offers a new perspective and a new paradigm for applying science in a geographical context.

Macmillan (1998) is much more accurate when he writes 'the claim I want to stake for geocomputation is that it is concerned with the science of geography in a computationally sophisticated environment. . . . It is also concerned with those computational questions . . . which are essential for the proper scientific use of our computational instruments' (p. 258). Later he adds '. . . the key feature of geocomputation . . . is the domain it belongs to – the domain of scientific research. Just as astronomy emerged with extraordinary vitality in the post-Galilean world, so geography can emerge from its post-modern slumbers in a geocomputational world' (Macmillan, 1998, p. 264). Let's hope so!

Longley writes 'The environment for geocomputation is provided by geographical information systems, yet what is distinctive about geocomputation is the creative and experimental use of GIS that it entails. The hallmarks of geocomputation are those of research-led applications which emphasize process over form, dynamics over statics, and interaction over passive response.' (Longley, 1998a, p. 3). Later he argues that 'geocomputation is the ways in which new computational tools and methods are used . . . also fundamentally about the depiction of spatial process.' (Longley, 1998a, p. 6). The important point here is that 'geocomputation is much more than GIS' (Longley et al., 1998, back cover).

Longley is right to point out the distinctiveness of GC but maybe it does not have to be so process and dynamics orientated and that this is only part of what GC is all about. Indeed Longley notes that GIS is but one of the many tools of 'geocomputation'. He argues that 'geocomputation' provides a framework within which those researching the development and application of GI technologies can address many of the important questions left unresolved by GIS. He writes very eloquently that 'The spirit of geocomputation is fundamentally about matching technology with environment, process with data model, geometry and configuration with application, analysis with local context, and philosophy of science with practice' (Longley, 1998a, p. 4).

So GC is not GIS and embraces a different perspective and set of tools. Longley says 'The data-rich environment of GIS today provides almost limitless possibilities for creating digital representations of the world, and the techniques of geocomputation provide better ways of structuring, analysing, and interpreting them than ever before' (Longley, 1998b, p. 83). There is a relationship with GIS but GC also has other relationships that may be just as important; for example, with computer science or numerical methods or statistics. Maybe also, from a geographical perspective, GC is what you do after GIS in that it does seek to make use of the data richness created by GIS and other developments in IT. If GIS is mainly about digital map information then GC is about using it in many different application areas within which the focus is not any longer particularly on the original GIS components. Nor is GC about evolving new or better data structures for use within GIS or about any of the GIS research agenda. To put it more bluntly, GIS is merely a database infrastructure which is nice to have but which is lacking in any science or theory other than the measurement science on which it is based. GC is not just an add-on to GIS; in fact it is not really part of it at all.

In essence GC is concerned with the application of a computational science paradigm to study all manner of geo-phenomena including both physical and human systems. It probably captures quite well the broad type of methodological approach that an informed computational physicist or chemist or aeronautical engineer would adopt if asked to suggest ways of studying subjects as diverse as river systems to human behaviour. It is not just data mining and it is not necessarily theory-free; indeed, both extremes of inductive and deductive approaches can be studied via a computational paradigm. GeoComputation is all about the use of relatively massive computation to tackle grand challenge (viz. almost impossible to solve) problems of immense complexity. However, a key feature is what is termed problem scalability. You start by tackling small and simpler versions of a more complex problem and then scaling up the science as and when either the HPC systems catch-up or knowledge of algorithms, models, and theory start to show signs of being able to cope. In many areas of social science and human geography, so great has been the scientific neglect, that we can no longer think about a 1 or 2 year time frame but need a 10 to 50 year period. Long time scales have not put off other sciences, e.g. mapping the human DNA or fusion reactor physics or laser physics. All you need is a bold but clear vision of what the end-goal is and then a path that connects where you are at present to where you want to be in some years time. However, not all GC need only be BIG science. This is fortunate because BIG science is still a wholly alien concept in the social sciences. Much progress can be made far more readily and with far less risk on small projects. The message is start small but think big.

However, not all researchers appear to agree with these definitions of GC. The problem appears to be that most commentators have focused on the content of the various conferences as a means of defining what it is that 'geocomputationalists' study and hence define the subject of GC. This is not a particularly good way of developing a definition; for example, the definition of geography based on the titles and content of the papers presented at the annual RGS/IBG or AAAG conferences would at best be confused and probably somewhat weird! It would be far better to think about the definition in a more abstract manner.

So far the most detailed study of the subject of GC is Couclelis (1998a, b). These two essays contain a delightful mix of fact, useful comment and suggestion blended with hints of confusion and flashes of future optimism. Couclelis has thought most about the meaning of the term. She starts by defining 'geocomputation' as 'the eclectic application of computational methods and techniques to portray spatial properties, to explain geographical phenomena, and to solve geographical problems' (Couclelis, 1998a, p. 17). She has observed from a study of the content of previous GC conferences that '. . . geocomputation is understood to encompass an array of computer-based models and techniques, many of them derived from the field of artificial intelligence (AI) and the more recently defined area of computational intelligence (CI)' (Couclelis, 1998a, p. 18). According to her, the key question now '. . . is whether geocomputation is to be understood as a new perspective or paradigm in geography and related disciplines, or as a grab-bag of useful computer based tools'. Longley also hints at a similar degree of confusion when he writes: '. . . geocomputation has become infinitely malleable . . .' (Longley, 1998b, p. 83). At issue here is whether or not there is something special and new to GC. The short answer is 'yes' for reasons that have already been stated. However, if you can accept that GC is a form of computation science applied to spatial or geo-problems (theory and data) then much of what Couclelis (1998a, b) is concerned about falls by the wayside.

Couclelis talks about an uneasy relationship with mainstream quantitative geography '. . . as evidenced by the relative dearth of geocomputation-orientated articles and topics in main quantitative geography journals and texts' (Couclelis, 1998a, p. 19). She also adds that 'geocomputation has thus far found only limited acceptance within the discipline'. However, this reflects her confusion as to what GC is all about and her mistaken belief that GC is not new but has been practised for at least a decade as a form of Dobson's automated geography; see Dobson (1983, 1993). Indeed one of the reasons for the author's insistence in this book on GeoComputation with a capital G and a capital C in the middle of the term rather than geocomputation, all in lower case, is to try and emphasize the newness and capture some of the excitement of what we understand GC to be about.

However, in writing about the nature of GC there is clearly a danger in associating it too closely with this or that exemplar technique. For example, Longley (1998a) and Macmillan (1998) both make several comments about the use of a highly automated form of exploratory geographical analysis in GC and, occasionally, they appear to think that GC is sometimes believed to be little more than this. The Geographical Analysis Machine (GAM) of Openshaw and associates (Openshaw, 1987; Openshaw and Craft, 1991) is the subject of this criticism but GAM was only really ever used as an illustration of one form or style of GC. Longley writes 'Geocomputation has been caricatured as uninformed pattern-seeking empiricism in the absence of clear theoretical guidance' (Longley, 1998a, p. 8). Maybe it should be added 'by the misinformed'! It was always intended to be more than this; indeed, this is an extremely biased, prejudiced, and blinkered view. Again Longley writes 'A central assumption of much of this work is that machine "intelligence" can be of greater import than *a priori* reasoning, by virtue of the brute force of permutation and combination – "might makes right" in this view of geocomputation' (Longley, 1998a, p. 12). This is a gross misunderstanding of the origins of GAM and also a reflection of a faith in theory and hypothesis that is quite unreal! It might help to know that GAM was developed for two main reasons: (1) knowledge of the data precluded proper *a priori* hypothesis testing; and (2) pre-existing hypotheses which could be legitimately tested reflected knowledge and theories that may well be wrong. For instance, one might speculate that disease rates will be higher within exactly 5.23 km of a specific point location. Suppose this general hypothesis is correct except the critical distance was 1.732 km! The hypothesis would be rejected and you would never be any the wiser about the form of the correct hypothesis! How silly! So why not use a GAM that would indicate the location of patterns treating all locations and distances equally. Of course, you would then have the problem of understanding and explaining the results but at least you would have found something if there was anything there that was sufficiently simple that GAM could find it. This does not reduce human thinking; it merely increases its utility.

There is nothing wrong with building pattern hunting machines that are able to be more successful at this task than we are, particularly in complex or multidimensional search spaces. Nor does it necessarily imply that pattern detection is sterile because there is no understanding. Any researcher with more than a modicum of intelligence or scientific curiosity will want to know 'why' a patterns exists here and not there. Pattern detection or the discovery of empirical regularities that are unusual or unexpected can be an important first step in scientific understanding. It does not have to be an end in itself! We should not be so ready to neglect inductive approaches based on data mining technologies. No one is insisting that GC has to be exclusively inductive, only that this is a useful technology in relevant circumstances. What is so wrong about building 'machines' dedicated to the inductive search for new theories or new models or new ideas? We would be daft to neglect any new opportunities to augment human reasoning, thinking, and deductive powers and processes by the use of machine based technologies. No one is yet suggesting that we relegate all thinking to machines; not yet anyway!

However it is hard for some to accept or appreciate what the possible benefits may be. Longley writes 'Technology empowers us with tools, yet conventional wisdom asserts that we need consciously and actively to use them in developing science without surrendering control to the machine' (Longley, 1998a, p. 5). Yes of course but when this comment is applied to the GAM then it shows an amazing naivety.

The meaning of GC is therefore no great mystery. It is essentially a computationally intensive science based paradigm used to study a wide range of physical and human geographical systems. It is neither a grab-bag set of tools, nor is it of necessity only rampant empiricism, nor must it be inductive, nor must it be without theory or philosophy! The distinctive features relate to its central emphasis on computation as a problem-solving device grounded in a scientific approach. It seeks to exploit the new opportunities for modelling, simulation, and analysis of human and physical systems that major new developments in high performance computing have created. In seeking to achieve this function, it is quite natural that GC should also seek to make good use of both old and new tools, particularly those emerging from artificial intelligence and computational intelligence backgrounds that are computationally based. However, it has not really been suggested anywhere that before you qualify as a 'geocomputationalist' you need simultaneously lots of data, a lack of theory, massive amounts of high performance computing, and heavy use of the latest artificial intelligence or computational intelligence tools.

To summarize, GeoComputation is:

    not another name for GIS;
    not quantitative geography;
    not extreme inductivism;
    not devoid of theory;
    not lacking a philosophy;
    not a grab-bag set of tools.

1.6 GeoComputation research

In many ways, the current GC research agenda reflects and evolves around that of its constituent parts. The most important of these are: high performance computing; artificial intelligence and its more generalized expression as computational intelligence; and a global GIS that has stimulated the appearance of many large spatial databases. However, there is no single dominant factor and others of more traditional importance probably need to be added, such as statistical techniques, mathematical modelling, and computer simulation relevant to a geographical context.

High performance computing is a most significant technological development. As computers become sufficiently faster and offer sufficiently large memories, HPC really does provide new ways of approaching geography based on a GC paradigm. GeoComputation encapsulates the flavour of a large-scale computationally intensive approach. It involves both porting and moving current computationally intensive activities onto HPC platforms, as well as the application of new computational techniques, algorithms, and paradigms that are dependent upon and can take particular advantage of supercomputing.

However, it is once again important to stress that it is much more than just supercomputing or high performance computing for its own sake. The driving factors are three-fold: (1) developments in HPC are stimulating the adoption of a computational paradigm to problem solving, analysis, and modelling; (2) the need to create new ways of handling and using the increasingly large amounts of information about the world, much of which is spatially addressed; and (3) the increased availability of AI tools and CI methods (Bezdek, 1994) that exist and are readily (sometimes instantly) applicable to many areas of geography suggesting better solutions to old problems and creating the prospect of entirely new developments. Geo-Computation also involves a fundamental change of style with the replacement of computationally minimizing technologies by a highly computationally intensive one. It also comes with some grand ambitions about the potential usefulness that may well result from the fusion of virtually unlimited computing power with smart AI and CI technologies that have the potential to open up entirely new perspectives on the ways by which we do geography and, indeed, social science. For instance, it is now possible to think about creating large-scale computing machine-based experiments in which the objects being modelled are artificial people living out their artificial lives as autonomous beings in computer generated artificial worlds (Dibble, 1996). High performance computing provides a laboratory within which many geographical and social systems can be simulated, studied, analysed, and modelled; see also Gilbert and Doran (1994); Gilbert and Conte (1995). A fusion of microsimulation and distributed autonomous intelligent agents is one way forward. The hardware, software, data and core algorithms largely exist. Perhaps the greatest obstacle is the difficulty of acquiring research funding for revolutionary ideas far beyond the conventional and then of gaining access to sufficiently powerful HPC to make it practicable.

GeoComputation may appear to some to be technique dominated; however, as previously discussed, the driving force is and has to be the 'geo' part, as it is not intended that GC becomes an end in itself. However, GC is unashamedly a problem-solving approach. One ultimate goal is an applied technology. Like GIS, it is essentially applied in character but this emphasis should in no way diminish the need for solutions that rest on a better theoretical understanding of how geographical systems work and of the processes that are involved. This focus on scientific understanding and theoretical knowledge provides a strong contrast with GIS. The challenge now is to create new tools that are able to suggest or discover new knowledge and new theories from the increasingly spatial data rich world in which we live generated by the success of GIS. In this quest for theory and understanding GC using HPC is a highly relevant technology.

There is an argument that GC would have developed sooner if the HPC technology had been more advanced. Indeed, until as recently as the early part of the 1990s, neither the power nor the memory capacities of the leading HPC machines were sufficient to handle many of the problems of immediate geographical interest. However, HPC is a relative concept. It is certainly true that most mathematical models developed by geographers made use of classical HPC hardware capable of a few thousand arithmetic operations per second. However, today the HPC hardware is many millions of times faster. It is still called HPC but it is like comparing the speed of a lame slug with a rocket!

Openshaw (1994a) suggested that by 1999 it was quite likely that HPC hardware available for use by geographers would be 10^9 times faster (and bigger in memory) than that available during the quantitative revolution years of the 1960s, 10^8 times faster than that available during the mathematical modelling revolution of the early 1970s, 10^6 times faster than that available during and since the GIS revolution of the mid-1980s, and at least a further 10^2 times faster than the current Cray T3D (in 1996, this was Europe's fastest civilian supercomputer located at the EPCC in Edinburgh yet by 1998 it is already more than 10 times slower than the fastest machines). Supercomputers capable of sustained speeds of 10 to 50 terraflops are promised within 5 to 10 years. One problem appears to be that most researchers in geography and the social sciences have failed to appreciate what these developments in HPC mean and they still think of a 450 MHz PC as being amazingly fast. The old Cray T3D with 512 processors had a peak theoretical performance of 76.8 gigaflops; but what does that mean? A gigaflop is 1000 million floating point operations per second but, again, what does this mean in a geographical context? One way of answering this question is to create a social science benchmark code that can be run on the widest possible range of computer hardware, ranging from PC to UNIX workstations, to massively parallel machines. The widely used science benchmark codes measure CPU performance in terms of fairly simple matrix algebra problems but it is not at all entirely clear what this means in a geographical context. Openshaw and Schmidt (1997) describe a social science benchmark code based on the spatial interaction model which can be run on virtually any serial or parallel processor. The benchmark is freely available from the WWW and Chapter 4 reports its use to measure the 'speed' of various HPC hardware available in the mid-1990s.

Another way of explaining what these changes in HPC hardware mean is to ask how would you do your research if that PC on your desk was suddenly 10 000 times faster and more powerful. It is likely that some researchers would not know what to do with it, some would not want it, but some would spot major new possibilities for using the computer power to do geography (and geo-related science) differently. It is this type of researcher who will switch to GC and be well placed to benefit from the next two or three generations of HPC. However, merely identifying applications that are by their nature potentially suitable for parallel hardware is not sufficient justification to invest in the necessary parallel programming effort. The applications also have to present a formidable computational challenge. What point is there in converting serial code that runs on a single CPU workstation in 30 minutes to run on a parallel supercomputer with 512 CPUs in 10 seconds? Certainly there is a software challenge, but the computational intensity of the task simply may not justify the effort involved. An additional criterion is that the parallel application should offer some significant 'extra benefit' that could not be realized without it. There should be some evidence of either new science or better science or of new results or improved results. The parallelization task is not an end in itself. In fact, it is totally irrelevant in the longer term. The biggest gains will come from those applications that were previously impossible but which can now be solved and, as a result, offer something 'worthwhile' knowing or being able to do.

What has changed dramatically during the 1990s is the maturity of parallel supercomputing, the continued speed-up of microprocessors, and the availability (after 20 years or so) of compilers that bring parallel computing within the existing skill domain of computationally minded geographers. The standardization of a highly parallel Fortran compiler and also of the message passing interface (MPI) eases the task of using parallel supercomputers in many areas of geographic application as well as producing reasonably future-proof portable codes (Openshaw and Turton, 1999). When viewed from a broader GC perspective, a major revolution in how geography and other spatial sciences may be performed is well underway; it is just that many researchers in these disciplines have not yet either realized it is happening or have not understood the possible implications for their interests.

The opportunities are essentially four-fold:

1. to speed up existing computer bound activities so that more extensive experimentation can be performed;
2. to improve the quality of results by using computational intensive methods to reduce the number of assumptions and remove shortcuts and simplifications forced by computational restraints that are no longer relevant;
3. to permit larger databases to be analysed and/or to obtain better results by being able to process finer resolution data;
4. develop new approaches and new methods based on computational technologies developed by other disciplines, particularly artificial intelligence and computer vision, new ways of solving optimization problems,

and generally to become opportunistic and entrepreneurial with a concern to tackle old problems using new technologies, and also to do new things that are relevant to geographical concerns but are currently limited by processor speed and perhaps also memory size.

All are important although some are much more readily attainable than others. In some applications, there are almost instant benefits that can be gained with a minimal degree of effort. Yet in others it could be 5–10 years before immature research blossoms into something useful. One problem for geographical HPC is that users in other areas of science have a 10–15 year head start in developing technology and raising awareness levels within their research communities and have research councils that now respond to their needs for HPC. Other problems are of their own making; for example, the various paradigm wars and artificially self-constructed philosophical and attitudinal barriers. Methodological pluralism is good but tolerance is also a necessary condition. Nevertheless, there is a growing belief that the time is ripe for HPC initiatives in geography and the social sciences and the international growth in popularity of GC is one indicator of this change.

1.7 Some examples of old and new GeoComputation

Even though GeoComputation is a new term, it is possible to recognize applications that today would be called GC but previously were regarded either as quantitative geography or GIS or spatial analysis. Some examples may help understand better the GC ethos or style and how GC fits in with what quantitative minded researchers have always done.

1.7.1 Parallel spatial interaction modelling and location optimization

One of the earliest uses of parallel computing in geography has concerned the parallelization of the spatial interaction model; see Harris (1985), Openshaw (1987). This model is central to several historically important areas of regional science, urban and regional planning and spatial decision support (Wilson, 1974; Birkin et al., 1996). For illustrative purposes the simplest spatial interaction model can be expressed as

$$T_{ij} = A_i O_i D_j B_j \exp(-b C_{ij}) \qquad (1.1)$$

where T_{ij} is the predicted flows from origin i to destination j, A_i is an origin constraint term, O_i is the size of origin zone i, D_j is the attractiveness of destination j, C_{ij} is the distance or cost of going from origin i to destination j, and b is a parameter that has to be estimated. This model was originally derived in a theoretically rigorous way by Wilson (1970) using an entropy maximizing method. Clearly this model is implicitly highly parallel since each T_{ij} value can be computed independently. Parallelization here is important because the model presents a computational challenge since computer times increase with the square of the number of zones (N). Small N values can be run on a PC but large N values need a supercomputer. The quality of the science reflects both the number of zones (more zones provide better resolution than few) and the specification of the model. Developments in information technology over the last decade have dramatically increased the availability and sizes of spatial interaction data sets. The 1991 census provides journey to work and migration data that contain 10 764 origin and destination zones. A parallel version of equation (1.1) has been run on the KSR parallel supercomputer at Manchester and later ported on to the Cray T3D (see Turton and Openshaw, 1996).

Scalability is a very important property in the world of parallel HPC. It creates new modelling opportunities applicable to the modelling of large-scale interaction data. Telephone traffic data exists for entire countries. In the UK, it is possible to imagine telephone call flow databases with between 1.6 million and 27 million zones in them. Equivalent data are generated by EFTPOS flows in the retail sector. These databases, currently being stored in data warehouses, are also of profound substantive interest since their data portrays the microfunctioning of selected aspects of the entire UK economic space. The daily trivia of a complete living nation is in there, just awaiting analysis. Retail catchments, changing network effects, space–time dynamics of individual behaviours are all in there, somewhere. The spatial interaction model could be scaled up to model only some of it and clearly entirely new modelling methodologies will be needed. Yet the possibilities are almost endless if we have the imagination to create them and the HPC hardware is sufficiently large and fast to meet the computational challenge. Modelling flow tables with 27 million rows and columns each with M attributes (one of which could be time) is incidentally beyond what even the largest parallel supercomputers can handle but for how much longer?

Computer technology able to model the behaviour of atoms will soon be able to model more and more of the behaviour in space and time of millions of individual people. As global resources become more limited, as environment concerns increasingly require behaviour modification, as governments aim at a lifelong equality consensus, so the task of people management will increase. However, better planning requires better prediction modelling. We need to be able to model people's behaviour if much progress is going to be made. The problem at present is that the science of human systems modelling (as it has been termed) is still at an extremely rudimentary stage of development; see Openshaw (1995a) for a brief review. Nearly all the existing models are aggregate rather than micro, static rather than dynamic, and insufficiently non-linear to be of much use. A start has been made but so much more is still needed.

1.7.2 New parameter estimation methods

Not all of GC needs the use of very large data sets or requires massive software investment or access to leading edge HPC. Diplock and Openshaw (1996) demonstrate some of the benefits of using genetic and evolutionary strategy-based parameter estimation methods compared with conventional non-linear optimization methods. Computer models (e.g. the spatial inter-action model in equation (1.1)) with exponential terms in them contain considerable opportunities for arithmetic instabilities to arise because the exponential deterrence function can readily generate very large and very small numbers depending on the parameter b. In fact, the numeric range where there are no arithmetic protection conditions being generated is ex-tremely small (typically plus or minus one depending on how the C_{ij} values are scaled) given that the parameter b could in theory range from minus infinity to plus infinity. The problem becomes worse when more parameters are used. Yet it is this function 'landscape' of flat regions, vertical cliffs, and narrow valleys leading to the optimal result, that conventional parameter optimization methods have to search. If they hit any of the barriers or the flat regions, they tend to become stuck and because it is dumb technology they have no way of telling you that this has happened. The implications are that potentially all statistical and mathematical models with exponential terms in them can produce the 'wrong' result because there is no assurance that the conventional non-linear optimizers in current use can safely handle the invisible arithmetic problems. There are newer methods which will func-tion well on these problems since they are more robust, they are not affected by floating point arithmetic problems and they can handle functions which are non-convex, discontinuous, and have multiple suboptima (see Diplock and Openshaw, 1996). The problem is that they require about 1000 times more computation. Once it was impossible to use this technology except on a small scale. Now it can be far more widely applied. This is a good example of one type of application where HPC (in the form of a fast workstation) can have an almost immediate impact. As HPC becomes faster so the scope for applying this robust parameter estimation technology will increase.

A related opportunity for a quick gain in benefit from HPC is the use of the bootstrap to estimate parameter variances. This is quite straightforward. You merely have to run the model of interest a few hundred or a few thousand times. It is naturally parallel because each run can be assigned to a different processor or else the code is left running on a workstation for a week or two. This raises another point of general significance. Research with a multiregion population forecasting model, that was used to make population forecasts for the European Union (EU), used this bootstrap approach to identify the error limits to forecasts for 2021 to 2051. This can be used to identify model data weaknesses. It also shows that currently there are no reliable long-term forecasts for the EU as the confidence limits are extremely wide. The problem appears to be due to uncertainty in the migration forecasting; see Turton and Openshaw (1998) for further details. Previously these error bands were unknown. Cross-validation using a jacknife is another useful computationally intensive tool. Here the additional computation is a factor of N times, where N is the number of observations.

It is important to use available HPC hardware but it is also important not to become too HPC crippled, where lack of access to leading-edge hardware equates with a total lack of progress. The basic performance equation of an HPC can be specified as follows. Imagine a parallel supercomputer with 256 processors. Typically each processor is about the same speed as a top-end workstation. So, on this assumption, one hour on this machine equals 256 hours (about 10 days) on your workstation (assuming you have enough memory to run the code of interest). However, parallel supercomputing is only 20–40% efficient and there are various overheads that further depress performance (network latency, bandwidth limitations, caching, etc.) none of which affect a workstation or slow it down to anything like the same extent. So maybe a workstation is able to do the run not in 10 days but in two to four days. Now, on a busy HPC platform you may easily wait a few weeks for your hour! A further possibility is to organize a set of unix workstations or PCs running NT into a local processor farm. Most workstations are idle, probably most of the time! If the parallelism in an application is sufficiently coarsely grained and the parallelism exploited using MPI (or PVM) message passing software (see Chapter 3) then the same code can be run on one or multiple local processors with the processors connected by a LAN. Obviously large amounts of network communications traffic will kill it but for coarsely grained applications (e.g. bootstrapping a complete model) it should run extremely well. There is the added benefit that the same code should run even better on a real HPC platform and also on future HPC platforms!

1.7.3 Network and location optimization

The basic spatial interaction model is often embedded in a non-linear optimization framework that can require the model to be run many thousands of times in the search for optimal locations, e.g. to determine the optimal network of shopping centres or car show rooms or good sites. There are many different types of important public and private sector location optimization problems of this sort. The quality of the final result is now critically dependent on the resolution of the data, the performance of the embedded model, and the quality of the optimization algorithm. The latter is, crudely put, usually related to how many million different candidate solutions can be evaluated in a fixed time period, because the problem can only be tackled by heuristic methods. The number of model evaluations per hour is dependent on processor speed, size of problem, granularity of the parallelism, and the skills of the programmer in teasing it out to ensure good performance on particular hardware; see Turton and Openshaw (1998) for an example. The problem here is that end-users (surprisingly) may be far more interested in a good solution than in obtaining an optimal solution, a view that is sometimes characterized by the dictum 'the best is the enemy of the good'. However, this is a distraction. The only way of determining whether a good result has been obtained is by knowing what the best attainable result is likely to be. Users will naturally assume that all of the results that they obtain are optimal, or nearly optimal, and it is a responsibility of the researcher to ensure that they are. It is not something that can be fudged but neither is the best result independent of the methodology used to find it especially in complex non-linear applications where optimality is computational-technology determined.

1.7.4 Automated modelling systems

There is also a need to improve the quality of the models being used in geographical research and not just speed up the time taken by legacy models or scale up the size of problem that can be tackled. The new computational technologies offer new ways of building models that either replace existing models based on mathematical and statistical approaches or else can be viewed as complementing them. The 'old' model shown in equation (1.1) assumes a single global deterrence function. This was quite reasonable when N was small and computer time was limited and without access to HPC not much more could be done. Yet building good models of many human systems is hard because of the complexity of the underlying processes, the lack of good relevant theory, and the seemingly chaotic non-linear behaviour of the systems of interest. It is important, therefore, to develop new ways of designing and building good performing models that can combine human intellect and modelling skills with HPC.

One approach is to create an automated modelling system that uses genetic algorithms and genetic programming techniques to suggest potentially useful models. The automated modelling system (AMS) method of Openshaw (1988) used a Cray Is vector supercomputer in an early attempt to define and then explore the universe of alternative spatial interaction models that could be built-up from the available pieces (e.g. variables, parameters, unary and binary operators, standard maths functions, and reverse polish rules for well formed equations) by using evolutionary programming algorithms to 'breed' new model forms. These methods are explicitly parallel (each member of a population of models is evaluated in parallel) and also implicitly parallel (the genetic algorithm's schemata theorem). The problem with AMS was the use of fixed length bit strings. Koza (1992, 1994) describes how this can be overcome by using what he terms genetic programming (GP). The AMS approach has been redeveloped in a GP format. Genetic programming is far more suitable for parallel rather than vector supercomputers.

The results from porting the genetic programming codes on to the Cray T3D suggest that not only can existing conventional models be 're-discovered' but that also new model forms with performance levels of two or three times better can be found (Turton *et al.*, 1996, 1997; Diplock, 1996, 1998, Chapter 6). Some of the GP runs reported in Turton *et al.* (1996, 1997) required over 8 hours on a 256 processor Cray T3D. It is likely that two week long runs on a 512 processor machine would yield even better results but this is seven times greater than the total ESRC allocation of Cray T3D time in 1996. In these complex search problems the quality of the results depends totally on the available HPC. In a decade or two's time, runs of this magnitude which are today barely feasible, will be considered trivial and historians will be amazed at how poor our current HPC hardware really is. If the new methods work well, then they would constitute a means of extracting knowledge and theories from the increasingly geography data rich world all around us. The key point to note here is that it is becoming increasingly possible to compute our way to better models.

Other new approaches to building new types of spatial models are described in Openshaw (1998). He compares the performance of a selection of genetic, evolutionary, neural net, and fuzzy logic spatial interaction models. In general, performance improvements of more than 200% over conventional models are possible and more than sufficient to justify the 10 000–100 000 times more computation that was involved. Some of these new models are purely black boxes (viz. the neural network models) but others are capable of plain English expression (the fuzzy logic models) or are in equation form (derived from AMS or GP).

Certainly there are problems that still need to be resolved but GC is about revolutionary technology. Old truisms may no longer hold good. Old barriers may have gone and have been replaced by others that are not yet understood. You have to believe that the impossible (i.e. previously the infeasible) is now possible else no progress will be made. However, put your GC spectacles on and suddenly the world is a different and more exciting place, but it still requires you to develop a degree of self-confidence that you can go safely and carefully where others have yet to tread.

1.7.5 Parallel zone design and optimal geographical partitioning

Some other GC applications involve applying existing methods that have been patiently waiting for increases in the speed of hardware and the provision of GIS data. Zone design is one of these. The basic algorithms were developed over 20 years ago (Openshaw, 1976, 1978, 1984) but until digital map boundary data became routinely available in the 1990s and computer hardware much faster, it was not a practical technology once N (the number of zones) exceeded a small number. The challenge now is to make routine access to the technology and make available the latest algorithms (Openshaw and Rao, 1995; Openshaw and Alvanides, 1999). If you can get that far then you have to start raising potential user awareness so that they realize what is now possible and start to use it. Most of the ground work has been done. Parallel zone design codes exist and a parallel simulated annealing algorithm has been developed; see Openshaw and Schmidt (1996). There are also potential applications, for example in designing census output areas for the 2001 census. In the UK this involves running a zone design algorithm on over 10 000 separate data subsets. But this can be easily solved as a parallel programming problem. Yet the principal barrier to application is not algorithmic or HPC aspects but awareness. It is unbelievable that in many countries the explicit and careful design of sensible census output areas is still not regarded as important, despite the likely role of census output areas in billions of pounds of public and private sector investment decisions, in the first decade of the 21st century. Surely this application is itself broadly equivalent in importance to many of the HPC projects in other areas of science yet because of the absence of a computational culture it is probably still regarded as being of the lowest priority and far too advanced for operational use. Yet we live in a world where computer speeds are doubling almost annually and the computer hardware likely to be considered state of the art in 2001 does not yet exist. User needs from the next census, or what is, indeed, feasible, will be much less conditioned by what was wanted from previous censuses and may well reflect what is now possible. Flexible output area definition is just one of these new needs and GC is one way of achieving it.

1.7.6 Parallel spatial classification methods

An obvious response to the spatial data explosion is to apply multivariate data summarizing tools, particularly classification, to the largest available databases. GeoComputation is also about rediscovering legacy methods and then scaling them up for a large data era. Twenty years ago the best (and most famous) cluster analysis package had an observation limit of 999. This would now be considered totally ridiculous, completely unnecessary, and a severe limitation. However, legacy methods can also usually be improved and replaced by more flexible and less assumption ridden more modern developments. The K-means technology of the 1970s that ran on a mainframe has now been replaced by unsupervised neural networks that run on parallel supercomputers and even workstations; see Openshaw (1994c), Openshaw et al. (1995), Openshaw and Turton (1996) for details. On the Cray T3D with 256 processors a single run takes 10 hours but the results are quite different from those produced by a more conventional method and may be substantially better and tell a very different story about the structure of Britain's residential neighbourhoods. At a time when 85% plus of the

UK's financial institutions use neighbourhood classifications, and there is a business worth £50 million per annum in selling them, one would have thought that the case for using HPC to generate 'better' and more advanced systems would be completely obvious, but currently it is not! No doubt this neglect will soon be rectified.

1.7.7 Parallel geographical pattern and relationship spotters

A major by-product of the GIS revolution of the mid-1980s has been to add geographic x, y co-ordinates to virtually all people and property related computer systems and to create multiple layers of other digital information that relate to the physical environment and which may be regarded as being related to it (as possible predictor variables). The success of GIS has created a growing imperative for analysis and modelling simply because the data exists. The problem now is how to do exploratory analysis on large databases, when there is little or no prior knowledge of where to look for patterns, when to look for them, and even what characteristics these might be based on. It goes without saying that the methods also have to be easy to use, automated, readily understood and widely available. A most difficult requirement but, nevertheless, a most important challenge for GC to consider.

One possible solution is Openshaw *et al.* (1987) who describe a prototype geographical analysis machine (GAM) able to explore a spatially referenced child cancer database for evidence of clustering. The GAM used a brute force grid search that applied a simple statistical procedure to millions of locations in a search for localized clustering. Fortunately, the search is highly parallel although it was originally run on a serial mainframe where the first run took 1 month of computer time. Subsequent work was done on Cray-XMP and Cray 2 vector supercomputer systems although the problem is not naturally a vectorizable one, see Openshaw and Craft (1991). A parallel version of the latest GAM/K code has been developed for the Cray T3D written in MPI but it will also now run on a PC in a few hundred seconds (if the Monte Carlo simulation option is not selected). More powerful computing is now needed only if the quality of the apparent results is of interest or concern. Previously it had to be used even to produce crude results. For many spatial analysis applications, the crude results may well be sufficient but if these are not, then it is now possible to use HPC to validate them (Openshaw, 1998b).

The same basic GAM type of brute force approach has been used to search for spatial relationships. The Geographical Correlates Exploration Machine (GCEM/1) of Openshaw *et al.* (1990) examines all 2^{m-1} permutations of m different thematic map layers obtained from a GIS in a search for localized spatial relationships. The GCEM was developed for a Cray Y-MP vector process. It is massively parallel because each of the 2^{m-1} map permutations are independent and can be processed concurrently. It too will now run on a PC in a few days. In both cases, the speed-up in computer hardware speeds has allowed very computationally intensive GC methods to filter down to the desk top. Yesterday's supercomputers are today's workstations and it is likely that this process will continue for at least a couple more decades. What it means is that you can develop and test new GC analysis tools using HPC and be fairly confident in the knowledge that soon it will run on far less powerful and far more available machines.

Another important development is to broaden the basis of the exploratory pattern search process to include all aspects of spatial data (e.g. location in space, location in time, and attributes of the space–time event) and to make the search intelligent rather than systematic. Indeed, the added complexity of additional data domains precludes a simple parallel brute force approach and emphasizes the importance of devising smarter search methods that can explore the full complexity of databases without being too restricted. What we now need are geographical data mining tools, a form of GC technology that is almost totally missing (at present). Only the most primitive of methods have so far been developed due to a seemingly widespread distaste for inductive analysis. That is a pity because this is exactly what the current era of massive data warehouses and unbelievable spatial data riches require.

Openshaw (1994d, 1995b) describes the development of space–time–attribute creatures, a form of artificial life that can roam around what he terms the geocyberspace in an endless hunt for pattern. The claim to intelligence results from the genetic algorithm used to control the search process and the use of computational statistics to reduce the dangers of spurious results. It is strongly dependent on having sufficient parallel computational power to drive the entire process. Openshaw and Perree (1996) show how the addition of animation can help users envisage and understand the geographical analysis. This type of highly exploratory search technology is only just becoming feasible with recent developments in HPC and considerable research is still needed to perfect the technology. The hope is that smart geographical analysis tools will run on PCs and workstations in tolerable elapsed times. More powerful computing is still needed but mainly in design and development of these methods where they can dramatically speed-up testing and be used to resolve design decisions via large-scale simulation and animation of the behaviour of alternative algorithms.

1.7.8 Building geographical knowledge systems

A final illustration describes HPC applications that are highly relevant to many areas of geography but which are probably not yet feasible but soon will be. All the components needed probably exist (a fairly common occurrence in GC research), usually in other contexts, and the trick is to find them, understand them sufficiently as to be able to safely use them, be bold enough to try them out and have access to a sufficiently fast HPC platform to permit experimentation. Creativity is the name of this game.

Consider the following view of the modern data landscape. Modern GIS have provided a microspatial data rich world but there are no non-cartographic tools to help identify in any scientific manner the more abstract recurrent patterns that may exist at higher levels of generalization if only we could 'see' them. Geography is full of concepts and theories about space that can be expressed as idealized two- and three-dimensional patterns that are supposedly recurrent. Traditionally, these concepts and theories have been tested using aspatial statistical methods that require the geography to be removed prior to analysis. For example, if you ask the question does the spatial social structure of Leeds as shown by the 1991 census conform to a broadly concentric ring type of pattern? This hypothesis can be tested by first defining a central point, then a series of three rings of fixed width, then a statistic of some kind is applied to census data to test the *a priori* hypo-thesized trends in social class. However, this clearly requires considerable precision and is not really an adequate test of the original hypothesis that specified no ring widths nor defined a central point nor defined at what level of geographic scale the pattern exists. A possible solution is to use pattern recognition and robotic vision technologies to see whether any evidence of a general concentric geographic structure exists in the census data for Leeds, after allowing for the distorting effects of scale, site and topography. If no idealized concentric patterns exist, then which of a library of different pat-tern types might be more appropriate?

The old pre-quantitative geographical literature of the 1940s and 1950s contains spatial patterns of various sorts, that could never really be tested using conventional statistical methods. Moreover, many of the models of geo-graphy also represent strongly recurring spatial patterns, e.g. distance decay in trip frequencies from an origin. The quantitative geographers of the 1960s and 1970s tried to develop precise mathematical models to describe these patterns but maybe they attempted too much precision and in the process lost the spatial dimension. The HPC revolution of the mid-1990s provides an opportunity to become less precise and more general, by developing pattern recognition tools that can build up recurring map pattern libraries of recur-rent idealized forms. Suppose you ask the question how many different spatial patterns do British cities exhibit? Currently this question cannot be answered but at least the tools exist to allow geographers (and others) to start to find out. Openshaw (1994e) argues that a more generalized pattern recognition approach provides the basis for a new look at geographical information with a view to extracting useful new knowledge from it. Turton (1997, 1999) provides some of the first computer experiments. But this will only become possible as HPC enters the terraflop era and it becomes feasible to apply pattern templates to many millions of locations at many different levels of resolution. There is, however, no reason not to start the research now.

1.8 GeoComputation and future geographers

Much of modern human geography is now in a mega-mess and is indefensible. Most is not geographical, much is little more than story telling, much is intensely theoretical and complex, nearly all is non-science based, there is little use of any of the world's databases, large volumes are anecdotal descriptions of the unique which are irrelevant to the needs of the modern world, there is little or no use of modern technology and no concern to meet either the needs of society or of commerce in a fast changing world; see Openshaw (1998a). Equally, quantitative geography is dead. It is no longer taught in many geography departments, it has failed to become properly spatial, it cannot cope with the needs of GIS, it emphasizes statistical and mathematical tools that are simultaneously too complex and too restrictive, it fosters the impression of being an unattractive, out of date, and old legacy based technology that is also out of touch with reality. Equally, there are problems with GIS, which is just about management and capture of mappable information and has steadfastly failed to develop beyond its original map based origins. As a result, GIS remains surprisingly narrowly focused, it is lacking in theory, it is largely devoid of much of the modelling and simulation relevant to the modern world and is limited in what it can deliver. The time is ripe for something new that can build on existing human and physical geographies, relates to GIS without being restrained by it, makes use of modern informatics, is exciting, relevant and applied but not devoid of theory, has depth but is not exclusive, and is acceptable to other sciences. It also needs to be able to invigorate and excite new generations of geographers with a development path that links past and present to the future and which also allows first year students to participate. It has to be both evolutionary and able to rapidly exploit new developments in a revolutionary manner when need be. However, if GC is to survive and expand then it also has to be no more difficult to the beginner than writing essays on the life and times of a particular gendered ethnic group with distinctive sexual practices in a specific rural village! The view here is that GC has the potential to meet these objectives and is likely to develop into a major paradigm in the new millennium.

It is argued that many areas within and outside of geography could benefit from the adoption of a GC paradigm. Couclelis writes '. . . geocomputation has the potential to influence a number of other spatial sciences, disciplines and application areas with a spatial component, but also to be noticed beyond the walls of universities and research centres. This potential is based on the fact that geocomputation blends well with several major trends in contemporary society. It is obviously in tune with the computer revolution, and capitalizes on the continuing dramatic expansion in computing power and the ubiquity of user-friendly, versatile machines. It has a better chance than stodgy quantitative geography to attract the interest of the coming generation of

researchers.' (Couclelis, 1998a, p. 25). Longley writes: '. . . geocomputation
. . . has become integral to our understanding of spatial structure' (Longley,
1998b, p. 83). However, there is no reason to assume that only quantitative
geography and GIS will benefit; indeed, those non-quantitative areas of geography which are concepts or theory rich but data poor may also have much
to gain; see Openshaw and Openshaw (1997) and Openshaw (1996, 1998a).

It is also argued that there will be no highly visible HPC revolution that
suddenly sweeps all before it. Instead the HPC revolution is silent and almost
invisible. Most of the potential users probably still read the wrong literature
and attend the wrong conferences to notice what is going on. A faster PC is
merely the sharpest tip of a massive iceberg of HPC developments. Yet in
those areas that need it and where a computational paradigm may be helpful,
then there is a way forward. If the current HPC machines are too slow then
be patient, soon there will be much faster ones, but you need to start developing the new approaches now and then safeguard your software investment by using portable programming languages and conforming to emerging
software standards. However, you do not need access to the world's fastest
HPC to start the process rolling. With modern parallel programming tools
you can now write portable scalable codes that can be developed and proven
to work on low-end HPC platforms (e.g. workstation farms) before moving
on to real-world large-scale applications. Indeed you can even assemble
your own workstation farms and test out your applications locally, secure in
the belief that if it works well on your workstation farm it will probably do
far better on a real HPC machine.

It is an interesting thought that GC could act as an attractor for
computationally minded scientists from other fields. It is becoming apparent
that the problems of using HPC are generic and discipline independent.
Cross-discipline research initiatives could be a useful way forward until
critical masses of users within specific disciplines appear. In a geographical
context, the combination of large amounts of data due to GIS, the availability of new AI and CI techniques, and other types of computer intensive
analysis and modelling technology, and the increasing accessibility to HPC
look set to create a new style of computational geography that in the longer
term will revolutionize many aspects of the subject by creating new ways
of doing nearly all kinds of geography. However, if this is to happen then
we need to attract computationally experienced researchers from outside.
GeoComputation has a most critical and focal role to play in this process.

The essential challenge is to use HPC to extend and expand our abilities
to model and analyse all types of geographical systems and not merely those
which are already quantitative and computerized. It would be a dreadful
waste if all they were used for was to make old legacy tools run faster
resulting in a kind of HPC based revival of old-fashioned quantitative geography. The opportunities are far broader than any backward looking view
would suggest. In some areas, almost instant benefits can be gained; for

example, by switching to computationally intensive statistical methods to reduce reliance on untenable assumptions or to discover new information about the behaviour of models. In other areas, whole new GC applications will emerge. In general it is likely that those with access to the biggest and fastest parallel hardware may well be best placed to develop leadership in this new form of internationally competitive computational-based geographical science. As HPC continues to develop it is likely that many subjects, not just geography, will have to undergo a major change in how they operate as HPC is more widely recognized as a paradigm in its own right.

In a world full of unimaginable data riches, maybe (just maybe) we can compute our way out of a massive data swamp fenced in by scores of traditional restrictions and discover how best to do more useful things with it. It is increasingly recognized that data are the raw materials of the information age. They are extremely relevant to commerce and the functioning of society. New scientific discoveries, new knowledge, new ideas, and new insights into the behaviour of complex physical and human systems will increasingly have to be created by a new kind of knowledge industry, something equivalent to a new knowledge manufacturing process. Maybe GC could become the geographer's own version of this knowledge processing industry of the new millennium. It will not be easy because many of the systems being studied are non-linear, chaotic, noisy, and extremely complex in ways we do not yet properly comprehend. Quite often all we have is masses of data that reflect the operation of as yet unidentified systems and processes about which we know next to nothing. It is all very complex, challenging, and exciting. Here lies one of the geocyberspace's research frontiers. It is easily reached but moving it on will be far harder. Those who readily deride data driven approaches as data dredging, data trawling, and data mining should appreciate how difficult it really is to apply science to these problems. It is far, far easier, and simpler, to be deductive than inductive; it is just that we no longer have the luxury of being able to do this. Immense complexity is the main reason why GC is needed and, once established, will last for a long time. However, it is also important not to neglect the new opportunities for building models of geo-systems, for understanding processes, for simulating new and old theories, and generally joining in the computational fun and games increasingly being enjoyed by most other sciences.

What is needed now are the new ideas, and young enthusiastic free-thinking spirits able to go and develop hitherto impossible or unthought of GC tools, the cleverest people from many different disciplines united by different aspects of the GC challenge and who believe it is both possible and worthwhile. We now know enough to start the process rolling (albeit slowly) but many others are now needed to develop the many threads and help guide it to a successful conclusion. The present is a very exciting time for computationally minded geographers and hopefully GC is a rallying call to which many more will feel able to respond to in the years ahead.

References

Bezdek, J. C. (1994) 'What is computational intelligence?', in Zurada, J. M., Marks, R. J., Robinson, C. J. (eds) *Computational Intelligence: Imitating Life*, New York: IEEE, pp. 1–12.

Birkin, M., Clarke, G., Clarke, M. and Wilson, A. G. (1996) *Intelligent GIS*, Cambridge: GeoInformation International.

Couclelis, H. (1998a) 'Geocomputation in context', in Longley, P. A., *et al.* (eds) *Geocomputation: A Primer*, Chichester: Wiley, pp. 17–30.

Couclelis, H. (1998b) 'Geocomputation and space', *Environment and Planning B, 25th Anniversary Issue*, pp. 41–47.

Dibble, C. (1996) 'Theory in a complex world: agent based simulation of geographical systems', in *Proceedings 1st International Conference on GeoComputation*, Leeds: Leeds University, September, Vol. 1, pp. 210–213.

Diplock, G. J. (1996) The Application of Evolutionary Computing Techniques to Spatial Interaction Modelling, Unpublished PhD thesis, University of Leeds.

Diplock, G. J. (1998) 'Building new spatial interaction models using genetic programming and a supercomputer', *Environment and Planning A*, 30, pp. 1893–1904.

Diplock, G. J. and Openshaw, S. (1996) 'Using simple genetic algorithms to calibrate spatial interaction models', *Geographical Analysis*, 28, pp. 262–279.

Dobson, J. E. (1983) 'Automated geography', *The Professional Geographer*, 35, pp. 135–143.

Dobson, J. E. (1993) 'The geographic revolution: a retrospective on the age of automated geography', *The Professional Geographer*, 45, pp. 431–439.

Gilbert, G. N. and Doran, J. (1994) *Simulating Societies: the Computer Simulation of Social Phenomena*, London: UCL Press.

Gilbert, N. and Conte, R. (ed.) (1995) *Artificial Societies*, London: UCL Press.

Harris, B. (1985) 'Some notes on parallel computing with special reference to transportation and land use modelling', *Environment and Planning A*, 17, pp. 1275–1278.

Koza, J. R. (1992) *Genetic Programming*, Cambridge, MA: MIT Press.

Koza, J. R. (1994) *Genetic Programming II: Automatic Discovery of Re-usable Programs*, Cambridge, MA: MIT Press.

Longley, P. A. (1998a) 'Foundations', in Longley, P. A. *et al.* (1998) *Geocomputation: A Primer*, Chichester: Wiley, pp. 1–16.

Longley, P. A. (1998b) 'Developments in geocomputation', *Computers, Environment and Urban Systems*, 22, pp. 81–83.

Longley, P. A., Brooks, S. M., McDonnell, R. and Macmillan, B. (1998) *Geocomputation: A Primer*, Chichester: Wiley.

Macmillan, B. (1998) 'Epilogue', in Longley, P. A. *et al.* (eds) *Geocomputation: A Primer*, Chichester: Wiley, pp. 257–264.

Openshaw, S. (1976) 'A geographical solution to scale and aggregation problems in region building, partitioning and spatial modelling', *Transactions of the Institute of British Geographers, New Series*, 2, pp. 459–472.

Openshaw, S. (1978) 'An empirical study of some zone design criteria', *Environment and Planning A*, 10, pp. 781–794.

Openshaw, S. (1984) 'Ecological fallacies and the analysis of areal census data', *Environment and Planning A*, 16, pp. 17–31.

Openshaw, S. (1987) 'Some applications of supercomputers in urban and regional analysis and modelling', *Environment and Planning A*, 19, pp. 853–860.

Openshaw, S. (1988) 'Building an automated modelling system to explore a universe of spatial interaction models', *Geographical Analysis*, 20, pp. 31–46.

Openshaw, S. (1994a) 'Computational human geography: towards a research agenda', *Environment and Planning A*, 26, pp. 499–505.

Openshaw, S. (1994b) 'Computational human geography: exploring the geo-cyberspace', *Leeds Review*, 37, pp. 201–220.

Openshaw, S. (1994c) 'Neuroclassification of spatial data', in Hewitson, B. and Crane, R. (eds) *Neural Nets: Applications in Geography*, Dordrect: Kluwer Academic, pp. 53–70.

Openshaw, S. (1994d) 'Two exploratory space-time attribute pattern analysers relevant to GIS', in Fotheringham, S. and Rogerson, P. (eds), *Spatial analysis and GIS*, London: Taylor & Francis, pp. 83–104.

Openshaw, S. (1994e) 'A concepts rich approach to spatial analysis, theory generation and scientific discovery in GIS using massively parallel computing', in Worboys, M. (ed.), *Innovations in GIS*, London: Taylor & Francis, pp. 123–138.

Openshaw, S. (1995a) 'Human systems modelling as a new grand challenge area in science', *Environment and Planning A*, 27, pp. 159–164.

Openshaw, S. (1995b) 'Developing automated and smart spatial pattern exploration tools for GIS applications', *The Statistician*, 44, pp. 3–16.

Openshaw, S. (1996) 'Fuzzy logic as a new scientific paradigm for doing geography', *Environment and Planning A*, 28, pp. 761–768.

Openshaw, S. (1998a) 'Towards a more computationally minded scientific human geography', *Environment and Planning A*, 30, pp. 317–332.

Openshaw, S. (1998b) 'Building automated geographical analysis and explanation machines', in Longley, P. A. *et al.* (1998) *Geocomputation: A Primer*, Chichester: Wiley, pp. 95–116.

Openshaw, S. (1998c) 'Neural network, genetic, and fuzzy logic models of spatial interactions', *Environment and Planning A*, pp. 1857–1872.

Openshaw, S. and Alvanides, S. (1999) 'Applying geocomputation to the analysis of spatial distributions', in Longley, P. A., Goodchild, M. F., Maguire, D. J. and Rhind, D. W. (eds) *Geographical Information Systems: Principles and Technical Issues*, Vol. 1, New York: Wiley.

Openshaw, S. and Craft, A. (1991) 'Using the geographical analysis machine to search for evidence of clusters and clustering in childhood leukaemia and non-hodgkin lymphomas in Britain', in Draper, G. (ed.) *The Geographical Epidemiology of Childhood Leukaemia and Non-Hodgkin Lymphomas in Great Britain, 1966–83*, London: HMSO, pp. 109–122.

Openshaw, S. and Openshaw, C. A. (1997) *Artificial Intelligence in Geography*, Chichester: Wiley.

Openshaw, S. and Perree, T. (1996) 'User centered intelligent spatial analysis of point data', in Parker, D. (ed.) *Innovations in GIS 3*, London: Taylor & Francis, pp. 119–134.

Openshaw, S. and Rao, L. (1995) 'Algorithms for re-engineering 1991 census geography', *Environment and Planning A*, 27, pp. 425–446.

Openshaw, S. and Schmidt, J. (1996) 'Parallel simulated annealing and genetic algorithms for re-engineering zoning systems', *Geographical Systems*, 3, pp. 201–220.

Openshaw, S. and Schmidt, J. (1997) 'A social science benchmark (SSB/1) Code for serial, vector, and parallel supercomputers', *Geographical and Environmental Modelling*, 1, pp. 65–82.

Openshaw, S. and Turton, I. (1996) 'A parallel Kohonen algorithm for the classification of large spatial datasets', *Computers and Geosciences*, 22, pp. 1019–1026.

Openshaw, S. and Turton, I. (1999) *An Introduction to High Performance Computing and the Art of Parallel Programming: for Geographers, Social Scientists, and Engineers*, London: Routledge.

Openshaw, S., Charlton, M., Wymer, C. and Craft, A. (1987) 'A mark I geographical analysis machine for the automated analysis of point data sets', *International Journal of GIS*, 1, pp. 335–358.

Openshaw, S., Cross, A. and Charlton, M. (1990) 'Building a prototype geographical correlates exploration machine', *International Journal of GIS*, 3, pp. 297–312.

Openshaw, S., Blake, M. and Wymer, C. (1995) 'Using neurocomputing methods to classify Britain's residential areas', in Fisher, P. (ed.), *Innovations in GIS 2*, London: Taylor & Francis, pp. 97–112.

Rees, P. and Turton, I. (1998) 'Geocomputation: solving geographical problems with computing power', *Environment and Planning A*, 30, pp. 1835–1838.

Turton, I. (1997) Application of Pattern Recognition to Concept Discovery in Geography, Unpublished MSc thesis, University of Leeds.

Turton, I. (1999) 'Application of pattern recognition to concept discovery in geography', in Gittings, B. (ed.), *Innovations in GIS 6*, London: Taylor & Francis.

Turton, I. and Openshaw, S. (1996) 'Modelling and optimising flows using parallel spatial interaction models', in Bouge, L., Fraigniaud, P., Mignotte, A. and Roberts, Y. (eds) *Euro-Par '96 Parallel Processing*, Vol. 2, Lecture Notes in Computer Science 1124, Berlin: Springer, pp. 270–275.

Turton, I. and Openshaw, S. (1998) 'High performance computing and geography: developments, issues and case studies', *Environmental and Planning A*, 30, pp. 1839–1856.

Turton, I., Openshaw, S. and Diplock, G. J. (1996) 'Some geographical applications of genetic programming on the Cray T3D supercomputer', in Jesshope, C. and Shafarenko, A. (eds) *UK Parallel '96: Proceedings of the British Computer Society Parallel Processing Specialist Group Annual Conference*, Berlin: Springer, pp. 135–150.

Turton, I., Openshaw, S. and Diplock, G. (1997) 'A genetic programming approach to building new spatial models relevant to GIS', in Kemp, Z. (ed.) *Innovations in GIS 4*, London: Taylor & Francis, pp. 89–102.

Wilson, A. G. (1970) *Entropy in Urban and Regional Modelling*, London: Pion.

Wilson, A. G. (1974) *Urban and Regional Models in Geography and Planning*, Chichester: Wiley.

Chapter 2

GeoComputation analysis and modern spatial data

A. Stewart Fotheringham

2.1 Introduction

Computation is a term which can take one of two possible meanings. In its broader sense, it refers to the use of a computer and therefore any type of analysis, be it quantitative or otherwise, could be described as 'computational' if it were undertaken on a computer. In its narrower and perhaps more usual sense, computation refers to the act of counting, calculating, reckoning and estimating – all terms which invoke quantitative analysis. This chapter will therefore restrict itself to this latter definition and uses the term GeoComputation (GC) to refer to the quantitative analysis of spatial data which is aided by a computer. Even more narrowly, I shall use the term GC to refer to quantitative spatial analysis *in which the computer plays a pivotal role*. This definition is still sufficiently vague though that fairly routine analyses of spatial data with standard statistical packages (for instance, running a regression programme in SAS) could be incorporated within it and I shall define such analyses where the computer is essentially a faster slide rule or abacus as *weak* GC. I will try to demonstrate in the examples below some of the ways in which spatial analysis is being extended through the use of computers well beyond that which standard statistical packages allow. What I define as *strong* GC analysis is where the use of the computer *drives* the form of analysis undertaken rather than being a convenient vehicle for the application of techniques developed independently of computers. Strong GC techniques are, therefore, those which have been developed *with the computer in mind* and which explicitly take advantage of large amounts of computer power.

Perhaps a simple example is useful to distinguish between weak and strong GC. Consider a correlation coefficient being calculated for two sets of spatially varying data, variable x and variable y. To assess the significance of the resulting correlation coefficient one could apply the standard formula for a t-statistic calculating the standard error of the correlation coefficient from a theoretical distribution – the procedure used in all standard statistical computer packages. I would term this weak GC because the computer is simply used to speed up a calculation developed well before computers were in use. A strong GC technique would be to derive an estimate of the standard error of the correlation coefficient by experimental methods. One such method would be to independently randomly permute the x and y variables across the spatial zones and calculate a correlation coefficient for each permutation. With a sufficiently large number of such correlation coefficients, an experimental distribution can be produced from which the significance of the observed correlation coefficient can be obtained. In this latter case, computational power is used to replace a theoretical distribution with the advantage being the avoidance of the assumptions underlying the theoretical distribution which may not be met, particularly with spatial data. More details on experimental significance tests, bootstraps and jacknifes are given by Diaconis and Efron (1983), Efron and Gong (1983), and Mooney and Duval (1993).

2.2 Philosophical issues in GeoComputation

While I have no time for those who are insistent that they have found the one true philosophical approach (usually something which avoids the use of quantitative methods in any form) to understanding geography (actually, just parts of human geography), there are nonetheless some interesting philosophical issues which underpin GC. It is, or will be, immediately condemned by those who espouse one of the myriad of anti-naturalism philosophies (Graham, 1997) as being a product of logical positivism. However, the anti-naturalists fail to realize that much of what GC has to offer is in fact the opposite of the 'Geography is Physics' approach which is such anathema to them and which in their minds is inextricably linked with quantitative geography. In what I would now regard as 'slightly old-fashioned' traditional spatial analysis, a key feature was indeed the attempt to establish global relationships – a questionable goal in terms of human behaviour (although not of course in terms of physical behaviour). A premise of much spatial analysis, however, is that things are *not* necessarily the same all over the map and it is of interest to describe such spatial variations as an aid to better understanding of spatial processes. Some of the techniques described below have been designed explicitly to investigate *local* rather than *global* relationships. Admittedly, the focus on local exceptions and anomalies can be used as a means of improving global models and so would not escape anti-naturalist critiques but, equally, local models can be used with the *a priori* assumption that there are intrinsically different types of behaviour over space and that no global model exists.

Besides the usual philosophical 'tug-of-war' between naturalists and anti-naturalists and between quantitative and non-quantitative geographers, it should be recognized that there are different views on analysis within quantitative geography. GeoComputation in fact touches upon two debates within quantitative analysis. In the first, that between *confirmatory* and *exploratory* techniques, GC is clearly seen as being in the latter camp. Confirmatory techniques emphasize hypothesis testing and include standard statistical techniques such as regression, *t*-tests, ANOVA, etc. However, the robustness of these methods can be greatly improved by replacing unrealistic distributional assumptions (that reflect a computational minimizing strategy) by computational alternatives (which are more computer intensive but far less assumption dependent). Exploratory techniques are data, and often computer, intensive and are used to *suggest* hypotheses from an analysis of the data. They are used relatively loosely to tease what appears to be interesting out of the data and are therefore often viewed as a precursor to more formal modelling and analysis although they do not have to be. However, exploratory techniques can also be used in a postconfirmatory role to examine the performance of a model, e.g. in what areas does a model appear to work better than others and what characteristics of such areas give rise to these differences?

Whichever way they are used, an important component of most exploratory spatial analysis is the map, the geographer's essential tool. Mapping residuals, or even data, for example, can be considered a relatively unsophisticated type of exploratory spatial data analysis. In more sophisticated types of exploratory analysis, interaction with the data being mapped is possible which is where GIS and exploratory graphics packages become useful (see below).

The other debate to which GC contributes is that which includes, on the one hand, those who support deductive reasoning and those, on the other, who believe in inductive techniques. Most quantitative geographers probably find a mixture of the two the most fruitful means of uncovering knowledge. GeoComputation techniques have a potential role to play in deductive reasoning by providing alternative means of model calibration (see, for example, the chapter on neurocomputing) and significance testing (an example being the development of experimental significance tests described above) which do not need to rely on untestable assumptions about the data. Their main role though is through induction: promoting and displaying aspects of the data to suggest new hypotheses concerning the spatial processes which have produced the data.

More controversially, GC techniques have been applied to model building (*inter alia* Openshaw, 1983, 1988; Diplock, 1996) as a means of by-passing traditional deductive logic. It is more difficult to support such applications because the data clearly drive the form of the models which result from these exercises and there is no guarantee that anything resembling the same model will result from a different data set although multiple data sets could be used in a hunt for more general empirically recurrent model forms. Even from the same data set, many different models could be produced which fit the data reasonably well and slight alterations in the goodness-of-fit criterion used to drive model selection may then produce very different models.

Clearly more research is needed to develop model breeding technologies that are robust and useful. For these reasons, data-driven methods of model derivation are ignored in what follows although they are described elsewhere in this book.

In the remainder of this chapter, four examples of GC are presented. The first example concerns GIS-based spatial analysis, the second describes computational issues surrounding the modifiable areal unit problem, the third covers examples of point pattern analysis using computational exploratory techniques, and the fourth describes a relatively new type of modelling known as *local modelling* in which the emphasis is on uncovering spatial variations in relationships.

2.3 GIS-based spatial analysis

Geographic information systems (GIS) provide potentially very powerful tools for spatial analysis. They provide a means of storing, manipulating and mapping large volumes of spatial data. Over the last two decades, various linkages have been established between GIS software and statistical analysis packages for GC analysis. As early as 1973 the Chicago Area Transportation Study used an IBM mainframe interactive graphics package called INTRANS to display and evaluate planning data generated by transportation models (Harper and Manheim, 1990). More recently, but very much in the same spirit, the Maryland National Capital Park and Planning Commission generated data from the SPANS GIS as input into EMME/2, a transportation modelling system. The former, and possibly also the latter, falls into what Anselin *et al.* (1993) refer to as one-directional or static integration, where the results of one operation are fed into the other with no feedback. More advanced integration between GIS and spatial analysis involves bidirectional connections where there is two-way interaction between the GIS and spatial analytical routines (such as where data are derived from a GIS to calculate a statistic and then the statistic is imported back into the GIS for mapping). An example of this type of integration is that of SpaceStat (Anselin, 1990) which is a software package for the statistical analysis of spatial data that can be hooked on to a variety of GIS packages. The most advanced coupling between GIS and spatial analysis comes through dynamic integration where movement between the two is continuous, an example being the brushing of data in a scatterplot in one window and the automatic referencing of those points on a map in another window. Examples of this type of integration are provided by the ESDA Arc/Info package developed by Xia and Fotheringham (1993) and the spatial analysis module (SAM) described in Ding and Fotheringham (1992). In the latter software, the user can open up to five linked windows for analysis and at least one window can contain a map of the study area. Observations brushed in any window will automatically be highlighted in the other four windows.

As a result of these initiatives and a multitude of calls for the development of more sophisticated spatial analytical tools to be incorporated within GIS (*inter alia* Goodchild, 1987; Rhind, 1988; Openshaw, 1990; Burrough, 1990; Fotheringham and Charlton, 1994), GIS vendors are now adding some of these tools to systems that previously only performed basic query, buffer and overlay routines. For instance: GIS-Plus has a suite of transportation-related models; SPANS has, amongst other things, a retail analysis package; SAS have developed their own GIS; and S-PLUS, an advanced statistical package for both exploratory and confirmatory analysis can now be run through AML commands under Arc/Info which gives the user access to over 1400 statistical functions although not many are relevant to spatial analysis.

In terms of new areas of GC, it is true that simply providing the means for greater linkages between mapping and analysis does not mean greater insights into spatial processes are guaranteed. In many cases it could be argued that it is not essential to integrate the statistical software with a GIS. However, for exploratory analysis, the ability to move seamlessly between the analytical and the mapping software and the ability to interrogate the mapped data produces a reasonably high probability of producing insights that would otherwise be missed if spatial data were not analysed within a GIS. Consequently, GIS-based computational analysis will continue to grow and is likely to become the preferred means of GC for many adherents. The development of new ways of interacting with and displaying spatial data and the results of spatial analyses provides a very fertile and exciting area into which GC will continue to expand.

2.4 The modifiable areal unit problem

GeoComputation frequently, although not always, involves the use of areal data, a common example being the analysis of census data reported at various spatial scales (such as enumeration districts and wards in the UK and block groups and census tracts in the US). One of the most stubborn problems related to the use of areal data is sometimes refered to as the zone definition problem or the modifiable areal unit problem and which relates to the sensitivity of analytical results to the definition of the spatial units for which data are reported (Openshaw, 1984; Openshaw and Rao, 1995). The policy implications of this problem are potentially severe because if the conclusions reached from an analysis of aggregate spatial data reported for one set of zones are different to those reached when data are reported for a different arrangement of zones, then how reliable can any one analysis be as a means of influencing policy?

One computationally intensive 'solution' to this problem demonstrated by Fotheringham and Wong (1991) is to provide analytical results not just for one set of zones but for a variety of zoning systems so that the stability or instability of a particular result can be assessed visually or statistically. Results (for example, parameter estimates) that are relatively stable to variations in reporting units are more reliable, *ceteris paribus*, than those which are relatively unstable. With data drawn from census units in the Buffalo Metropolitan Area, the relationship between mean family income and a series of independent variables is examined within both a linear and non-linear modelling framework. The relationship with the percentage of elderly in a spatial unit is most interesting: by varying the spatial scale at which the data are collected, the parameter estimate for the elderly variable is consistently insignificant when the data are obtained from 800 zones but is consistently significant and negative when the data are aggregated to 200 or fewer zones. Thus, two very different interpretations can be drawn from the same underlying data.

A similar inconsistency is found even when scale (i.e. the number of zones) remains constant but the arrangement of these zones is allowed to vary. An examination of the parameter estimate for the elderly variable estimated with 150 different spatial arrangements of the data at the same spatial scale reveals that while the majority of the zoning systems yield the conclusion that there is no significant relationship between mean family income and the proportion of the elderly, a substantial number of zoning systems yield the conclusion that there is a significant *negative* relationship between the two variables, while two zoning systems yield the conclusion that there is a significant *positive* relationship between income and the elderly! It should be noted that the results reported by Fotheringham and Wong (1991), as well as being computationally intensive, rely heavily on the ability to combine large numbers of zones large numbers of times based on their topological relationships and that this is greatly facilitated by having the data stored in a GIS.

A similar computationally intensive sensitivity analysis is described by Fotheringham *et al.* (1995) for a set of techniques for locational analysis known as location-allocation modelling. These techniques provide information not only on the optimal locations for a set of facilities but also on the demand that is served by each facility. Common to almost all applications of location-allocation modelling is that demand is computed for a set of aggregate zones and an issue analogous to that investigated by Fotheringham and Wong (1991) is to what extent the outcome of a location-allocation procedure is affected by the particular way in which demand is aggregated. That is, to what extent is the set of locations and the allocation of demand to these locations simply a product of the way in which data are aggregated and how sensitive are these results to variations in the zoning system? This is a particularly relevant question in location-allocation modelling because the outputs from such a procedure are taken to be the *optimal locations* for a set of facilities and the *optimal allocation* of demand to those facilities and as such often carry a great deal of weight in decision-making. However, if the results can be varied substantially simply by varying the scale of the analysis or by altering the arrangement of the reporting units for which the data are collected, the word 'optimal' in this situation would need to be used with a great deal of caution.

To examine this issue in a computationally intensive manner, Fotheringham *et al.* (1995) examined the location of 10 senior citizen centres in Buffalo, NY, with the objective of minimizing the aggregate distance travelled to the nearest facility. An initial zoning system is formed by 871 demand zones and these are aggregated to six spatial scales with 20 different zoning systems at each scale. The optimal locations of the 10 facilities are then mapped for all 120 aggregate zoning systems. The degree of fuzziness on the resulting map depicts the sensitivity of the model results to the way in which the data are aggregated and gives a strong visual impression of the somewhat arbitrary nature of the results using any one zoning system. Whilst some locations are more consistent targets than others, optimal sites are dispersed throughout the study region indicating the locations of the facilities are highly dependent on the way in which the data are aggregated.

Given that policy decisions are sometimes guided by the analysis of spatial data in aggregate zones and that the results of such analysis appear to be dependent on the nature of the zoning system used, there is an increasing need for computationally intensive techniques such as those described above. In order to provide a convincing set of results for any spatial analysis using aggregated data, it is necessary to demonstrate that the results are likely to hold regardless of the type of zoning system used. If this cannot be demonstrated then the results may be mere artefacts of the particular zoning system used and might not necessarily reflect any underlying process.

2.5 Point pattern analysis

There are important questions to be answered in many types of spatial point patterns. Do significant clusters of diseases occur around pollutant sources, toxic waste dumps and nuclear power stations? Are certain types of crime clustered in certain areas? Can we identify significant clusters of plants which may give clues to the underlying soil quality? Until relatively recently, answers to these questions were only given on a 'global' or 'whole-map' scale. That is, a point pattern was declared 'clustered', 'random' or 'dispersed' based on its whole map distribution using a single statistic such as a nearest neighbour statistic. Such an analysis was computationally easy but hopelessly flawed in many ways, not least of which was that it ignored potentially interesting subregions within the point pattern where different processes might be at work. For instance, it is quite conceivable that local areas of clustering could be hidden in the calculation of a global statistic by broader areas of randomness or dispersion.

The recognition of such problems with traditional methods of point pattern analysis and the increasing availability of georeferenced point data in digital form led to the development of new and computationally intensive forms of point pattern analysis where the emphasis is not on whole map statistics but on identifying local areas of the map in which interesting things appeared to be happening. These new exploratory techniques were developed by researchers across several disciplines, not least of which was geography (Clayton and Kaldor, 1987; Openshaw *et al.*, 1987; Doll, 1990; Besag and Newall, 1991; Marshall, 1991; Fotheringham and Zhan, 1996).

To demonstrate the GC aspects of these new forms of point pattern analysis, consider the techniques described by Openshaw *et al.* (1987) and by Fotheringham and Zhan (1996). Both techniques can be described in terms of the Geographical Analysis Machine for detecting spatial point clusters developed by Openshaw *et al.* (1987) and which consists of an efficient method for accessing large amounts of spatial data, a method for assessing cluster significance, and a method for returning information on significance for mapping. The basic steps in this type of computationally intensive analysis are as follows:

1. define a point on a map and draw a circle of random radius r around that point;
2. count the number of observations within this circle;
3. compare this count with an expected value (usually assuming a random generating process);
4. draw the circle on the map *if* it contains a significantly higher number of observations than expected;
5. select a new point on the map and repeat the procedure.

In Openshaw *et al.* (1987), points are selected through systematic sampling on a regular grid and then Monte Carlo techniques are used to assess significance. This process does not demand knowledge of the underlying, at-risk, population but is computationally intensive in terms of significance testing. In Fotheringham and Zhan (1996), points are selected randomly and a Poisson probability function with data on the at-risk population is used to assess significance. The latter method is shown to be more discerning in identifying significant clusters although it does demand geo-coded data for the at-risk population which is not always available. An empirical comparison of the two techniques plus another by Besag and Newall (1991) is given in Fotheringham and Zhan (1996).

Both techniques are computationally intensive because they apply a routine statistical analysis at many different points in space. Both are clearly exploratory in nature because they identify potentially interesting clusters of points in space around which further investigations can be concentrated. Also, by investigating the geography of significant clusters, hypotheses may be suggested concerning the processes producing the clusters.

Essentially these newer, more computationally intensive, forms of point pattern analysis are examples of a broader and growing concern in the literature for *local statistics* and this forms the basis of the last set of examples.

2.6 Spatial non-stationarity and local modelling

A recent and potentially very powerful trend within spatial analysis concerns the identification and understanding of *differences* across space rather than *similarities*. This trend is termed local modelling and it encompasses the dissection of global statistics into their local constituents, the concentration on local exceptions rather than the search for global regularities, and the production of local or mappable statistics rather than 'whole-map' values. By its nature, local modelling is much more computer intensive than traditional global modelling. This shift in emphasis from global to local reflects the increasing availability of large and complex spatial data sets, the development of faster and more powerful computing facilities, and the recognition that relationships might not be stationary but might vary significantly across space (i.e. they might exhibit *spatial non-stationarity*).

The *raison d'être* for local modelling is that when analysing spatial data, whether by a simple univariate analysis or by a more complex multivariate analysis, it might be incorrect to assume that the results obtained from the whole data set apply equally to all parts of the study area. Interesting insights might be obtained from investigating spatial variations in the results. Simply reporting 'average' sets of results and ignoring any possible spatial variations in those results is equivalent to reporting the mean value of a spatial distribution without providing a map of the data.

In order to appreciate the GC aspects of local versus global analysis, consider a frequently encountered aim of data analysis: that of understanding the nature of the distribution of one variable in terms of the distributions of other variables through multiple linear regression (Draper and Smith, 1981; Berry and Feldman, 1985; Graybill and Iyer, 1994). If the data are drawn from spatial units, the output from regression is a single set of parameter estimates which depict relationships assumed to be stationary over space. That is, for each relationship a single parameter estimate is implicitly assumed to depict the nature of that relationship for all points in the entire study area. Clearly, any relationship which is not stationary over space will not be modelled particularly well and indeed the global estimate may be very misleading locally.

There are at least three reasons to question the assumption of stationarity in spatial data analysis. The first is that there will inevitably be spatial variations in observed relationships caused by random sampling variations. The contribution of this source of spatial non-stationarity is generally not of interest but it needs to be accounted for by significance testing. That is, it is not sufficient to describe fluctuations in parameter estimates over space as they may simply be products of sampling variation. It is necessary to compute whether such variations are statistically significant. The second, which is in accord with postmodernist views on the importance of localities, is that, for whatever reasons, some relationships are intrinsically different across space. Perhaps, for example, there are spatial variations in attitudes or preferences or there are different contextual issues that produce different responses to the same stimuli. The third reason why relationships might exhibit spatial non-stationarity, more in line with positivist thinking, is that the model from which the relationships are measured is a gross misspecification of reality and that one or more relevant variables are either omitted from the model or are represented with an incorrect functional form. This view assumes that ultimately a global model might be attainable but currently the model is not correctly specified. In this case, local modelling is very much an exploratory tool which can guide the development of more accurate spatial models. The local statistics are useful in understanding the nature of the misspecification more clearly: in what parts of the study region does the model replicate the data less accurately and why?

Within the last few years there has been a mini-explosion of academic work reflecting the calls of Fotheringham (1992), Fotheringham and Rogerson (1993) and Openshaw (1993) for greater attention to be given to local or mappable statistics. The style of point pattern analysis described above is one area in which local mapping of statistical analysis has come to the fore. Much of the work undertaken in exploratory graphical analysis (*inter alia* Haslett *et al.* (1991)) is essentially concerned with identifying local exceptions to general trends in either data or relationships. Computational techniques such as linking windows and brushing data allow the data to be examined interactively and linked to their locations on a map. Usually, this type of interaction is undertaken within relatively simple univariate or bivariate frameworks.

More formally, although again with univariate statistical analysis, local spatial association statistics have been developed by Getis and Ord (1992), Anselin (1995) and Ord and Getis (1995). Brunsdon *et al.* (1998) show how local spatial autocorrelation statistics can be derived in a multivariate framework. These statistics show how data cluster in space: to what extent are high values located near to other high values and low values located near to other low values? Previously, standard global spatial autocorrelation statistics such as Moran's I only produced a global value assumed to apply equally to the clustering of all data values across space. Local values of this, and other, statistics indicate how data values cluster locally around each point in space. Some of these statistics have been implemented in a GIS environment by Ding and Fotheringham (1992).

The increasing availability of large and complex spatial data sets has led to a greater awareness that the univariate statistical methods described above are of limited application and that there is a need to understand local variations in more complex relationships. In response to this recognition, several attempts have been made to produce localized versions of traditionally global multivariate techniques, with the greatest challenge being that associated with developing local versions of regression analysis. Perhaps the best-known attempt to do this is the expansion method (Casetti, 1972; Jones and Casetti, 1992) which attempts to measure parameter 'drift'. In this framework, parameters of a global model are expanded in terms of other attributes. If the parameters of the regression model are made functions of geographic space, trends in parameter estimates over space can then be measured (Eldridge and Jones, 1991; Fotheringham and Pitts, 1995). While this is a useful and easily applicable framework, it is essentially a trend-fitting exercise in which complex patterns of parameter estimates can be missed.

More computationally complex procedures for local regression modelling include geographically weighted regression (GWR) which has been developed specifically for spatial data (Brunsdon et al., 1996; Fotheringham et al., 1997a, b); spatial adaptive filtering (Foster and Gorr, 1986; Gorr and Olligshlaeger, 1994); random coefficients modelling (Aitkin, 1996); and multilevel modelling (Goldstein, 1987). Problems with the latter three techniques for handling spatial processes have been noted by Fotheringham (1997) and the remainder of this discussion concentrates on GWR.

Geographically weighted regression is a relatively simple, although computationally complex, procedure which extends the traditional global regression framework by allowing local rather than global parameters to be estimated. The model has the general form:

$$y_i = a_{i0} + \sum_k a_{ik} x_{ik} + \varepsilon_i$$

where y represents the dependent variable, x_k represents the kth independent variable, ε represents an error term and a_{ik} is the value of the kth parameter at location i. In the calibration of this model it is assumed that observed data near to point i have more influence in the estimation of the a_{ik}s than do data located further from point i. In essence, the equation measures the relationship inherent in the model *around each point i*. Calibration of the model is more complex than with ordinary regression because, although a weighted least squares approach is used, the data are weighted according to their location with respect to point i and therefore the weights vary with each point i rather than remaining fixed. The estimator for the parameters in GWR is then:

$$a_i' = (x^t w_i x)^{-1} x^t w_i y$$

where w_i is an $n \times n$ matrix whose off-diagonal elements are zero and whose diagonal elements denote the geographical weighting of observed data for point i. It should be noted that as well as producing localized parameter estimates, GWR also results in localized versions of all standard regression diagnostics such as R^2 and standard errors of the parameter estimates. The former can be particularly useful as an exploratory tool when mapped to suggest additional explanatory models to add to the model.

Computationally, the GWR framework described above incorporates some interesting issues connected with spatial processes. The first is the definition and calibration of the spatial weighting function. A weighting function has to be defined which weights data in close proximity to point i more than data which are farther away. One possibility is to set the weights equal to one within a prespecified distance of point i and zero beyond this distance. This is relatively easy to compute but as it assumes a discrete spatial process, it is perhaps not very realistic for most processes. An example of this type of weighting function is given in Fotheringham *et al.* (1996) and in Charlton *et al.* (1997). A more realistic, but more computationally intensive, weighting function is one which is a continuous decreasing function of distance so that a distance-decay parameter is calibrated for the function. This can be done by a cross-validation goodness-of-fit criterion (see Brunsdon *et al.* (1996) and Fotheringham *et al.* (1997a, b) for examples of this type of spatial weighting function and its calibration).

Given that a weighting function and, where necessary, a calibration procedure have been selected, a further element of realism and computational complexity can be added by allowing the weighting function to vary spatially. That is, in what is described above, a global weighting function is calibrated but it is possible to allow the function to vary across space, presumably with the kernel becoming larger in sparser areas and smaller in more populated areas. There are several ways in which a spatially adaptive kernel can be computed. One is to allow a point-specific distance-decay parameter to be calibrated. A second is to base the weighting function on the x nearest neighbours of point i and so a continuous distance-based measure is replaced with an ordinal topological measure. A third is to set a constraint so that every point has the same sum of weighted data. This creates a weighting function with the added attraction of having a constant number of 'virtual' degrees of freedom. This constant could be given exogenously or, with yet another computational complexity, it could be calibrated within the GWR routine.

To this point it is assumed that all the parameters in the regression model are allowed to vary spatially and have the same weighting function. A final computational complexity is to allow the weighting functions to vary across the parameters and even to allow 'mixed' models in which some of the parameters are fixed over space whilst the others are allowed to vary spatially.

Clearly, this is an exciting frontier of spatial analysis because it allows us to input the types of spatial processes we think operate in a given situation. It also provides us with a great deal of additional *spatial* information in the form of maps of parameter estimates, goodness-of-fit statistics and other regression diagnostics. It brings issues of space to the fore in the previously aspatial, yet very powerful, statistical technique of regression.

2.7 Summary

There are at least two constraints on undertaking empirical research within spatial analysis. One is our ability to think about how spatial processes operate and to produce insights which lead to improved forms of spatial models. The other is the tools we have to test these models. These tools might be used for data collection (e.g. GPS receivers, weather stations, stream gauges, etc.) or they might be used for data analysis (computers). In the early stages of computer use, it was relatively easy to derive models which could not be implemented because of the lack of computer power. This was an era when the second constraint was more binding than the first: the level of technology was behind our ability to think spatially. We are now no longer in this era. We are now in a situation where the critical constraint is more likely to be our ability to derive new ways of modelling spatial processes and analysing spatial data. The increases in computer power within the last 20 years have been so enormous that the technological constraint is much less binding than it once was. The challenge now is to come up with ideas which make full use of the technology to improve our understanding of spatial processes. Hopefully, the examples given above suggest ways in which this is being done. However, we are just at the beginning of this new era of virtually unconstrained computing power and there is still much to be done in revamping our analytical methods to take advantage of the new situation. In many instances the change is so profound that it can alter our whole way of thinking about issues. The development of experimental significance testing procedures and the subsequent decline in the reliance on theoretical distributions is a case in point. The movement from global modelling to local modelling is another. Who knows what changes the next decade will bring?

References

Aitkin, M. (1996) 'A general maximum likelihood analysis of overdispersion in generalised linear models', *Statistics and Computing*, 6, pp. 251–262.

Anselin, L. (1990) 'SpaceStat: a program for the statistical analysis of spatial data', NCGIA Publications, Department of Geography, University of California at Santa Barbara.

Anselin, L. (1995) 'Local indicators of spatial association – LISA', *Geographical Analysis*, 27, pp. 93–115.

Anselin, L., Dodson, R. F. and Hudak, S. (1993) 'Linking GIS and spatial data analysis in practice', *Geographical Systems*, 1, pp. 3–23.

Berry, W. D. and Feldman, S. (1985) *Multiple Regression in Practice*, Sage Series in Quantitative Applications in the Social Sciences 50, London: Sage.

Besag, J. and Newall, J. (1991) 'The detection of clusters in rare diseases', *Journal of the Royal Statistical Society*, 154A, pp. 143–155.

Brunsdon, C., Fotheringham, A. S. and Charlton, M. E. (1996) 'Geographically weighted regression: a method for exploring spatial nonstationarity', *Geographical Analysis*, 28, pp. 281–298.

Brunsdon, C., Fotheringham, A. S. and Charlton, M. E. (1998) 'Spatial non stationarity and autoregressive models', *Environment and Planning A*, 30, pp. 957–973.

Burrough, P. A. (1990) 'Methods of spatial analysis in GIS', *International Journal of Geographical Information Systems*, 4, pp. 221–223.

Casetti, E. (1972) 'Generating methods by the expansion method: applications to geographic research', *Geographical Analysis*, 4, pp. 81–91.

Charlton, M. E., Fotheringham, A. S. and Brunsdon, C. (1997) 'The Geography of Relationships: An Investigation of Spatial Non-Stationarity', in Bocquet-Appel, J.-P., Courgeau, D. and Pumain, D. (eds) *Spatial Analysis of Biodemographic Data* Montrouge: John Libbey Eurotext.

Clayton, D. and Kaldor, J. (1987) 'Empirical Bayes' estimates of age-standardized relative risks for use in disease mapping', *Biometrics*, 43, pp. 671–681.

Diaconis, P. and Efron, B. (1983) 'Computer intensive methods in statistics', *Scientific American*, 248, 5, pp. 116–130.

Ding, Y. and Fotheringham, A. S. (1992) 'The integration of spatial analysis and GIS', *Computers, Environment and Urban Systems*, 16, pp. 3–19.

Diplock, G. (1996) The Application of Evolutionary Computing Techniques to Spatial Interaction Modelling, Ph.D. Thesis, School of Geography, University of Leeds.

Doll, R. (1990) 'The epidemiology of childhood leukaemia', *Journal of the Royal Statistical Society*, 52, pp. 341–351.

Draper, N. and Smith, H. (1981) *Applied Regression Analysis*, New York: Wiley.

Efron, B. and Gong, G. (1983) 'A leisurely look at the bootstrap, the jacknife and cross-validation', *American Statistician*, 37, pp. 36–48.

Eldridge, J. D. and Jones, J. P. (1991) 'Warped space: a geography of distance-decay', *Professional Geographer*, 43, pp. 500–511.

Foster, S. A. and Gorr, W. L. (1986) 'An adaptive filter for estimating spatially varying parameters: application to modeling police hours spent in response to calls for service', *Management Science*, 32, pp. 878–889.

Fotheringham, A. S. (1992) 'Exploratory spatial data analysis and GIS', *Environment and Planning A*, 24, pp. 1675–1678.

Fotheringham, A. S. (1997) 'Trends in quantitative methods I: stressing the local', *Progress in Human Geography*, 21, pp. 88–96.

Fotheringham, A. S. and Charlton, M. E. (1994) 'GIS and exploratory spatial data analysis: an overview of some research issues', *Geographical Systems*, 1, pp. 315–327.

Fotheringham, A. S. and Pitts, T. C. (1995) 'Directional variation in distance-decay', *Environment and Planning A*, 27, pp. 715–729.

Fotheringham, A. S. and Rogerson, P. A. (1993) 'GIS and spatial analytical problems', *International Journal of Geographical Information Systems*, 7, pp. 3–19.

Fotheringham, A. S. and Wong, D. W.-S. (1991) 'The modifiable areal unit problem in multivariate statistical analysis', *Environment and Planning A*, 23, pp. 1025–1044.

Fotheringham, A. S. and Zhan, F. (1996) 'A comparison of three exploratory methods for cluster detection in point patterns', *Geographical Analysis*, 28, pp. 200–218.

Fotheringham, A. S., Curtis, A. and Densham, P. J. (1995) 'The zone definition problem and location-allocation modeling', *Geographical Analysis*, 27, pp. 60–77.

Fotheringham, A. S., Charlton, M. E. and Brunsdon, C. (1996) 'The geography of parameter space: an investigation into spatial non-stationarity', *International Journal of Geographic Information Systems*, 10, pp. 605–627.

Fotheringham, A. S., Charlton, M. E. and Brunsdon, C. (1997a) 'Measuring spatial variations in relationships with geographically weighted regression', Chapter 4 in Fischer, M. M. and Getis, A. (eds) *Recent Developments in Spatial Analysis, Spatial Statistics, Behavioral Modeling and Computational Intelligence*, London: Springer-Verlag, pp. 60–82.

Fotheringham, A. S., Charlton, M. E. and Brunsdon, C. (1997b) 'Two techniques for exploring non-stationarity in geographical data', *Geographical Systems*, 4, pp. 59–82.

Getis, A. and Ord, J. K. (1992) 'The analysis of spatial association by use of distance statistics', *Geographical Analysis*, 24, pp. 189–206.

Goldstein, H. (1987) *Multilevel Models in Educational and Social Research*, London: Oxford University Press.

Goodchild, M. F. (1987) 'A apatial analytical perspective on geographic information systems', *International Journal of Geographical Information Systems*, 1, pp. 327–334.

Gorr, W. L. and Olligshlaeger, A. M. (1994) 'Weighted spatial adaptive filtering: Monte Carlo studies and application to illicit drug market modeling', *Geographical Analysis*, 26, pp. 67–87.

Graham, E. (1997) 'Philosophies underlying human geography research', in Flowerdew, R. and Martin, D. (eds) *Methods in Human Geography: A Guide for Students Doing a Research Project*, Harlow: Addison Wesley Longman.

Graybill, F. A. and Iyer, H. K. (1994) *Regression Analysis: Concepts and Applications*, Belmont, CA: Duxbury.

Harper, E. A. and Manheim, M. L. (1990) 'Geographic information systems in transportation planning: a case for a geographic-based information services strategy', *Regional Development Dialogue*, 11, pp. 188–212.

Haslett, J., Bradley, R., Craig, P., Unwin, A. and Wills, C. (1991) 'Dynamic graphics for exploring spatial data with applications to locating global and local anomalies', *The American Statistician*, 45, pp. 234–242.

Jones, J. P. and Casetti, E. (1992) *Applications of the Expansion Method*, London: Routledge.

Marshall, R. J. (1991) 'A review of methods for the statistical analysis of spatial patterns of disease', *Journal of the Royal Statistical Society*, 154A, pp. 421–441.

Mooney C. Z. and Duval, R. D. (1993) *Bootstrapping: A Nonparameteric Approach to Statistical Inference*, Sage Series in Quantitative Applications in the Social Sciences, Newbury Park, CA: Sage.

Openshaw, S. (1983) 'From data crunching to model crunching: the dawn of a new era', *Environment and Planning A*, 15, pp. 1011–1013.

Openshaw, S. (1984) 'The modifiable areal unit problem', *CATMOG*, 38, Norwich: GeoAbstracts.

Openshaw, S. (1988) 'Building an automated modelling system to explore a universe of spatial interaction models', *Geographical Analysis*, 20, pp. 31–46.

Openshaw, S. (1990) 'Spatial analysis and GIS: a review of progress and possibilities', in *Geographical Information Systems for Urban and Regional Planning*,

Scholten, H. S. and Stillwell, J. C. H. (eds), Dordrecht: Kluwer Academic, pp. 53–163.

Openshaw, S. (1993) 'Exploratory space-time attribute pattern analysers', in *Spatial Analysis and GIS*, Fotheringham, A. S. and Rogerson P. A. (eds). Taylor & Francis: London.

Openshaw, S. and Rao, L. (1995) 'Algorithms for re-engineering 1991 census geography', *Environment and Planning A*, 27, pp. 425–446.

Openshaw, S., Charlton, M. E., Wymer, C. and Craft, A. W. (1987) 'A mark I geographical analysis machine for the automated analysis of point data sets', *International Journal of Geographical Information Systems*, 1, pp. 359–377.

Ord, J. K. and Getis, A. (1995) 'Local spatial autocorrelation statistics: distributional issues and an application', *Geographical Analysis*, 27, pp. 286–306.

Rhind, D. (1988) 'A GIS research agenda', *International Journal of Geographical Information Systems*, 2, pp. 23–28.

Xia, F. F. and Fotheringham, A. S. (1993) 'Exploratory spatial data analysis with GIS: the development of the ESDA module under Arc/Info', *GIS/LIS '93 Proceedings*, 2, pp. 801–810.

Chapter 3

Parallel processing in geography

Ian Turton

3.1 Introduction

This chapter is about parallel processing, or parallel computing; the terms are used synonymously. It will focus on ways to produce real applications not computer science abstractions. It will start out by describing what parallel computing is and why as a geographer you should even care. It will then give a brief historical overview of supercomputing and the rise of parallel computers. It will then attempt to set out what parallel computing is good for and what it is not good for and then finish up by showing you how you might get started with parallel computing.

3.2 What is parallel computing?

Parallel processing at its simplest is making use of more than one central processing unit at the same time to allow you to complete a long computational task more quickly. This should not be confused with so called multitasking where a single processor gives the appearance of working on more than one task by splitting its time between programs; if both the programs are computationally intensive then it will take more than twice the time for them to complete, nor are we concerned here with specialized processors (e.g. graphics controllers or disk managers) that work in parallel with a processor.

Some tasks are easy to parallelize. One example often used is building a brick wall. If it would take one person four days to build, it would probably take a well-organized team of four bricklayers one day to build it. However some tasks are clearly unsuited to parallelization; for example, if it takes a woman nine months to have a baby, it will still take nine women nine months! But if the task was to produce nine babies then getting nine women to help would provide a considerable speedup. Speedup is a term we will return to later which is often used to describe how well (or badly) a parallel program is working; it can be defined as the time taken to run the program on a single processor divided by the time taken to run it on a larger number of processors (N). The closer the speedup is to N the better the program is performing.

Parallel processing is often felt to be the preserve of large number crunching engineering disciplines; however geography has many large complex problems that require the use of the largest supercomputers available. It also has many smaller problems that can still benefit from parallelism on a smaller scale that is easily available to geographers.

3.3 Types of parallel computer

There are several different types of parallel computer, some of which are better than others for different types of task. Flynn (1972) proposed a classification of computers based on their use of instructions (the program) and their use of data (Table 3.1). He divided computers into four possible groups formed by the intersection of machines that used single steams of data and multiple streams of data, and machines that used single streams of instructions and multiple streams of instructions.

A SISD processor works one instruction at a time on a single piece of data; this is the classical or von Neuman processor. The operations are ordered in time and are easily traced and understood. Some would argue that the introduction of pipelining in modern processors introduces an element of temporal parallelism into the processor. This is, however, not true parallelism and can be ignored for our purposes as it is not completely within the control of the programmer.

A MISD processor would have to apply multiple instructions to a single piece of data at the same time. This is clearly of no use in the real world and therefore is seen by many to be a serious failing of Flynn's classification since it classifies non-existent processor types.

The last two classes of processor are of more interest to parallel programmers. Firstly, SIMD machines have a series of processors that operate in exact lockstep, each carrying out the same operation on a different piece of data at the same time. In the 1980s, Thinking Machines produced the massively parallel connection machine with many thousands of processors. These machines were useful for some limited types of problem but proved to be less than useful for many types of real problem and Thinking Machines went bankrupt and has ceased to make hardware.

Secondly, MIMD machines are more generally useful having many processors performing different instructions on different pieces of data at the same time. There is no need for each processor to be exactly in step with each

Table 3.1 Flynn's taxonomy of parallel computers

	Single instruction stream	Multiple instruction streams
Single stream of data	SISD	MISD
Multiple streams of data	SIMD	MIMD

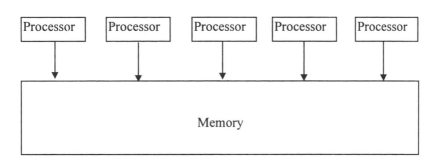

Figure 3.1 A shared memory machine.

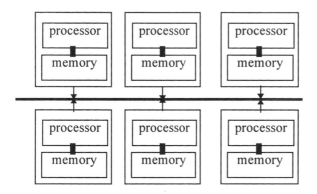

Figure 3.2 A distributed memory machine.

other processor or even to be carrying out a similar task. This allows the programmer much greater flexibility in programming the machine to carry out the task required as opposed to coercing the algorithm to fit the machine.

Parallel machines can also be divided into shared memory and distributed memory machines. In a shared memory machine, all processors have access to all of the memory in the machine (Figure 3.1). This can lead to problems; for instance, if all the processors want to write a value to the same variable at the same time, does the first to write succeed and the remainder fail or does the last to write succeed and the remainder fail? For different runs of the machine different processors can be involved, so it is exceedingly hard to debug a program with this problem if the compiler allows you to write code that contains this sort of contention. There are also problems in scaling shared memory machines with large numbers of processors since the amount of memory becomes unmanageable and there is inevitably congestion as many processors attempt to read and write to the memory.

Distributed memory machines are built from processing elements (PEs) which contain a processor and some memory (Figure 3.2). Each processor can only access the memory in its own PE. So if a data value is required by another processor an explicit message must be sent to request it and another to return the value required.

Modern machines like the Cray T3D are distributed memory machines that emulate shared memory (virtual shared memory) by hiding the explicit message passing between processors from the user. This can be very successful if handled carefully but can also lead to very poor program performance if handled naively.

3.4 Historical perspective

Supercomputer development has always been driven by the concerns of the military, but the need to count and tabulate people has also been a driving force. The US Census Bureau is widely believed to have been the first user of mechanical computers to produce censuses in the 1800s and were the first buyer of a UNIVAC (UNIVersal Automatic Computer) in 1951, which was the first commercially available electronic computer. However the continuing growth of supercomputing has been shaped by the US military for nuclear bomb design and simulation and the NSA (National Security Agency) for encryption and decryption. This has led to the design of ever larger and faster computers. Initially it was possible to make a computer go faster by building better and smaller chips; for instance, the improvements from the Cray-1 to the Cray X-MP came mostly from better chip technology. Unfortunately there are several physical barriers to continuing along this path. Electrons are limited to the speed of light so a smaller chip runs faster because the electrons have less distance to travel. Unfortunately there is a limit to how fine a mask can be used in the manufacture of chips due to diffraction effects. Also at very small length scales it becomes impossible to contain the electrons in the 'wires' due to quantum tunnelling effects. Barring any significant breakthroughs in quantum or DNA computing, parallel computing is the only way to continue to push machine performance upwards.

With the end of the Cold War, the supercomputer market is no longer large enough or rich enough to support the design of new custom chips and the remaining manufacturers in the market are now moving to the use of highly parallel arrays of cheap consumer chips with fast interconnections between them. These chips are the same ones as are used in workstations and personal computers. Likewise, physical limitations will eventually prevent chip designers from making these chips any faster. The speed limit will probably be reached in the early years of the 21st century and we will then see the descent of parallel computing from the elevated heights of supercomputing to the desks of users.

3.5 Supercomputing in geography: grand challenges

3.5.1 What is a grand challenge?

In 1987 and subsequent years the US Office of Science and Technology (OST) has set out a series of grand challenges, which are defined as:

> 'a fundamental problem in science or engineering, with broad applications, whose solution would be enabled by the application of high performance computing resources that could become available in the near future'.

(Office of Science and Technology, 1987)

OST then went on to define several areas of science and engineering that they felt to be grand challenges:

1. computational fluid dynamics;
2. electrical structure calculations for the design of new materials;
3. plasma dynamics;
4. calculations to understand the fundamental nature of matter (including quantum chromodynamics (QCD) and condensed matter theory);
5. symbolic computation (including speech recognition, computer vision, tools for the design and simulation of complex systems).

This list can be summarized in a graph showing memory size and computational speeds required for each challenge (Figure 3.3).

3.5.2 Should geography and the social sciences have any?

Parallel computing does not have a long history in geography or the social sciences. Fogarty (1994) pointed out that this list of grand challenges lacked any reference to the social sciences and argued for the inclusion of more research on the use of supercomputing in GIS. This continued the trend of the preceding years where the only interest geographers had shown in parallel computing was as a means of making GIS go faster (Faust *et al.*, 1991; Kriegel *et al.*, 1991; Costanza and Maxwell, 1991). In this view Fogarty was mistaken as to what a grand challenge was; while building a faster GIS is important as a tool for geography it lacks the novelty and size of the challenges laid out above. It was left to Openshaw (1995) to successfully propose human systems modelling as a new grand challenge area to place geography and the social sciences firmly in the supercomputing arena. He argues that knowledge and the ability to model human systems, such as cities, is of vital importance since the majority of the world's population live in cities. Human influences on the climate are also poorly understood or modelled and that

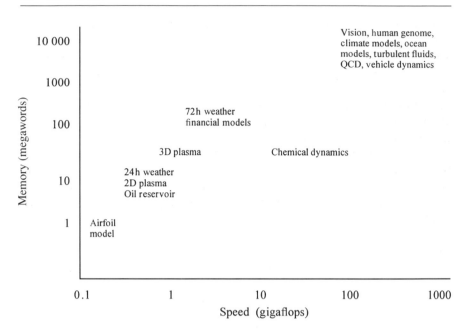

Figure 3.3 Some grand challenge problems with speed and memory requirements.

with the size and speed of the new supercomputers becoming available in the mid-1990s, combined with the ever larger amounts of data being produced by the spatial data revolution, it is becoming possible for geography and the social sciences to start to model these phenomena. It can be argued that human systems modelling is covered in Item 5 of the list above as a simulation of a complex system and does not so much represent a new grand challenge but the fleshing out and focusing of an existing one.

Progress continues to be made in the field of GIS but this is largely in experimental systems with few if any of the vendors of GIS systems showing any interest in developing parallel systems commercially. The only exception to this are the developments in parallel databases that are being developed for the business market but do have obvious benefits to the GIS community. In general, it is the database operations that make a GIS slow; with the development of graphics co-processors, visualizing even very large datasets is no longer a problem. However, carrying out large and complex queries of the database can still be very slow and is inherently parallel.

3.5.3 Should we worry about it anyway?

Openshaw (1995) may be mistaken in his call for an immediate jump into massively parallel processing. Geography and the social sciences may benefit from a slower but steadier move to parallel processing which starts out using small machines to produce easily achievable successes in the early stages. In part this is due to the lack of historical involvement in the computational environment by the social sciences. This leads more established sciences to consider the encroachment of geography as a threat to their hard won rights to supercomputing time, which is a scarce resource so every new user community must be prepared to justify preventing an established user from making use of the machines.

So it is probably best for geography to consider small and achievable problems to start out. This allows the necessary training and development work to take place in an easier environment and allows geographers to develop a track record to carry into the supercomputing arena. Obviously parallel programmers should keep in the back of their minds the possibility that one day their code will be run on a MPP system and avoid making any assumptions about the underlying number of processors and the specific hardware they are using. But it is just common sense for any programmer to think first and last of portability.

There may well be real advantages to building small-scale systems of a few processors which are loosely coupled to existing GIS, allowing enhanced modelling capabilities to be added. An example is ZDES (Openshaw and Rao, 1995; Openshaw and Schimdt, 1996). This system allows users to design new zoning systems from smaller regions and is highly computationally intensive. By breaking the system out of the GIS framework into a loosely coupled system, they allow the use of available parallel resources to speed up the problem. This, however, should not be considered to be a parallel GIS since Arc/Info remains firmly tied to a single processor.

3.6 When not to use parallel processing

It may seem odd in a chapter promoting the use of parallel computers to include a section about when not to use them. However it will save a lot of time if, before rushing headlong into parallel programming, you stop and think if your problem really requires you to put so much effort into it. Some questions to ask are: How often is this code to be run in production? How long does it take to run at the moment? If the answers to these questions are not often and not too long then don't bother to think parallel! It will take at least twice as long and probably a lot longer to write, test and debug the parallel version of your code than it would to write (or optimize) some serial code. If you only need to run the model once or twice then even runtimes of a week or two are still quicker than building a parallel version of your code. However you will also need to consider the memory requirements of your problem; if they are many times the real memory available on your serial machine then parallel may well be the right route. If you have a program that takes a month or more to run then parallelism may be right and if you need to run a program or model many thousands or millions of times then again parallel is the correct path to consider.

It is also necessary to consider the size of the task that you want to parallelize. If the problem is a small kernel which can not be easily sub-divided and it depends on the previous result then it is unlikely to parallelize, since you can not split the job between processors and you can not start computing the next part of the problem until the current step is complete. Problems of this type are, however, rare in geography.

3.7 When to use parallel processing

First re-read the section above, which will attempt to talk you out of parallelization. If you can pass all the tests above or are persistent in the face of them then now is the time to think parallel. First you need to decide what sort of parallelism you want to use, and then consider the language you want to use. Some of these choices will be dictated by what sort of parallel machine you have access to and which parallel languages it supports.

3.8 How to use parallel programming

If you are certain that you need to go parallel then the next question is: Do you have access to a parallel computer? Even if at first glance the answer to this seems to be no, do not be disheartened as it is possible with some extra software to turn a room (or building) full of workstations into a virtual parallel computer. There are free versions of MPI available that provide exactly the same functionality on a network of workstations as MPI found on large supercomputers. Obviously this is not as fast as a dedicated parallel machine but it is a lot cheaper. This approach can also be used for development work on parallel software since you will only be allocated a limited amount of time on a large machine for your work, which can often be used up very quickly during development.

Parallel problems can be broken down into several types. The first group is known as trivially parallel or embarrassingly parallel, and as the name suggests they are the easiest to handle. The example of producing nine babies given above is an example from the real world. There is no need for the women involved to be in constant communication with each other or even to be in the same place. For certain types of computing problem, this is also the case; for instance, if you want to carry out an error simulation on a model that involves running the model one hundred times and comparing the spread of the results from slightly different starting points then the easiest way to make this run faster is to run each model on a separate processor and collate the results at the end of the runs. This kind of problem can almost always produce speedups of near N (where N is the number of processors) without much work on the part of the programmer. Harder forms of parallelism are where there is insufficient room in the memory of each processor to store the problem or where the model is only to be run once or a few times. Examples of this are often found in physical geography, such as climate models which cover the whole of the earth, or in human geography with large spatial interaction models where the cost matrix and the trip matrix must be stored. In this case it becomes necessary to divide the problem up between different processors. This nearly always involves the program in communicating between the processors which takes up time when the processor could be performing 'real' work. In problems of this type it is quite possible to achieve speedups of significantly less than N. If the number of processors becomes too large compared to the size of the program task then communication completely swamps the problem run time. For instance, in the example of the brick wall mentioned earlier, if a team of 1000 bricklayers was put to work then it would be very unlikely that any bricks would have been laid at the end of the first day since the foreman would still be trying to sort out the teams.

How to divide up a problem is also an important part of the parallelization task. In a simple model, it may be possible to simply divide the problem up evenly between the processors. This is often the approach carried out in computational physics and chemistry where problem shapes are often rect-angular and evenly spaced. Unfortunately, in geography this is rarely the case. The spatial interaction problem can be divided up in this way since journeys to work are usually local and it is possible to split the rows of the matrix in such a way as to spread out the major conurbations. However if you wished to carry out a pattern recognition task on a population raster map of Great Britain then one approach might be to divide the map up into as many rectangles as there are processors and assign one rectangle to each processor (Figure 3.4(a)). However the processor that was allocated to the top right-hand rectangle would finish almost immediately since it contains mostly sea, whereas the one that was allocated the bottom right rectangle would take a long time since it had the whole of London to deal with. In this case, it would be better to divide the map up into smaller regions (Figure 3.4(b)) and to allocate several regions to each processor, or to hand out the next region to each processor as it finished. Care has to be taken not to divide the problem into regions too small for the processors to work on efficiently or more time will be spent allocating areas than will be spent working on them.

3.9 Different ways of parallel programming

There are two very different views of parallel computing in the world. There are computer scientists who spend their time thinking about parallelism and designing new and better languages to allow the natural expression of parallel constructs and who feel that parallel programs are exciting because they are

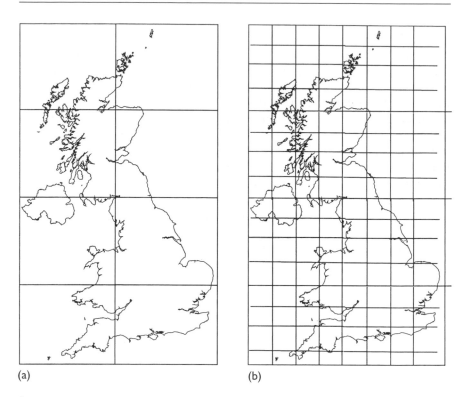

(a)   (b)

Figure 3.4 Possible ways of dividing up a raster population map of the UK.

parallel. There are also application scientists who need their programs to run faster and will use a parallel machine only if it can provide this. These scientists tend to have existing code or a tight deadline in which to get new code working; therefore they usually program in languages such as Fortran and C which they already know and avoid worrying about the niceties of the underlying hardware. In fact, they tend to believe in portability; since this means not having to rewrite all of your programs when the next machine is purchased. In general, if you want to make use of parallel processing in geography (or any other field) it is necessary to write your own software. This is not as daunting as it sounds. In many cases, a serial program that already exists can be adapted to make use of many processors. In some cases, this can be a distinct advantage as it allows you to test the new parallel version by comparing it to the results of the serial version.

There are two main ways of programming parallel computers which both have their supporters and detractors. The first method, High Performance Fortran (HPF), was developed on shared memory machines but is slowly moving to distributed memory machines. In this method, HPF allows the programmer to define the parallelism of their code in a very high-level manner using compiler directives (small messages to the compiler that look like comments to compilers that do not understand parallelism). These directives allow the programmer to define how data is to be shared between the different processors and also to describe which loops within the program can be carried out safely in parallel. The compiler is then responsible for converting these directives as best it can into a parallel executable that can be run on a parallel computer. Unfortunately, at the time of writing, there is still some way to go before code written specifically for one HPF compiler can be compiled using another as many vendors have yet to implement the full standard but many have added extensions that they hope to get the standards committee to accept in the future.

Supporters of this method will say that since HPF requires few changes from the serial code, it allows you to get a good start in parallel programming. It also allows the programmer to produce very poor programs. This is because HPF hides the issues of data distribution from the programmer. This can lead to communication between the processors dominating the performance of the program. With care it is possible to produce efficient code using HPF but this requires rethinking the existing serial code thus undermining some of the benefits claimed for HPF.

The second main method of producing a parallel program is to use message passing. In this case, the programmer uses a standard FORTRAN (or C) compiler and places explicit subroutine calls to pass messages from one processor to another which either contain data or instruct a processor what to do next. The message passing implementation (MPI, 1994) is a standard backed by a majority of parallel computer vendors and by many software developers. This approach seems like more work but tends to produce better code since the programmer is in control of the process and is aware of which pieces of data are being passed from processor to processor. At its simplest level, a programmer has two commands to use: send and receive, used as a pair with one processor instructed to start sending to another which in turn initiates a receive. All other parts of the standard are essentially built from this combination of send/receive events. There are also some utility commands which allow the programmer to discover which processor is which and to group processors. This is the method that we will concentrate on for the remainder of this chapter.

3.10 Message passing

If you have a simple parallelization task, such as error simulation in a model, then the task is straightforward. First you develop a serial version of your model that is structured in such a way that there is, if possible, a single routine that assigns the input variables and a single routine that outputs the result(s). This program should be tested thoroughly on known data and on any special cases that might apply, for example zeros in the input and very

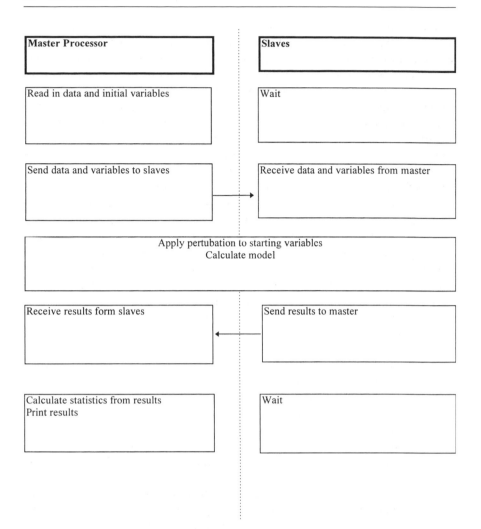

Figure 3.5 How to parallelize a simple model.

large numbers. You then need to write a parallel wrapper for the model; this should read in the initial values on one processor and then pass them out to the other processors. Each processor should then apply a random error of some predetermined size and compute the model. In the results routine, you now need to collect all the results on to one processor and calculate some statistics. If necessary this process can be repeated until sufficient model runs have been completed. Finally the master processor should write out the results. This is summarized in Figure 3.5. This is probably as easy as parallel programming gets. The only possible problem is in the way that you send and receive the required information between the processors. Each process is running its own copy of the same program. However, a common mistake is to think that because a variable is set on one processor that it must also be set on another. Each processor has its own separate version of the variable space, so you must pass a copy of the variable from processor to processor. This is what allows you to be running the same model on each processor without them interfering. So once the master processor (this is the processor numbered 0, since this will always exist no matter how many processors you are using) has read in some data, only this processor 'knows' what these values are, the next step is to use a broadcast to send the data to all the other processors. The model can then be run and at the end each processor has a different result, since they had different starting values. You can either loop through all available processors on the master, requesting the data values for the other processors, or make use of a higher level routine called a reduce operator which may be more efficient in collecting the results into an array on the master processor. Once this is done, the master can calculate any statistics necessary and either start the process again in which case you will need to make sure your model keeps copies of any variables that are changed during a run or print out the results and exit. For an example of this method in use, see Turton and Openshaw (1999) who use this technique in a location optimization problem.

The communication overheads of this type of program are very low (assuming the model is fairly substantial) with only the initial data and the final results being passed between processors. This allows the problem to scale well as more processors are added to solve the problem. There is, of course, the need to increase the number of models to be run so that it is always at least as large as the number of processors and for best results several times larger than that, but still an exact integer multiple of, the number of processors. If the number of models does not divide evenly by the number of processors the final trip through the model loop will only be using some of the processors. Obviously if this involves 3 processors out of 5 for 10 000 parallel model runs it can be ignored, but if it is 2 out of 2000 processors for 2 parallel model runs then it is a problem.

A more complex problem is when you do not know or cannot predict how long an individual task will take to complete. For instance, if you are interested in optimizing some parameters for a series of models, it is sometimes difficult to predict how long each individual model will take to run. So if you followed the strategy outlined above there could be times when all but one of your processors are waiting for the last one to finish and this could be a significant amount of time if one of the models happened to take twice as long as the remainder. In this case, it is usually best to set up a list of jobs to be completed on the master processor and then to have it loop around this list handing out the next job to each slave processor as it finishes. This approach is summarized in Figure 3.6. Here it is necessary to remove the master from the processing section of the code to allow it to be responsive

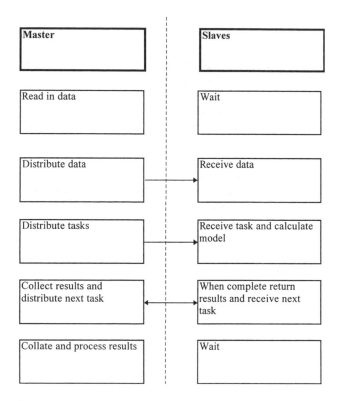

Figure 3.6 Task farming.

to the needs of the slaves. The master spends most of its time waiting for a result to be sent from a slave. When it receives one, it should immediately send the slave a new task and then store the result somewhere for later processing. It should then return to waiting for the next message from a slave. Great care must be taken in this approach to make sure that the master knows which slave is working on which problem so that it can match the result sent back to the task allocated. It is also important that the master sends out a close down signal to each slave in turn once there are no further tasks since the program cannot finish until all processors call the finalize routine. An example of this method is presented in Turton *et al.* (1996) who present a parallel genetic programming system that is used to 'breed' better spatial interaction models.

Again, in this method, there is often little communication between the processors; however you must decide whether you can afford the loss of one processor from the calculation to a supervisory mode. If you have 512 or 1024 processors then this is clearly not a problem. However if you only have 4 processors available then the gain in load balancing may be outweighed by the loss of one processor's power. In this case, you would be wise to consider a modified version where the master executes a model as well as controlling the allocation of tasks. This may lead to some of the slaves waiting longer than before but it will provide more computing power to the task in hand.

The most complex type of parallelization task is where a single model is to be run in parallel either because of excessive run times or memory size. This case cannot be handled in either of the two ways discussed above. The choice really depends upon the complexity of the model to be evaluated. In each case, instead of passing whole models out to the slaves, the master passes out part of the data to each processor. The slaves then compute their part of the model on the data provided. There is usually a need to stop and synchronize the processors while intermediate results are calculated or passed between processors. For example, in a spatial interaction model, it is necessary for each slave to calculate the A_i value based on its share of the rows and then combine all of these values to produce a global A_i value which must then be passed out to the other processors. This is demonstrated by Turton and Openshaw (1996) for a large (10 764 zones) spatial interaction model.

In this case, there can be very large communication overheads, especially if many stops are required in order to pass variables around. It becomes important to plan out exactly what data are required for each part of the computation and where they will be required. Some variables that are required by all the processors should be replicated across all of the processors so that every part of the program can access them quickly. If possible when global communication is unavoidable you should attempt to store up the variables to be passed and then send them all in one long message. This reduces some of the overheads involved in setting up the messages.

This type of parallel program is the hardest to write, since you need to have a very clear view of which variables are needed where, and which are local and which are global in your head as you write the code. However, it also leads to better structured and often faster code than can be produced using High Performance Fortran. In general, this is not a case where having existing serial code helps, except as a test to check that the parallel version gives the same answer as the serial code on some small test problem.

It is also important that you consider what sort of parallelism is important to your problem and decide if this is appropriate to the machine that you are using. For instance, if you have a problem that requires a lot of communication and a network of workstations on a fairly slow ethernet then you are unlikely to gain much benefit from a parallel program. This would be a good way to develop the code for later use on a purpose-built parallel computer. However, if this was to be the production machine, it would be better to attempt to rethink the parallelism to avoid as much communication as possible or to restructure the algorithm to use asynchronous communications so that a processor can start a message and then continue computing before completing the message sometime later.

The secret, if there is one, in efficient parallel programming is to maximize the parallel sections of the task whilst minimizing the amounts of inter-processor communications. How this is achieved in practice is entirely dependent on the application, your ability to re-think the task in a parallel way, and often a degree of fortune. It can be a major challenge but it is often also loads of intellectual fun and excitement on the world's fastest and biggest computing hardware.

3.11 Conclusions

Parallel computing can be expected to be the future for high and medium performance computing, barring some unforeseen physical breakthrough in uniprocessor design. This is simply due to the fact that processor chips cannot get too small without quantum effects becoming a problem. It must also be recognized that automatic parallelization is a long way into the future. So there is a need for geographers and social scientists to learn to program the present generation of parallel machines, even if you suspect that next year's machines will not look like today's machines. It does not seem likely that there will be any significant changes in the type of parallel machines being produced in the near term. However, many people said that in the 1980s when the trend was to ever larger massively parallel machines. They were nearly all proved wrong!

There are benefits to learning parallel programming as an abstract concept without becoming too tied to one particular architecture as the company that makes the machine you use may go bankrupt and you will be left to move to the next machine unaided. So it is best to use an abstracted model of programming, such as message passing or shared data with the parallelism embedded in localized subroutines, to make it easy to change the particular library calls without needing to change whole sections of the code if necessary. The future of high performance computing is parallel and you can start writing parallel code secure in the belief that in 10 years time it will probably still run on whatever hardware exists at that time.

Further reading

Electronic sources
The usenet news groups comp.parallel and comp.parallel.mpi are both useful. New users should especially read the frequently asked questions list of comp.parallel. The world wide web also provides many useful resources on parallel programming. The following is not an exhaustive list but should provide pointers to some useful sites.

http://www.epcc.ed.ac.uk: The Edinburgh Parallel Computer Centre has documents about HPF and MPI as well as a freely available implementation of MPI. There are also links to other useful parallel sites.

http://www.mcs.anl.gov/mpi: The Argone National Laboratory in the United States also has a freely available MPI implementation and other useful parallel computing documents.

http://www.hensa.ac.uk/parallel/: is an archive site with links to a large collection of documents related to parallel computing.

Printed sources
This is not an exhaustive list but merely indicates some of the range of parallel programming books now available.

Practical Parallel Computing, H. S. Morse, Academic Press, London, 1994.
Introduction to Parallel Computing, T. Lewis and H. El-Rewini, Prentice Hall, New Jersey, 1992.
Using MPI, W. Gropp, E. Lusk and A. Skjellum, MIT Press, Cambridge, MA, 1994.
Practical Parallel Processing, A. Chalmers and J. Tidmus, Intl. Thompson Computer Press, London, 1996.
Parallel Programming with MPI, P. Pacheco, Morgan Kaufmann Publishing, San Francisco, 1997.
An Introduction to Parallel Computing, K. Chandy and S. Taylor, Jones and Bartlett, Boston, MA, 1992.

References

Costanza, R. and Maxwell, T. (1991) 'Spatial ecosystem modelling using parallel processors', *Ecological Modelling*, 58, pp. 159–183.
Faust, N. L., Anderson, W. H. and Star, J. L. (1991) 'Geographic information systems and remote-sensing future computing environment', *Photogrammetric Engineering and Remote Sensing*, 57,6, pp. 655–668.
Flynn, M. J. (1972) 'Some computer organisations and their effectiveness', *IEEE Trans. Computers*, 21,9, pp. 948–960.
Fogarty, B. W. (1994) 'Grand Challenges for GIS and what's really relevant to social-science', *Social Science Review*, 12,2, pp. 193–201.
Kriegel, H. P., Brinkhoff, T. and Schneider, R. (1991) 'The combination of spatial access methods and computational geometry in geographic database-systems', *Lecture Notes in Computer Science*, Vol. 525, pp. 5–21.
MPI Forum (1994) *MPI: A Message-passing Interface Standard*, Knoxville: University of Tennessee.
Office of Science and Technology (1987) *The Federal High Performance Computing Program*, Washington DC: Office of Science and Technology.
Openshaw, S. (1995) 'Human systems modelling as a new grand challenge area in science, what has happened to the science in social science?', *Environment and Planning A*, 27, pp. 159–164.
Openshaw, S. and Rao, L. (1995) 'Algorithms for re-engineering 1991 census geography', *Environment and Planning A*, 27, pp. 425–446.
Openshaw, S. and Schimdt, J. (1996) 'Parallel simulated annealing and genetic algorithms for re-engineering zoning systems', *Geographical Analysis*, 3, pp. 201–220.

Turton, I. and Openshaw, S. (1996) 'Modelling and optimising flows using parallel spatial interaction models', in Bougé, L., Fraigniaud, P., Mignotte, A. and Roberts, Y. (eds) *Euro-Par '96 Parallel Processing*, Vol. 2, Lecture Notes in Computer Science 1124, Berlin: Springer, pp. 270–275.

Turton, I. and Openshaw, S. (1999) *High Performance Computing in geography and GIS*, London: Routledge.

Turton, I., Openshaw, S. and Diplock, G. (1996) 'Some geographical applications of genetic programming on the Cray T3D supercomputer', in Jesshope, C. and Shafarenko, A. (eds) *UK Parallel '96: Proceedings of the British Computer Society Parallel Processing Specialist Group Annual Conference*, Berlin: Springer-Verlag, pp. 135–150.

Chapter 4

Evaluating high performance computer systems from a GeoComputation perspective

Joanna Schmidt and Stan Openshaw

4.1 Introduction

The mid-to-late 1990s have witnessed a dramatic speed-up in the performance of parallel supercomputers. The 1990s started with high performance computing (HPC) based mainly on vector supercomputing hardware offering about one gigaflop of arithmetic computing power (about 10 times more than a decade earlier). Parallel computers existed but were based on many different architectures. They were often not a practical proposition from a reliability perspective; the software was lagging far behind the hardware and most of the programs written in that period will now be either defunct (like the hardware) or else have been rewritten in one of the newly-emerged standard parallel programming languages. Yet by the end of the 1990s it is quite clear that the parallel supercomputing era is here. The nineties decade is ending with the rapid disappearance of HPC based on vector processors and the emergence of mature, reliable, affordable and highly efficient parallel supercomputers offering between 150 and 1000 times more computational power. The early years of the new millennium are confidently expected to witness the appearance of multi-teraflop machines that will slowly percolate into the scientific infrastructures accessible to GeoComputation. For those applications that require unbelievably vast amounts of computational power then HPC hardware offers a way forward that is now within the reach of the skill base of many geographers and social scientists. HPC is thus a key tool for those researchers who need to perform large-scale analysis of big datasets, to create numerical experiments of complex processes, or to construct computer models of many physical and human systems. For example, in order to model the largest public domain 'flow' dataset that is available in the UK, the journey to work flows from the 1991 census for all wards in Britain, there is a requirement to process a table comprising 10 764 columns by 10 764 rows. The biggest flow dataset (i.e. telephone flows between houses) that may well exist could well have between 1.7 and 28 million rows and columns. These data can be disaggregated in many different ways (e.g. by time or age or sex or purpose), creating dataset sizes of gigabytes maybe even terabytes. This complements the flood of geo-information being generated by remote sensing devices. It is quite clear that data mining and modelling of these, and other even larger datasets, can (or could) now be undertaken, and the results analysed and mapped by applying HPC technologies. HPC, however, also has the potential to open up entirely new areas for research.

Many GC problems are naturally data intensive and place heavy demands on storage resources as they require large amounts of memory and disk space. Ideally, the data should be stored in computer memory removing the need for large numbers of input and output operations. Given the increased processing capabilities and memory resources of contemporary parallel systems, the modelling and analysis of virtually all of our very large datasets is now within reach. The principal remaining difficulties are largely self-imposed. The low level of usage of HPC by the non-traditional user communities in areas such as GC creates a self-reinforcing pattern of little usage. There is a widespread lack of awareness of what HPC can deliver in terms of computational speed, and a failure to appreciate that the top-end machines really are offering or promising much more computing power and memory than most PC based systems. If GC is really a geo-version of other computational sciences then it is utterly inconceivable that they too will not soon develop into major users of future HPC systems. If the quality of the science you are using to solve your GC problems is dependent on the speed of the best available hardware or its memory size then sooner or later you will be wanting to move into the HPC world. Those cellular automata models that run on a PC with large cell numbers will also run on HPC machines. The difference being you now have available 1000 to perhaps 30 000 times more computing power and the results should be improved. The problem that GC faces in starting to develop its HPC usage is that of creating its own vibrant HPC user community, training its researchers in the basic skills of parallel programming, and trying to acquire the research resources needed to sustain major grand challenge problems in this area. Chapters 1 and 16 have many suggestions that may help identify some of the key researchable topics for the near future.

The objective of this chapter is to try and explain how to measure the performance of HPC hardware on a type of problem that is clearly relevant to GC. Seemingly most of the current researchers involved in GC do not use HPC much or at all and appear to have no good understanding of what is involved or what is on offer. Openshaw and Turton (1999) provide an introduction to HPC programming and try to demonstrate that it is more unfamiliar than difficult. However, this still leaves unresolved the question as to why you should bother. Are these machines really that much faster than a PC when running a practical application related to a GC topic compared to solving some problem in theoretical physics? Everyone knows that comparing processor clock speeds provides little guidance to useful computing power and that quoting MIPS (million instructions per second, often rephrased as meaningless instructions per second) is unhelpful as a guide to what HPC can deliver. So if you were interested in measuring the speed of your PC or UNIX workstation or file server or this or that chunk of HPC metal, how would you set about it?

The usual solution is to select a representative application that is relevant to GC, write it in a portable way, and then run it on different hardware making a note of how long each run takes. These timing results provide a basis for comparing different machines, assessing the effects of different compilers or languages on the same machine, identifying the performance and efficiency of different architectures, and for researching scalability issues. The most difficult design question concerns the choice of representative problem and then coding it in such a way that it will run on virtually any and every computer that ever existed or may soon exist. Yet the availability of a portable and representative benchmark code is a crucial requirement for raising awareness of the capabilities of current HPC systems.

To date many vendors and NASA researchers have reported their results in terms of the so-called NAS parallel benchmark (Hwang and Xu, 1998). But this is representative only of certain types of aerophysics applications. Likewise, the GENESIS benchmark for distributed-memory machines re-lates mostly to problems from physics and theoretical chemistry (Hey, 1991). The SPEChpc96 suite (SPEC, 1997), which incorporates two application areas, the seismic industry computational work and computational chemis-try, has not yet been widely adopted by the industry but is equally irrelevant to GC. Perhaps the most commonly used is Dongarra's Linpack suite based on solving dense systems of linear equations (see Dongarra, 1995). None of these benchmarks are particularly relevant to GC (see Brocklehurst, 1991; Lewis et al., 1992). It is for this reason that Openshaw and Schmidt (1997) developed their social science benchmark (SSB/1) code which is in fact based on a classical GC application. The benchmark code described in this chap-ter can be run on a wide diversity of target architectures ranging from desktop computers to the most powerful systems available anywhere in the world. It is scalable and offers portability across a large number of comput-ers ranging from a network of workstations to massively parallel systems. The benchmark provides a means to measure compute speeds in terms of an understandable and problem-specific measurement-metric which may well be helpful to researchers wondering what HPC is capable of doing. The bench-mark results may also be useful to other users who have computational projects with similar types of operations. There are many potential parallel applications that are memory intensive rather than pure number crunchers; thus the benchmark described here adds to a pool of tools for evaluating parallel hardware from the perspective of this type of application. Finally, a GC benchmark code will help attract the attention of others to this area of application. Vendors will probably use it to measure the speed of their newest machines, and it is likely to be far more meaningful than many of the benchmarks that currently exist.

4.2 Spatial interaction models

4.2.1 Brief description

The SSB/1 benchmark code is based on two types of spatial interaction model (SIM). The models were defined a long time ago but are still of considerable value; see Wilson (1970, 1981), Birkin *et al.* (1996). They are used to simulate the flow of people, goods or information between geographical areas that act as origin and destination zones. The computational complexity of these models is related to their levels of disaggregation and the number of zones (origins and destinations) used. Note that the quality of the science is also related to both factors; indeed, as computers become faster both should be able to increase (Turton and Openshaw, 1996). Spatial interaction models are also used within certain types of spatial optimization problem where these items are embedded in a location optimization framework and have been applied to retail site evaluation, to transportation planning and to healthcare planning (Birkin *et al.*, 1996). The two different types of SIMs used here are an origin constrained model and a doubly constrained model.

4.2.2 An origin constrained model

The equations for an origin (singly) constrained (SC) model are given below:

$$T_{ij} = D_j O_i A_i f(C_{ij}) \qquad i = 1, \ldots, N \qquad j = 1, \ldots, M \tag{4.1}$$

with:

$$A_i = 1 \Big/ \sum_{j=1}^{M} D_j f(C_{ij}) \qquad i = 1, \ldots, N \tag{4.2}$$

to ensure that:

$$\sum_{j=1}^{M} T_{ij} = O_i \qquad i = 1, \ldots, N \tag{4.3}$$

where T_{ij} is the number of trips (flows) between zone i and zone j, O_i is the number of trips starting in zone (origin) i, D_j is the number of trips ending in zone (destination) j, C_{ij} is the distance between origin i and destination j, N is the number of origin zones, and M is the number of destination zones.

The $f(C_{ij})$ function represents the trip-deterring effect of travel cost. The modelling process requires the computation of all elements of $\{T_{ij}^{predicted}\}$. Typically an error measure such as the sum of errors squared:

$$F = \sum_{i=1}^{N} \sum_{j=1}^{M} (T_{ij}^{predicted} - T_{ij}^{observed})^2 \tag{4.4}$$

would be used to assess the model's goodness of fit.

4.2.3 A doubly constrained model

The equations for a doubly constrained (DC) model, are more complex than those for the singly constrained model, and are given below:

$$T_{ij} = O_i D_j A_i B_j f(C_{ij}) \qquad i = 1, \ldots, N \qquad j = 1, \ldots, M \qquad (4.5)$$

where:

$$A_i = 1 \bigg/ \sum_{j=1}^{M} D_j B_j f(C_{ij}) \qquad i = 1, \ldots, N \qquad (4.6)$$

$$B_j = 1 \bigg/ \sum_{i=1}^{N} O_i A_i f(C_{ij}) \qquad j = 1, \ldots, M \qquad (4.7)$$

This model is constrained at both ends: viz.

$$\sum_{j=1}^{M} T_{ij} = O_i \qquad i = 1, \ldots, N \qquad (4.8)$$

and

$$\sum_{i=1}^{N} T_{ij} = D_j \qquad j = 1, \ldots, M \qquad (4.9)$$

where T_{ij}, O_i, D_j, C_{ij}, N, M and the $f(C_{ij})$ function have the same meaning as described for the SC model.

4.2.4 Computations in SIMs

These models are useful for benchmarking purposes because the GC form of HPC tends to be much more data intensive than that found in many other areas of science. In a large number of geographical applications, there is typically little computation being executed in comparison to the number of memory references performed. Spatial interaction models are good examples of this class of problem that makes heavy demands on memory resources. Their performance will in part measure a machine's ability to execute a small ratio of arithmetic operations to memory references. Furthermore, the SIM benchmark also has a reasonably high level of non-floating point integer arithmetic, as it uses sparse matrix methods to store the observed flow data for T_{ij}. The need to perform large amounts of integer arithmetic is often overlooked in more conventional HPC benchmarks and in hardware designed solely to optimize pure number crunching throughput on data stored in a highly optimized cache with a large amount of arithmetic being performed in proportion to each memory access. In other words, benchmark codes based on, for example, physics applications or matrix operations are not particularly appropriate for far less computationally intensive applications that characterize many areas of GC, e.g. those naturally applying neural networks, genetic algorithms, or certain geographic models. High performance computer hardware that can do billions of dot products per second may not do nearly so well on running SIMs.

Finally, it is noted that in both models the quality (i.e. spatial resolution) of the results depends on both the number of origins (N) and the number of destination-ends (M). Increasing the values of N and M for a given geographical area leads to more realistic modelling at a finer spatial grain resulting in arguably better science. However, this increases the demands placed on memory, as there is a need to store large arrays. Models with small N and M values (i.e. 1000 or less) can be run on a PC. Large values (i.e. 25 000) need a parallel system with many processing nodes (e.g. Cray T3D), and the bottom end of the maximum possible values (i.e. 1.6 million) may need the next generation or two of highly parallel teraflop computers before they can be processed.

So the benchmark is built using the two types of spatial interaction model expressed by equations (4.1)–(4.9). For both models it computes the elements of the $\{T_{ij}^{predicted}\}$ matrix and a set of error measures. The distance deterrence function is defined in the benchmark as:

$$f(C_{ij}) = \exp\,(-\beta C_{ij})C_{ij}^{\alpha}$$

where α and β are parameters fixed at some non-trivial values. The use of this function is deliberate as it involves both *exp* and *log* function calls from a mathematical library. These functions are not always optimized by vendors for maximum performance nor are they often included in megaflop counting, but they are very typical of GC work involving mathematical and statistical models. Finally, the benchmark stores the correct values for equation (4.4), and uses them to verify results for different hardware that may well use different arithmetic models.

The benchmark includes its own data generator. This creates artificial spatial interaction data but it is realistic in terms of both the sparsity pattern and general structure of the flows. This is useful because it makes the benchmarking software self-contained, it permits the generation of virtually any size of dataset, it avoids the need for obtaining permission to access copyrighted data and it also allows for simulation of flow data of much greater size and volume than the largest currently available public datasets. The data generator always creates the same data, regardless of hardware or floating point precision, for the same problem size. As such the data generator is an excellent substitute for multiple benchmark datasets but without the practical inconvenience of having to port them around the world or to store them on a local machine.

4.3 Characteristic features of the benchmark

In designing a good benchmark a number of factors need to be considered. The key features of the spatial interaction model benchmark make it an excellent means for quantifying the performance of a range of computer systems. These are as follows:

1. The benchmark is easily portable to a broad range of hardware. It contains only a few hundred lines of code, therefore it is easy to understand and port to a new platform. A useful benchmark must require the minimum of vendors' time and effort to apply it. Porting large code is a non-trivial task and the simplicity of the model and code is important.
2. The code is available in the public domain and can be easily downloaded from the World Wide Web.
3. A good benchmark has to be representative of the area of science being studied and the SSB/1 is certainly representative of some of the HPC uses made by GC but there are others that could be developed in a similar fashion.
4. The benchmark has a built-in data generator so there is no need to store any data on disk or perform large amounts of input and output during benchmarking. This neatly avoids the problem of shipping large volumes of data around and yet permits the benchmark to be run on realistic and variable volumes of data. The self-contained nature of the benchmark code is another important ease-of-use feature.
5. The executable time and memory requirements for the benchmark are easily adjusted by altering the problem size, i.e. N, M values. A set of ten standard (N, M) values has been suggested.
6. The benchmark permits a wide range of dataset sizes to be investigated, providing a platform for various numerical experiments and allows scaling experiments with datasets of virtually any size that may be considered important in the future.
7. The ratio of computation to memory references in the benchmark is typical of GC, geography, and most social science problems that may require the use of high performance computers; typically, there is a high ratio of memory access to computation. In many statistical models (and also here) there is a memory access for each floating point operation performed. This places considerable stress on memory bandwidth.
8. The performance indicators can be readily interpreted because it is easy to establish a model of the computational load being generated by the benchmark.
9. The benchmark has a built-in results verifier that checks, against the reference values, whether or not the numerical results are acceptable for a standard set of problem sizes.

4.4 Measuring performance

The aim of benchmarking is to compare the performance of different computer systems in relation to the representative application. The performance indicator used here is the elapsed execution time for one or a specified number of evaluations of the model. This is an absolute metric. It indicates which hardware performs best for solving this type of application. In other words, it allows the user to identify the platform (with the smallest value of performance indicator) that is seemingly most suitable for running the code and which performs similar kinds of computations to that of the benchmark. Note that, as in all benchmarks, problems of a dissimilar computational structure may perform very differently. High performance computer hardware is notoriously quirky and the selection of hardware is nearly always going to favour some applications more than others.

The benchmark code can be retrieved from the web pages at: *http://www.leeds.ac.uk/iss/projects/benchmarks/*. It is also useful as a guide on how to write portable code with the same models being written in two different parallel programming styles.

4.5 Computer systems under evaluation

A number of computer systems have been evaluated with this benchmark, ranging from PCs through workstations to vector processors and massively parallel systems. Table 4.1 contains a short description of all computers

Table 4.1 Computer systems under evaluation

Type of system	Name	System description	Location
Vector processor	Fujitsu VPX	VPX240/10 Fujitsu vector processor with one vector unit with a clock of 3.2 ns and one scalar unit with a clock speed of 6.4 ns and with the main memory of 1 Gb	Manchester (MC)
MPP	Cray T3D	512 processor system built using 150 MHz 21064 DEC Alpha RISC chips; each with 64 Mb of local memory	Edinburgh (EPCC)
MPP	IBM SP2	16 RS/6000 nodes, including 14 IBM thin 2 nodes; each node runs at 67 MHz and has 64 Mb of local memory	Daresbury Laboratory
MPP	KSR 1	A system with 32 cells; each 20 MHz super scalar RISC chip with 32 Mb of local memory	Manchester (MC)
S2MP (MPP)	SGI Origin	14 processors running at 195 MHz; system with scalable shared memory of 1.8 Gb	Reading, SGI

Table 4.1 (cont'd)

Type of system	Name	System description	Location
PVP	Cray J90	32 cpu machine running at 100 MHz with 4 Gb main memory	Rutherford (RAL)
SMP	DEC 8400	2 Gb shared memory 6 processor system; each PE is 300 MHz DEC EV5 (21164) chip	Rutherford (RAL)
SMP	Cray CS6400	A superserver with 12 SuperSPARC processors running at 60 MHz and 768 Mb of shared memory	Manchester (MC)
SMP	SGI Power Challenge	A system with 4 MIPS R8000 processors running at 75 MHz and 256 Mb of memory	Leeds University, Computer Studies
SMP	SGI Onyx	A system with 4 MIPS R10000 processors running at 194 MHz and 256 Mb memory	Daresbury workshop
SMP	SGI Challenge	8 MIPS R4400 processors running at 150 MHz and 256 Mb memory	Leeds University
Workstation	SGI Indy	Indy workstation with a single 180 MHz MIPS R5000 processor with 160 Mb memory	Daresbury workshop
Workstation	DEC Alpha 600	Model 600/5/333 – with a single 333 MHz EV5 processor and 512 Mb memory	Daresbury workshop
Workstation	Sun Ultra-2	2×200 MHz UltraSPARC-2 processors and 256 Mb of memory	Daresbury workshop
Workstation	Sun Ultra-1	1 UltraSPARC-1 processor running at 150 MHz with 64 Mb memory	Leeds University, Geography
Workstation	Sun 10/41	1 SPARC processor running at 40 MHz with 64 Mb memory	Leeds University, Geography
Workstation	HP9000-K460	Model K460 with a single 160 MHz PA8000 processor and 512 Mb memory	Daresbury workshop
Workstation	HP9000-C160	Model C160 with a single 160 MHz PA8000 processor and 512 Mb memory	Daresbury workshop
PC	Pentium PC	Gateway 133 MHz Intel Pentium with triton chip and 16 Mb memory	Leeds
PC	486 DX PC	Elonex 66 MHz Intel processor with 16 Mb memory	Leeds University

NB: MPP stands for Massively Parallel Processing system
SMP stands for Symmetrical Multiprocessor
PVP stands for Parallel Vector Processor
S2MP stands for Scalable Shared-memory Multiprocessor

with their locations where the benchmarking has been carried out. However, very often this benchmarking was performed within the limits set for guest users on the use of available resources. Limits were often applied to memory sizes and the number of parallel processing nodes that could be used.

4.5.1 Parallel processor benchmark performance

Table 4.2 presents the results (the elapsed execution time for one evaluation of the model) for the origin (singly) constrained model with 1000 origins and 1000 destinations for different parallel platforms. The first column in this table (and also in Tables 4.3, 4.4 and 4.5), contains the name of the system under evaluation. The second column shows the number of processors used. The third column contains the execution time of the benchmark and the fourth column the relative performance of the system, which has been computed as the execution time of the benchmark on the system under evaluation divided into the execution time of the benchmark on a 133 MHz Pentium PC. In the last column, the command used for benchmarking and the version of the environment under which the benchmarking was performed, are given. This clarifies which compiler options were used to optimize the performance of the benchmark code. Table 4.3 presents the results for the doubly constrained model with 1000 origins and 1000 destinations for different parallel platforms.

The relative performance measured suggests that the singly constrained model runs 137 times faster on the 256 processor Cray T3D than on a PC, but the doubly constrained model runs only 86 times faster in comparison to a PC run. This may be due to the extensive amount of message-passing that occurs between the processors in the doubly constrained model owing to the estimation of the balancing factors. For the singly constrained model it would appear that the SGI SMP Onyx system is 8.8 times faster than the SGI Challenge but only three times faster than the SGI Power Challenge system. For the doubly constrained model the SGI Onyx system is 11.5 times faster than the SGI Challenge but only four times faster than the SGI Power Challenge system.

Note that the 1000 origins by 1000 destinations SC and DC models were the largest that could be run on all available platforms. Benchmark results for other problem sizes and systems are available on the World Wide Web (WWW). The problem size with $25\,000 \times 25\,000$ origin-destination pairs computed very fast (13 seconds for the singly constrained model and 570 seconds for the doubly constrained model) on the 512 processor Cray T3D. This sized model could not be computed on the 4-processor SGI Onyx system with 256 megabytes of memory as the benchmark's memory requirements were larger than the available real memory. This illustrates another benefit of HPC, namely offering large real memory options.

Table 4.2 Results for the parallel SC model with 1000 origins and 1000 destinations

System name	Number of processors	Time (ms)	Relative performance	Compiler version/reported by[1] and command used for compilation[2]
SGI Origin	14	46.2	180.8	v 7.0 under IRIX 6.2/ J.Barr- SGI f77 -mips4 -O3 -r10000 −lfastm
SGI Origin	12	52.1	160.3	v 7.0 under IRIX 6.2/ J.Barr- SGI f77 -mips4 -O3 -r10000 −lfastm
Cray T3D	256	61	136.9	v 6.0.4.3 cf77 -Cray-t3d -Oscalar3
SGI Origin	10	62	134.7	v 7.0 under IRIX 6.2/ J.Barr- SGI f77 -mips4 -O3 -r10000 −lfastm
SGI Origin	8	78.5	106.4	v 7.0 under IRIX 6.2/ J.Barr- SGI f77 -mips4 -O3 -r10000 −lfastm
Cray T3D	128	129	64.7	v 6.0.4.3 cf77 -Cray-t3d -Oscalar3
SGI Origin	4	154	54.2	v 7.0 under IRIX 6.2/ J.Barr- SGI f77 -mips4 -O3 -r10000 −lfastm
SGI Onyx	4	173	48.3	v 7.0 compiler running under IRIX 6.2 f77 -mips4 -O3 -r10000 −lfastm
Cray T3D	64	202	41.3	v 6.0.4.3 cf77 -Cray-t3d -Oscalar3
SGI Origin	2	323	25.9	v 7.0 under IRIX 6.2/ J.Barr- SGI f77 -mips4 -O3 -r10000 −lfastm
Cray T3D	32	439	19.0	v 6.0.4.3 cf77 -Cray-t3d -Oscalar3
SGI Power Challenge	4	530	15.8	v 6.0.1 f77 -O3 -64 -mips4 −lfastm
IBM SP2	8	560	14.9	AIX XL v03.02.0001.0000 mpxlf -O3
Cray T3D	16	796	10.5	v 6.0.4.3 cf77 -Cray-t3d -Oscalar3
Cray J90	8	847	9.9	v 6.0.4.3(6.62) cf77
IBM SP2	4	1060	7.9	AIX XL v03.02.0001.0000 mpxlf -O3
SGI Challenge	4	1479	5.6	v 5.2 f77 -O2 -mips2 −lfastm
Cray T3D	8	1580	5.3	v 6.0.4.3 cf77 -Cray-t3d -Oscalar3
Cray T3D	4	3186	2.6	v 6.0.4.3 cf77 -Cray-t3d -Oscalar3

[1] Where benchmarking was not performed by the authors the name of the person who submitted the results is given.
[2] In addition the MPI library must be invoked.

Table 4.3 Results for the parallel DC model with 1000 origins and 1000 destinations

System name	Number of processors	Time (s)	Relative performance	Compiler version/reported by[1] and command used for compilation[2]
SGI Origin	14	1.57	155.5	v 7.0 under IRIX 6.2/ J.Barr- SGI f77 -mips4 -O3 -r10000 −lfastm
SGI Origin	12	1.76	138.8	v 7.0 under IRIX 6.2/ J.Barr- SGI f77 -mips4 -O3 -r10000 −lfastm
SGI Origin	10	2.08	117.4	v 7.0 under IRIX 6.2/ J.Barr- SGI f77 -mips4 -O3 -r10000 −lfastm
SGI Origin	8	2.59	94.3	v 7.0 under IRIX 6.2/ J.Barr- SGI f77 -mips4 -O3 -r10000 −lfastm
Cray T3D	256	2.83	86.3	v 6.0.4.3 cf77 -Cray-t3d -Oscalar3
Cray T3D	128	4.41	55.4	v 6.0.4.3 cf77 -Cray-t3d -Oscalar3
SGI Origin	4	5.12	47.7	v 7.0 under IRIX 6.2/ J.Barr- SGI f77 -mips4 -O3 -r10000 −lfastm
SGI Onyx	4	5.39	45.3	v 7.0 compiler running under IRIX 6.2 f77 -mips4 -O3 -r10000 −lfastm
Cray T3D	64	6.67	36.6	v 6.0.4.3 cf77 -Cray-t3d -Oscalar3
SGI Origin	2	10.5	23.3	v 7.0 under IRIX 6.2/ J.Barr- SGI f77 -mips4 -O3 -r10000 −lfastm
Cray J90	8	14.5	16.8	v 6.0.4.3(6.62) cf77
Cray T3D	32	14.5	16.8	v 6.0.4.3 cf77 -Cray-t3d -Oscalar3
IBM SP2	8	20.8	11.7	AIX XL v 03.02.0001.0000 mpxlf -O3
SGI Power Challenge	4	22.4	10.9	v 6.0.1 f77 -O3 -64 -mips4 −lfastm
Cray T3D	16	25.7	9.5	v 6.0.4.3 cf77 -Cray-t3d -Oscalar3
IBM SP2	4	41.2	5.9	AIX XL v 03.02.0001.0000 mpxlf -O3
Cray T3D	8	51.4	4.8	v 6.0.4.3 cf77 -Cray-t3d -Oscalar3
SGI Challenge	4	67.9	3.6	v 5.2 f77 -O2 -mips2 −lfastm
Cray T3D	4	102.7	2.4	v 6.0.4.3 cf77 -Cray-t3d -Oscalar3

[1] Where benchmarking was not performed by the authors the name of the person who submitted the results is given.
[2] In addition the MPI library must be invoked.

Table 4.4 Results for the serial SC model with 1000 origins and 1000 destinations

System name	Number of processors	Time (ms)	Relative performance	Compiler version/reported by[1] and command used for compilation
Fujitsu VPX	S	170	49.1	fortran 77 ex/vp v12110 frt -Pd -Nf -Wv,-md,-px240 −Ianulib
DEC Alpha 600	S	729	11.5	v 4.0 f77 -fast -tune ev5
SGI Origin	S	730	11.4	v 7.0 under IRIX 6.2/ J.Barr- SGI f77 -mips4 -O3 -r10000 −Ifastm
SGI Onyx	S	900	9.3	v 7.0 compiler running under IRIX 6.2 f77 -mips4 -O3 -r10000 −Ifastm
DEC 8400	S	988	8.5	v 3.8-711 f77 -fast -tune ev5
SGI Power Challenge	S	1 744	4.8	v 6.0.1 f77 -O3 -64 -mips4 −Ifastm
Sun Ultra-2	S	2 144	3.9	v4.2 f77 −fast
Sun Ultra-1	S	2 491	3.4	v4.2 f77 −fast
HP9000/K460	S	3 110	2.7	rel 10.0 f77 +Oall +U77
HP9000/C160	S	3 180	2.6	rel 10.0 f77 +Oall +U77
SGI Indy	S	3 645	2.3	v 7.0 f77 -O3 -64 -mips4 −Ifastm
IBM SP2	S	4 743	1.8	AIX XL v 03.02.0001.0000 mpxlf -O3
Cray CS6400	S	5 122	1.6	SUNWspro/SC3.0.1 f77 -fast -O4
Cray J90	S	6 085	1.4	v 6.0.4.3(6.62) cf77
SGI Challenge	S	6 360	1.3	v 5.2 f77 -O2 -mips2 −Ifastm
Pentium PC	S	8 351	1.0	NAGWARE f90 v 2.05 ftn90
Sun 10/41	S	8 459	1.0	v SC3.0.1 f77 -fast -O4
Cray T3D	S	12 832	0.7	v 6.0.4.3 cf77 -Cray-t3d -Oscalar3 -X1
KSR1	S	22 379	0.4	v 1.2.1.3-1.0.2 f77 -O2
486 PC	S	33 242	0.3	NAGWARE f90 v 2.05 ftn90

[1] Where benchmarking was not performed by the authors the name of the person who submitted the results is given.

Table 4.5 Results for the serial DC model with 1000 origins and 1000 destinations

System name	Number of processors	Time (s)	Relative performance	Compiler version/reported by[1] and command used for compilation
Fujitsu VPX	S	6.56	37.2	fortran 77 ex/vp v12110 frt -Pd -Nf -Wv,-md,-px240 −lanulib
SGI Origin	S	19.2	12.7	v 7.0 under IRIX 6.2/ J.Barr- SGI f77 -mips4 -O3 -r10000 −lfastm
SGI Onyx	S	19.5	12.5	v 7.0 compiler running under IRIX 6.2 f77 -mips4 -O3 -r10000 −lfastm
DEC Alpha 600	S	24.0	10.2	v 4.0 f77 -fast -tune ev5
DEC 8400	S	34.3	7.1	v 3.8-711 f77 -fast -tune ev5
SGI Power Challenge	S	72.0	3.4	v 6.0.1 f77 -O3 -64 -mips4 −lfastm
Sun Ultra-2	S	73.5	3.3	v 4.2 f77 −fast
Sun Ultra-1	S	87.6	2.8	v 4.2 f77 −fast
Cray J90	S	88.0	2.8	v 6.0.4.3(6.62) cf77
HP9000/K460	S	102.5	2.4	rel 10.0 f77 +Oall +U77
HP9000/C160	S	103.3	2.4	rel 10.0 f77 +Oall +U77
SGI Indy	S	129.3	1.9	v 7.0 f77 -O3 -64 -mips4 −lfastm
IBM SP2	S	166.9	1.5	AIX XL v 03.02.0001.0000 xlf -O3
Cray CS6400	S	167.5	1.5	SUNWspro/SC3.0.1 f77 -fast -O4
SGI Challenge	S	196.0	1.2	v 5.2 f77 -O2 -mips2 −lfastm
Pentium PC	S	244.2	1.0	NAGWARE f90 v 2.05 ftn90
Sun 10/41	S	277.7	0.9	v SC3.0.1 f77 -fast -O4
Cray T3D	S	457.0	0.5	v 6.0.4.3 cf77 -Cray-t3d -Oscalar3 -X1
KSR1	S	677.5	0.4	v 1.2.1.3-1.0.2 f77 -O2
486 PC	S	1045.8	0.2	NAGWARE f90 v 2.05 ftn90

[1] Where benchmarking was not performed by the authors the name of the person who submitted the results is given.

4.5.2 Single processor benchmark performance

The performance of serial systems is reported by presenting the elapsed time required for one evaluation of the model. Table 4.4 presents the results for the singly constrained model with 1000 origins and 1000 destinations for different single processor runs. Table 4.5 presents the results for the doubly constrained model with 1000 origins and 1000 destinations for different serial runs. The Fujitsu/VPX240 vector supercomputer reliably takes top place in all our tables with the single processor runs. The SSB benchmark 1 is easily vectorizable and allows full advantage to be taken of the vector processing capabilities of the VPX without restructuring the serial code. Yet vector processors seem to be disappearing as available computing platforms due to the high cost of megaflop computation, as compared with the parallel alternative based on multiple cheaper commodity processors, linked via fast and high bandwidth networks.

The spatial interaction model performs better (1.54 faster for the SC model and 1.87 for the DC model) on the Intel 133 MHz Pentium processor than on a single Cray T3D processor. None of this is particularly surprising as parallel HPC systems gain their speed from having several hundred processors operating in parallel. The faster the speed of an individual processor, the faster the speed of the entire system, if other aspects scale to a similar degree. The seemingly sluggish performance of a single processor Cray T3D should not detract from the immense power of an MPP system comprising 512 processors. Nor should the benefits of having 512 times more memory than a single PC processor be neglected. On the other hand, the results do suggest that the historic speed advantage between UNIX RISC chips and Pentium type CISC machines is narrowing.

4.5.3 Scalability issues

The scalability analysis studies the degree to which benchmark performance changes as either more processors are added and/or the problem size changes. Many factors may affect the scalability of codes and one of the most basic issues is the communication overhead.

4.5.3.1 Communication patterns for the SC model

This description of the communication costs (measured as time delay) does not include data generation and initialization phases and it refers only to the computation of the SIM model implementing a message-passing paradigm. The communication costs of the parallel implementations for the SC model are related to the costs of a global sum reduction operation and the costs of a broadcast operation. Each processor is the owner of the data and computes partial values of the model without the need to communicate with other processors. In the final stage, each processor communicates a few real numbers to the master node where the reduction of global error sums takes place. The master node then reduces all error sums and broadcasts the final error sum to all processors. The SC benchmark generates the following communications:

$$C_{max} = 3 * \text{reduction}(p) + \text{broadcast}(1)$$

4.5.3.2 Communication patterns for the DC model

The DC model encompasses the SC model and includes additional computations and interprocessor communications. The overheads of the parallel implementation include the following costs of communication. In the horizontal sweep, after computing a block of the $\{T_{ij}^{predicted}\}$ matrix, each processor sends one of the partial sums (one number or word) to the master processor (master node) for reduction. This is followed by a message of length n/p words, where n is the number of origins and p the number of processors, that is sent to the master node. Now the master node broadcasts a message comprising n words to all processors. In effect, the set of values, of $\{A_i\}$, which are computed on the individual nodes, are gathered by the master node and then the full $\{A_i\}$ array is broadcast to all processors. This communication scheme could have been replaced by the collective communication gather-to-all, where a block of data sent from each process is received by every process and placed in the correct place in the buffer. At the design stage, it was thought that this implementation provided the user with more explicit information about communication patterns.

Similarly, in the vertical sweep, each processor computes a set of values and sends at first a single value (one word) to be reduced by the master, followed by m/p words (m is the number of destinations and p the number of processors) to the master node. Now the master node broadcasts a message comprising m words followed by a one word in length message to each processor. This process is iterated 20 times. In the last step, the final model is computed and the communication costs are the same as those described for the SC model. Each processor sends three messages (each message one word in length) to the master node where the reduction of global error sums takes place. The master node then computes all error sums and broadcasts the final error sum (one word) to all processors. The total number of messages sent and received is quite large in relation to the amount of computation performed. The DC benchmark generates the following communications:

$$C_{max} = 20 * \{ \ 2 * \text{reduction}(p) + \text{broadcast}(m) + \text{broadcast}(n)$$
$$+ \text{broadcast}(1) + \max[(p-1)*\text{receive } (n/p), \text{send}(n/p)] + \max[(p-1)$$
$$* \text{receive } (m/p), \text{send}(m/p)]\} + 3*\text{reduction}(p) + \text{broadcast}(1)$$

where broadcast(m) means to broadcast a message of length m words.

4.5.3.3 Workload

To determine scalability of a parallel system it is important to establish the workload associated with the problem size (see Hwang, 1993). The size of a job should vary according to the size of the parallel system being benchmarked. Usually, a larger parallel system should run a larger parallel application, in order to avoid the implications of Amdahl's law (Gustafson et al., 1991). Therefore, it is important to understand how the workload changes with changes in the size of the problem being benchmarked. The amount of workload for both models is proportional to the problem size (n,m). The workload per processor can be measured experimentally by the elapsed execution time, which for both models can be expressed as: $t(p,n,m) = n*m/p*K + h(n,m,p)$ where $h(n,m,p)$ is the overhead, p is the number of processors and K is constant. The overhead is due to the costs of communications and synchronization; other factors are not considered here.

4.6 Comparison of performance of HPF and MPI codes

4.6.1 HPF implementation

There are two versions of the benchmark code. Each of them employs a conceptually different parallel computational model. As SIMs contain inherently data-parallel computations, the second version of the benchmark code uses a data-parallel paradigm, which is implemented through a high-level data-parallel language, High Performance Fortran (HPF) (see Koelbel et al., 1994). This version of the benchmark was not used for benchmarking computer systems but was implemented with the aim of benchmarking the performance of the HPF compiler. It is important to know whether or not HPF provides a fast and efficient alternative to a message passing interface (MPI) (see Gropp et al., 1994). Current HPF compilers target a single program multiple data (SPMD) programming concept and implement the owner-computes rule; each processor executes the same program but operates on a local portion of the distributed data. Although HPF does not provide as much control over data distribution as MPI, it does offer a dramatic simplicity of approach and implementation at a higher level of abstraction. Most importantly, HPF implementation does not require a serial code to be restructured for parallel processing provided the algorithm is already in a data-parallel or vectorizable form.

For HPF, the cost array is distributed blockwise by rows. Another copy of the array is made and distributed blockwise by columns across parallel processors. The following statements are added to the serial code to explicitly map the data on to parallel processors:

```
!HPF$ DISTRIBUTE cij(BLOCK,*)
!HPF$ DISTRIBUTE cijc(*,BLOCK)
!HPF$ DISTRIBUTE rn(BLOCK,*)
!HPF$ DISTRIBUTE mcol(BLOCK,*)
```

This mapping has to reflect the way in which the data are to be accessed otherwise performance will be degraded. The serial data generator was converted to a data-parallel version. The subroutine that generates the data was modified in order to avoid side effects and converted into what is called a pure subroutine. In the serial code the $\{T_{ij}^{predicted}\}$ matrix is stored in sparse format in a one-dimensional array with additional integer arrays for indexing. This approach was beneficial in terms of memory requirements. However, it could not be directly implemented in HPF which lacks a convenient way for the indirect addressing of data. Instead the sparse matrix was substituted by a regular matrix which loses the benefits of sparse storage. High Performance Fortran implementation of the benchmark contains a number of !HPF$ INDEPENDENT statements in front of the DO loops in the source code. Examples of statements converted into a parallel form by inserting the !HPF$ INDEPENDENT directive are shown in Appendix 2. Finally, HPF allows a programmer to map and tune the code on to a target architecture. Preliminary experiments with tuning did not deliver any significant performance improvement for code running on the Cray T3D.

4.6.2 SC benchmark results

The performance indicators which are presented in this chapter are the elapsed execution time of one model evaluation, the speedup, the relative efficiency, and the relative difference in execution time for both implementations of the model. A whole set of these performance metrics is included here in order to allow the user to make a comparative analysis of both implementations of the benchmark.

The *speedup* factor has been computed as: $S(p) = T(1)/T(p)$ where $T(1)$ is the execution time of the benchmark on one processor and $T(p)$ is the execution time on p processors. Calculating this factor we used as $T(1)$ the elapsed execution time of one model evaluation for each implementation on the Cray T3D. The speedup metric shows how well both implementations performed with the increase in the number of processors used.

Table 4.6 Results for the SC model

No of processors	SC 1000×1000 problem size						
	HPF execution time (ms)	HPF speedup	Relative efficiency of HPF	MPI execution time (ms)	MPI speedup	Relative efficiency of MPI	Relative difference for HPF and MPI
1	13 697	1.00	1.00	11 351	1.00	1.00	20.67%
4	3 422	4.00	1.00	2 943	3.86	0.96	16.28%
8	1 709	8.01	1.00	1 294	8.77	1.10	32.07%
16	864	15.85	0.99	644	17.63	1.10	34.16%
32	440	31.13	0.97	327	34.71	1.08	34.56%
64	221	61.98	0.97	165	68.79	1.07	33.94%
128	112	122.29	0.96	83	136.27	1.07	34.94%
256	57	240.30	0.94	43	263.98	1.03	32.56%
512	30	456.57	0.89	22	515.95	1.01	36.36%

The speedup metric is augmented by the efficiency indicator which is a relative measure. It allows the user to formulate a picture of how well both implementations perform in relation to a single processor run on the same system. The efficiency has been calculated using the formula: $E(p) = S(p)/p$ This indicator shows the actual degree of speedup performance achieved in relation to the execution time of a job on one processor. This factor usually takes values between 0 and 1 unless there are some other factors (for example, the cache effects) that affect the performance of a one-processor run.

The *relative difference* for both implementations was computed as follows: $D(p) = (| T_{MPI}(p) - T_{HPF}(p) | / T_{MPI}(p))$ where $T_{HPF}(p)$ is the execution time for the HPF implementation and $T_{MPI}(p)$ is the execution time for the MPI implementation. These values show the degree to which one implementation outperforms the other.

None of the above metrics shows the effects of the performance change with a change in the problem size. This analysis was carried out for the SC model with the following metrics: the *predicted execution time* $T_{pred}(p,n_1,m_1) = T(p,n,m)*(n_1*m_1/n*m)$ and the *relative error* $Er(p,n_1,m_1) = | T_{pred}(p,n_1,m_1) - T(p,n,m) | / T(p,n,m)$ where $T(p,n,m)$ is the execution time for the benchmark of problem size (n,m) executed on p processors.

The MPI implementation and the HPF implementation of the benchmark were run on the Cray T3D. The MPI results differ from the results presented in Table 4.2 because they were obtained with a later version of the compiler (i.e. the Cray 6.0.5.1 compiler). The HPF version was compiled with the Portland Group's High Performance Fortran (pghpf) compiler.

Table 4.6 contains results for the singly constrained model for the 1000×1000 problem and Table 4.7 for the 5000×5000 problem. The

Table 4.7 Results for the SC model

No of processors	SC 5000 × 5000 problem size						
	HPF execution time (ms)	HPF speedup	Relative efficiency of HPF	MPI execution time (ms)	MPI speedup	Relative efficiency of MPI	Relative difference for HPF and MPI
16¹				15 997	16.07	1.00	
32	10 937	32.00	1.00	8 032	32.00	1.00	36%
64	5 506	63.56	0.99	4 036	63.68	0.99	36%
128	2 786	125.62	0.98	2 038	126.12	0.98	37%
256	1 395	250.88	0.98	1 022	251.49	0.98	36%
512	699	500.69	0.98	515	499.08	0.97	36%

¹ There was not enough memory to run this problem with 16 processors.

efficiency of parallel jobs was computed with reference to the one-processor job for the 1000×1000 problem. There was not enough memory to run on one node the 5000×5000 problem; thus the relative efficiency is calculated with reference to a 32-processor job.

The benchmark achieves close to linear speedup for both 1000×1000 and 5000×5000 problem sizes. For the 1000×1000 problem the relative efficiency for HPF implementation decreases from 1.00 for a one-processor job to 0.89 for a 512-processor run. For the same problem size with MPI implementation it ranges between 1.00 to 1.10 for varying numbers of processors (Table 4.6). The values over 1.00 may be due to the effect of the cache sizes. With an increase in the number of processors the amount of data stored per processor reduces, more of the data can be kept in the cache giving a superlinear speedup.

For the 5000×5000 problem the relative efficiency was computed with reference to a 32-processor run. When the number of processors is increased to 64, the efficiency lowers slightly to 0.99, and decreases to 0.97 for a 512-processor job. The overall impression gained here is that performance scales linearly, viz. double the number of processors and the clock time halves. This is observed for both the MPI and HPF implementations.

The MPI implementation of the SC model performed better, in terms of execution time, than the HPF implementation for both problem sizes on the Cray T3D. For the 1000×1000 problem the relative difference for a four-processor job is approximately 16% (Table 4.6). With an increase in the number of processors to eight, the relative difference doubles to 32% and then slightly increases up to 36% for different numbers of processors up to 512. The 5000×5000 HPF implementation produces on average 36% worse results than the MPI implementation. This high relative difference between both implementations is due to the type of model under consideration in which the ratio of memory references to computations is very high. The computation of this model involved very few inter-process communications so the main overheads were due to the creation of parallel threads and the building of the communications schedule in the HPF implementation. Nevertheless, given the simple and highly data-parallel nature of the singly constrained SIM this is a very disappointing result for HPF and must cast doubts over its usefulness in those applications where maximum performance is required.

As the SC benchmark scales linearly it can be used to predict performance of the code run across larger number of processors. For example, Table 4.8 contains the results for the SC model together with the predicted values and the relative errors.

Table 4.8 Results for the SC model

No of processors	The SC model							
	HPF 1000 execution time (ms)	HPF 5000 execution time (ms)	Predicted HPF 5000 execution time (ms)	Relative error	MPI 1000 execution time (ms)	MPI 5000 execution time (ms)	Predicted MPI 5000 execution time (ms)	Relative error
32	440	10 937	11 000	0.58%	327	8 032	8 175.00	1.78%
64	221	5 506	5 525	0.35%	165	4 036	4 125.00	2.21%
128	112	2 786	2 800	0.50%	83	2 038	2 082.50	2.18%
256	57	1 395	1 425	2.15%	43	1 022	1 075.00	5.19%
512	30	699	750	7.30%	22	515	550.00	6.80%

The results in Table 4.8 were obtained by multiplying the execution time for the 1000×1000 problem by 25 as the 5000×5000 problem contains 25 times more workload than the 1000×1000 problem. The relative error for the predicted performance is small and fluctuates between 0.58% to 7.30%.

4.6.3 DC benchmark results

Table 4.9 contains results for the doubly constrained model for the 1000×1000 problem for the 5000×5000 problem and Table 4.10. The efficiency of parallel jobs was computed with reference to the one-processor job for the 1000×1000 problem. There was not enough memory to run on one node the 5000×5000 problem thus the relative efficiency is calculated with reference to a 32-processor job. The DC 1000×1000 model shows approximate linear scalability up to 64 processors. However, with more processors, the benchmark does not scale. This is more evident for the HPF implementation, which achieves the scalability of 51 for 64 processors, and this then

Table 4.9 Results for the DC model

No of processors	DC 1000 × 1000 problem size						
	HPF execution time (s)	HPF speedup	Relative efficiency of HPF	MPI execution time (s)	MPI speedup	Relative efficiency of MPI	Relative difference for HPF and MPI
1	433.07	1.00	1.00	441.41	1.00	1.00	1.81%
4	108.61	3.97	0.99	124.35	3.55	0.89	12.38%
8	54.40	7.96	0.99	54.29	8.12	1.02	0.20%
16	27.65	15.63	0.98	27.17	16.23	1.01	1.95%
32	14.51	29.86	0.93	13.93	31.73	0.99	4.32%
64	8.45	51.30	0.80	7.13	61.85	0.97	18.37%
128	8.18	52.93	0.41	3.94	111.93	0.87	107.61%
256	19.08	22.79	0.09	2.86	154.20	0.60	564.34%
512	65.12	6.66	0.01	3.02	146.03	0.29	2052.3%

Table 4.10 Results for the DC model

No of processors	DC 5000 × 5000 problem size						
	HPF execution time (s)	HPF speedup	Relative efficiency of HPF	MPI execution time (s)	MPI speedup	Relative efficiency of MPI	Relative difference for HPF and MPI
16[1]				672.09	21.65		
32	353.72	32.00	1.00	454.83	32.00	1.00	22%
64	179.79	62.95	0.98	208.99	69.67	1.09	14%
128	96.06	117.17	0.92	87.90	165.57	1.29	10%
256	64.20	176.85	0.69	44.49	324.13	1.27	43%
512	90.73	124.79	0.24	29.98	486.74	0.95	203%

[1] There was insufficient memory to run this size of problem with 16 processors.

falls to 6.7 for 512 processors. The MPI implementation, in contrast, achieves the scalability of 62 for 64 processors and this then falls to 146 for 512 processors. With a large number of processors, the amount of workload per processor is very low and thus the overheads of parallelization and communications stop the benchmark from scaling. For the 128-processor run each processor worked only on eight rows before communicating the values and then worked on eight columns which was again followed by communication. This process was repeated for 20 iterations before the model was computed. The DC benchmark results demonstrate that the user must be aware of the type of operations in an application when converting it to a parallel version. The ratio of communication to computation in an application must be known in order to choose the right number of processors for the job. The HPF implementation brings additional overheads, which are easily recovered, on the small number of processors where the amount of work is larger per processor.

The HPF implementation for the 5000×5000 problem obtains a scalability of 117 for 128 processors and 124 for 512 processors. The MPI implementation scales linearly up to 512 processors. However, it is possible that with a further increase in the number of processors this implementation will not scale any further. The solution is to scale up the problem size as the number of processors increases. The problem seen here would then disappear. Finally, the poor performance of the HPF implementation for the large number of processors (over 128) is noted. It is questionable whether the gain in ease of programming using HPF is worth the loss of performance. After all, HPF should be most useful for those applications on the edge of what is computable and it is here where squeezing the last drops of speed out of the code, compiler, and hardware, is likely to be most worthwhile.

4.7 Conclusions

This chapter has described a social science benchmark suitable for measuring the performance of the widest possible set of HPC hardware on a representative social science computational problem. The benchmark is thought to be a useful means for geographers and other HPC users to quantify the performance of the diversity of computational systems they have or could have access to. A large number of performance results are collected and reported here; these are also available on the WWW (*http://www.leeds.ac.uk/ iss/projects/benchmarks*). The performance tables express HPC developments in terms of hardware that social scientists and others engaged in GC activities have access to or can relate to, so that they may more readily appreciate the potential gains in speed that are now available on various HPC platforms for those applications that need it. However, such is the current rate of improvement in hardware speeds, that it is important for these performance results to be regularly updated. Knowing how fast a computer runs is fundamental to much of GC as well as being interesting in its own right.

Acknowledgements

The work was supported by the EPSRC grant number GR/K40697 under the PAGSIM project comprising part of the Portable Software Tools for Parallel Architecture projects.

References

Birkin, M., Clarke, M., George, F. (1995) 'The use of parallel computers to solve non-linear spatial optimisation problems', *Environment and Planning A*, 27, pp. 1049–68.

Birkin, M., Clarke, G., Clarke, M., Wilson, A. (1996) *Intelligent GIS: Location Decisions and Strategic Planning*, Cambridge: GeoInformation International.

Brocklehurst, E. R. (1991) *Survey of Benchmarks*, NPL Report, DITC 192/91.

Dongarra, J. J. (1995) *Performance of Various Computers Using Standard Linear Equations Software*, CS-89-85, Computer Science Department, University of Tennessee, Knoxville, TN, July.

Gropp, W., Lusk, E., Skjellum, A. (1994) *USING MPI: Portable Parallel Programing with the Message-Passing Interface*, Cambridge, MA; London: MIT Press.

Gustafson, J., Rover, D., Elbert, S. and Carter, M. (1991) 'The design of a scalable, fixed-time computer benchmark', *Journal of Parallel and Distributed Computing*, 12, pp. 388–401.

Hey, A. J. G. (1991) 'The GENESIS distributed memory benchmarks', *Parallel Computing*, 17, pp. 1275–1283.

Hwang, K. (1993) *Advanced Computer Architecture: Parallelism, Scalability, Programmability*, McGraw-Hill.

Hwang, K. and Xu, Z. (1998) *Scalable Parallel Computing*, Boston-Toronto: McGraw-Hill.

Koelbel, C., Loveman, D. B., Schreiber, R. S., Steele Jr, G. L., Zozel, M. E. (1994) *The High Performance Fortran Handbook*, Cambridge, MA; London: MIT Press.

Lewis, T. G. and El Rewini, H. (1992) *Introduction to Parallel Computing*, Englewood clalls, NJ: Prentice Hall.

Openshaw, S. and Schmidt, J. (1997) 'A Social Science Benchmark (SSB/1) Code for Serial, Vector, and Parallel Supercomputers', *Geographical and Environmental Modelling*, 1,1, pp. 65–82.

Openshaw, S. and Turton, I. (1999) High Performance Computing and the Art of Parallel Programming, London: Routledge.

SPEC (1997) SPEChpt96 Results: http://www.specbench.org/hpg/results.html

SSB/1 (1997) *A Social Science Benchmark (SSB/1) Code for Serial, Vector, and Parallel Supercomputers*: http://www.leeds.ac.uk/ucs/projects/benchmarks/

Turton, I. and Openshaw, S. (1996) 'Modelling and optimising flows using parallel spatial interaction models', in Bougé, L., Fraigniaud, P., Mignotte, A. and Roberts, Y. (eds) *Euro-Par '96 Parallel Processing*, Vol. 2, Lecture Notes in Computer Science 1124, Berlin: Springer, pp. 270–275.

Wilson, A. G. (1970) *Urban and Regional Models in Geography and Planning*, London: Wiley.

Wilson, A. G. (1981) *Geography and the Environment, System Analytical Methods*, London: Wiley.

Appendix 1: Parallelization of a doubly constrained SIM

Master node communicates to each processor: istart,iend; jstart,jend; n, m, nproc, alpha, beta

data generation on each (rth)processor: C_{kj}, k=istart,. . .,iend; j = 1, . . . ,m

$T_{kj}^{observed}$, k=istart,. . .,iend; j = 1, . . . ,m

O_i, i = 1, . . . ,n

partial values of D_j^r, j = 1, . . . ,m

Master node reduces and broadcasts: $D_j = \sum_{r=1}^{P} D_j^r; j = 1, . . . ,m$

Each processor communicates to others: C_{kl}, k = istart,. . .,iend; l = jstart, . . . ,jend

Each processor computes: $T^r = \sum_{k=istart}^{iend} \sum_{j=1}^{M} T_{kj}^{observed}; O^r = \sum_{k=istart}^{iend} O_k;$

$$Z^r = \sum_{k=istart}^{iend} \sum_{j=1}^{M} T_{kj}^{observed} f(C_{kj})$$

Master node reduces: $T = \sum_{r=1}^{P} T^r; O = \sum_{r=1}^{P} O^r; Z = \sum_{r=1}^{P} Z^r$

Master node computes and broadcasts: K = Z/O

Loop for 20 iterations:

each processor (rth)computes: $A_k = \sum_{j=1}^{m} B_j f(C_{kj})$, k=istart, . . . ,iend;

$$SA^r - \sum_{k=istart}^{iend} A_k$$

each processor communicates to master: A_k, k = istart,. . .,iend

master broadcasts: A_i, i = 1, . . . ,n

master node reduces received partial SA^r: $SA = \sum_{r=1}^{P} SA^r$

each processor computes: $B_i = \sum_{i=1}^{n} A_i f(C_{il})$, l=jstart, . . . ,jend;

$$SB^r = \sum_{l=jstart}^{jend} B_l$$

| master node reduces received partial SB^r: | $SB = \sum_{r=1}^{P} SB^r$ |

master node computes and broadcasts: $\quad$ f = SB/SA

each processor computes: $\quad$ $B_1 = D_1 B_1 / f$, $1 = $ jstart, . . . ,jend;

each processor communicates to master: $\quad$ B_1, $1 = $ jstart, . . . ,jend

Each processor computes: $\quad$ $F^r = \sum_{k=istart}^{iend} \sum_{j=1}^{m} (T_{kj}^{predicted} - T_{kj}^{observed})^2;$

$$G^r = \sum_{k=istart}^{iend} \sum_{j=1}^{m} |T_{kj}^{predicted} - T_{kj}^{observed}|;$$

$$H^r = \sum_{k=istart}^{iend} \sum_{j=1}^{m} T_{kj}^{predicted} \, f(C_{kj})$$

Master node reduces partial errors: $\quad$ $F = \sum_{r=1}^{P} F^r; \; G = \sum_{r=1}^{P} G^r; \; H = \sum_{r=1}^{P} H^r$

Master node computes: $\quad$ $E = (K-H/O)^2$

Master node broadcasts the error sums: $\quad$ E, F; G

Appendix 2: HPF code fragments

```
!HPF$ INDEPENDENT, NEW (k1,k2,t,j,icol,aa,pred,sum,b,addon),          &
!HPF$&              REDUCTION (f1,f2,c1)
        DO 33000 i = 1, n
          k1 = ip (i)
          k2 = num (i)
          DO 11 j = 1, m
            t (j) = 0.0
 11     END DO
        IF (k2.ge.k1) then
          DO 12 j = k1, k2
            icol = mcol (i,j)
            t (icol) = rn (i,j)
 12     END DO
        ENDIF
        sum = 0.0
        DO 14 j = 1, m
          aa = beta * cij (i, j)
          IF (aa.gt.prec) aa = prec
          IF (aa.lt. - prec) aa = - prec
          pred = exp (aa) * d (j)
```

```
        addon = cij (i, j) **alpha
        pred = pred * addon
        sum = sum + pred
        b (j) = pred
14      END DO
        IF (sum.lt.small) sum = small
        sum = o (i) / sum
        DO 384 j = 1, m
          pred = b (j) * sum
          f1 = f1 + (pred – t (j)) **2
          f2 = f2 + abs (pred – t (j))
          c1 = c1 + pred * cij (i, j)
384     END DO
33000   END DO
```

Chapter 5

GeoComputation using cellular automata

Michael Batty

5.1 Cellular automata as GeoComputation

Cellular automata (CA) are computable objects existing in time and space whose characteristics, usually called states, change discretely and uniformly as a function of the states of neighboring objects, i.e. those that are in their immediate vicinity. The objects are usually conceived as occupying spaces which are called cells, with processes for changing the state of each cell through time and space usually articulated as simple rules which control the influence of the neighborhood on each cell. This formulation is quite general and many systems can be represented as CA but the essence of such model-ling consists of ensuring that changes in space and time are always gener-ated locally, by cells which are strictly adjacent to one another. From such representation comes the important notion that CA simulate processes where local action generates global order, where global or centralized order 'emerges' as a consequence of applying local or decentralized rules which in turn embody local processes. Systems which cannot be reduced to models of such local processes are therefore not likely to be candidates for CA, and although this might seem to exclude a vast array of geographical processes where change seems to be a function of action-at-a-distance, this criterion is not so restrictive as might appear at first sight.

Formally, we can restate the principles of CA in terms of four elements. First there are *cells*, objects in any dimensional space but manifesting some adjacency or proximity to one another if they are to relate in the local manner prescribed by such a model. Second, each cell can take on only one *state* at any one time from a set of states which define the attributes of the system. Third, the state of any cell depends on the states and configurations of other cells in the *neighborhood* of that cell, the neighborhood being the immediately adjacent set of cells which are 'next' to the cell in question where 'next' is defined in some precise manner. Finally, there are *transition rules* which drive changes of state in each cell as some function of what exists or is happening in the cell's neighborhood. There are further assump-tions and conditions. It is assumed that the transition rules must be *uniform*, i.e. they must apply to every cell, state and neighborhood at all times, and that every change in state must be *local*, which in turn implies that there is no action-at-a-distance. There are conditions which specify the start and end points of the simulation in space and time which are called *initial* and *boundary conditions* respectively. Initial conditions apply to the spatial configuration of cells and their states which start the process, as well as the time at which the process begins. Boundary conditions refer to limits on the space or time over which the CA is allowed to operate.

To illustrate these principles, consider an elementary example. The most usual configuration of cells comprising a cellular automata is based on a regular two-dimensional (2D) tessellation, such as a grid where the array of square cells are contiguous to one another. The simplest categorization of states is that each cell can be either alive or dead, active or inactive, occupied or empty, on or off, true or false, while the neighborhood within which any action changes the state of a cell is composed of the eight adjacent cells in the band around the cell in question, at the eight points of the compass. This is the so-called Moore neighborhood. A very basic rule for changes from cells which are 'off' to 'on' might be: *if* any cell in the neighborhood of any other cell in question is 'on', *then* that cell becomes 'on'. In this way, cells which are off are turned 'on', those that are 'on' remain 'on'. To show how this automata might change the state of an array of cells, we need an initial condition, a starting point for the configuration of cells, and also a stopping rule which in spatial terms is the boundary condition. We will assume that the temporal conditions are dictated by the spatial conditions in that once the process begins from time zero, it finishes when the spatial boundary is reached.

In this case, we fix the initial configuration as one active or live 'on' cell in the center of a 100×100 square cellular array and start the process. Note how easy and natural it is to represent this using pixels on a computer screen. In Figure 5.1(a), we show a magnification of the grid of cells, a typical Moore neighborhood around an arbitrarily chosen cell, and the central live cell which starts the simulation. The process is particularly simple. At every time period, each cell in the array is examined and if there is a live cell in its neighborhood, then that cell is made live or switched on. Here, on the first time cycle, the cells in the band around the center cell each have a live cell in their Moore neighborhoods, and thus they are switched on. In the next iteration, the band around this first band all have live cells in their neighborhoods and the same occurs. A process of growth begins in regular bands around the initial seed site, with the spatial diffusion that this growth implies clearly originating from the operation of the rules on a system with a single seed site. In Figure 5.1(b), we show this process of growth up to time $t = 40$ when the 40th band around the seed becomes live. We take this as the spatial boundary condition which in this case is coincident with the temporal. Clearly the process could continue indefinitely if the array of cells were infinite.

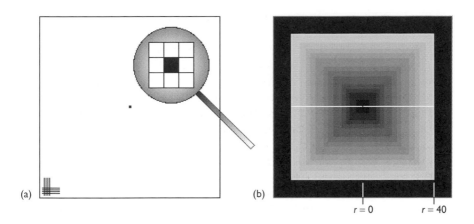

(a)

(b)

$r = 0$ $r = 40$

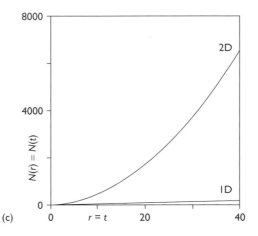

(c)

Figure 5.1 Neighborhoods, cellular growth and growth paths for the simplest automata: (a) origin of growth, 100×100 grid and neighborbood; (b) 1D (line) and 2D (square) cellular automata: (c) 1D and 2D cellular growth.

This kind of growth and diffusion is an analog to many systems. For example, consider the cellular array as a grid of light bulbs all wired to those in their immediate (Moore) neighborhood. The rule is that we switch one on when one of those to which it has been wired has been switched on. If we begin by switching the central bulb on, the process whereby all the bulbs are lit follows the diffusion shown in Figure 5.1(b). If the central seed were a population which grew in direct response to the immediate space around it, like a city, then the process illustrated in Figure 5.1(b) might mirror urban development. These kinds of example can be multiplied indefinitely for any simple growth process from crystals to cancers. The morphology produced is very simple in that it is one based on an entirely filled cluster whose form is dictated by the underlying grid and by the influence of the neighborhood. We should also look at the model's dynamics. The number of cells occupied in this model can be predicted either as a function of time or space. Calling the number of cells at the horizontal or vertical distance r from the seed $N(r)$, the cells occupied can be predicted as $N(r) = (2r + 1)^2$. As r and time t are completely synchronized in this automata, then $N(t) = (2t + 1)^2$. In Figure 5.1(c), we plot the number of cells which are live at each distance and time, $r = t = 0, 1, 2, 3, \ldots$, which follows the progression $1, 9, 25, 49, \ldots$, and so on.

It is also useful at this point to examine the dynamics of an equivalent 1D CA where the neighborhood is now a 3×1 set of cells, i.e. each cell east and west of the cell in question. We also show this in Figure 5.1(a) and using the same rules as in the 2D CA, the resulting pattern is shown in Figure 5.1(b). This is a horizontal line of cells diffusing from the central cell in exactly the same manner as the 2D automata. It is now easy to guess the dynamics which give the number of cells produced at distance r from the central cell and at time t as $N(r) = (2r + 1)$ and $N(t) = (2t + 1)$, also plotted in Figure 5.1(c). For now, simply note that the exponent on the general equations for $N(r)$ and $N(t)$ is the dimension of the system, 2 for the 2D automata, 1 for the 1D. We will return to this, but all the elements for our study of CA as a basis for GeoComputation (GC) have now been assembled. Although we will explore how these features can be adapted to generate many different types of spatial system in the rest of this chapter, we can anticipate some of this before we digress back into the history and origins of this powerful approach.

With these definitions and principles in mind, it is worth demonstrating just how flexible the CA framework is at representing and simulating very diverse types of system. Readers who require a thorough discussion with many examples, especially from physics, are referred to Toffoli and Margolus (1987). Although CA give equal emphasis to objects and their relations in space and time, the focus on cells means that the framework is organized from the spatial rather than the temporal viewpoint. However, 2D arrays of cells, although the most usual, are simply one case, for CA can exist in any number of dimensions, and all the principles generalize accordingly. One-dimensional models can be used to model relations on a line but space need not be real, it might simply be used as an ordering principle; in this context, the cells might be, say, time cells. Three-dimensional automata might be used to represent the explicit 3D world of terrain and built form, but in our exposition, we consider the 2D world the most appropriate in that this is the usual world of GC. States can take on any value in a range of discrete values while in geographical applications there is usually some argument as to whether the concept of neighborhood should be relaxed. What we refer to as *strict CA* are those automata where there is no action-at-a-distance, i.e. where the neighborhood of interest is entirely local, being composed of those cells which are topologically nearest neighbors to each cell in question. Geographical systems however are often characterized by action-at-a-distance, and if this is to be represented, then neighborhoods must be defined to reflect it. Such variants are better called cell-space or *CS models* (after Albin, 1975) but here we will restrict our focus to CA. In many instances, action-at-a-distance is in fact the product of the system's dynamics – it is a consequence of local actions through time – and thus it is eminently feasible to generate morphologies and their dynamics which display such properties using strict CA.

The major representational issue in CA modelling involves the extent to which the discreteness that the framework demands matches the system's elements, relations and behavior. In principle, with any continuous system that can be made discrete, thus assuming that local action and interaction characterize the system, this type of modelling is applicable. However, in practice, it is often difficult or even impossible to associate cells and states of the model to those of the real system. For example, consider a town whose basic elements are buildings. Within each building, there may be several distinct activities and thus cells cannot be buildings; they must be parts of buildings disaggregated to the point where each distinct activity – state – is associated with a single cell. Often this is impossible from the available data with this problem usually being compounded at higher levels of aggregation such as the census tract. The same problem may exist in defining states. No matter how small a cell, there may always be more than one state associated with it in that the elemental level may not be a state *per se* but some object that can take on more than one state simultaneously. Sometimes, redefinition of the system can resolve such ambiguities but often to use CA at all certain approximations have to be assumed; for example, a cell which may be a geographical location, say, can have only one land use or state and thus this may have to be the dominant land use. We have said little about time – that much underworked dimension in GC – but similar problems emerge where different temporal processes requiring synchronization in complex ways characterize the automata.

5.2 The origins of cellular automata

Cellular automata date back to the very beginnings of digital computation. Alan Turing and John von Neumann who pioneered the notion that digital computation provided the basis for the universal machine, both argued, albeit in somewhat different ways, that digital computation held out a promise for a theory of machines that would be self-reproducible (Macrae, 1992), that computers through their software could embody rules that would enable them to reproduce their structure, thus laying open the possibility that digital computation might form the basis of life itself. This was a bold and daring speculation but it followed quite naturally from the philosophy of computation established in the 1920s and 1930s. Von Neumann perhaps did most to establish the field in that up until his death, in 1956, he was working on the notion that a set of rules or instructions could be found which would provide the software for reproducibility. The idea of automata flowed quite easily from this conception, and the notion that the elements of reproducibility might be in terms of 'cells' was an appeal more to the 'possibility' of using computers as analogs to create life than any actual embodiment of such functions through computer hardware.

Von Neumann worked on many projects, cellular automata being only one. His work was published posthumously by his student and colleague Arthur Burks. Burks carried on this work at the University of Michigan in the 1960s and 1970s (Burks, 1970) where, through his Logic of Computers Group, he kept the field alive until the glory years began. The early years were fallow in that although von Neumann's insights marked his usual genius, computer hardware and software had not reached the point where much could be done with CA. In fact, progress came from a much simpler, more visual approach to automata. Von Neumann had drawn some of his inspiration from Stanislav Ulam, the mathematician who worked with him on the Manhattan Project. Ulam had suggested to him as early as 1950 that simple cellular automata could be found in sets of local rules that generated mathematical patterns in two- and three-dimensional space where global order could be produced from local action (Ulam, 1962, 1976). It was this line of thinking that was drawn out, as much because in 1970, John Conway, a mathematician in Cambridge, England, suggested a parlor game called 'Life' which combined all the notions of CA into a model which simulated the key elements of reproduction in the simplest possible way. 'Life' has become the exemplar *par excellence* of cellular automata, but its popularity rests on the fact that a generation of hackers took up Conway's idea and explored in countless ways the kinds of complexity which emerge from such simplicity.

It is probably worth stating the elements of 'Life' for it is a more general embodiment of the key elements of CA than our first example. In essence, 'Life' can be played out on any set of cells which exist in any space but it is most convenient to think of this space as being a regular tessellation of the two-dimensional plane such as the usual cellular grid. Any cell can be 'alive' or 'dead', 'on' or 'off', and there are two rules for cells becoming alive/giving birth, or dying/not surviving. The rules are simplicity itself. A cell which is not alive becomes alive if there are exactly three live cells immediately adjacent to it in its Moore neighborhood. A cell remains alive if there are two or three live cells adjacent to it, otherwise it dies. Less than two adjacent cells implies the cell dies from isolation, more than three from overcrowding. The event that set the field humming in 1970 was John

Conway's challenge, reported by Martin Gardner (Gardner, 1970) in his recreational mathematics column in *Scientific American*, that he, Conway, would give a prize of 50 dollars to the first person who could unequivocally demonstrate that certain configurations of 'Life' could be self-perpetuating. The challenge was won within the year by Bill Gosper and his group at MIT, who showed that a particular configuration of cells and their dynamics called a 'glider gun' would, under these rules, spawn live cells indefinitely (Poundstone, 1985).

However suggestive the game of 'Life' might be, the original logic of automata eventually came to fruition. Burk's group produced some of the basic ideas which now serve to underpin complexity theory. Work on genetic algorithms associated with Holland (1975, 1992), and with cooperative evolutionary game theory (Axelrod, 1984) has come from Michigan as have new developments of CA-like models in the fast growing field of artificial life (Langton, 1989). Much of this work has been quickly disseminated, linked to more general approaches to systems theory, and joined to new developments in morphology and nonlinear dynamics such as fractal geometry, chaos, and bifurcation theory. The field itself has also been the subject of more fundamental theoretical exploration particularly by Wolfram (1984) who has classified CA in terms of four varieties of system stability, and there have been various attempts to consider CA as parallel computation (Toffoli and Margolus, 1987). Applications now abound in many fields which have a spatial bias and involve the evolution of populations, from ecology to astrophysics, but all are marked by a strong pedagogic flavor. It would appear that CA are most useful in simplifying phenomena to the point where the kernel of any local-global interaction is identified, and this has meant that full-scale system simulations based on CA are still rare, which is surprising given that their charm and attraction lie in their ability to reduce systems to their barest essentials.

5.3 Neighborhoods, transitions, and conditions

We will begin by examining the range of different CA that might be constructed by varying neighborhoods, transition rules, initial conditions, and system states. Figure 5.2 illustrates four different types of neighborhood which are all based on the notion that the neighborhood around any cell is composed of cells which are geometrically contiguous. It is entirely possible to consider neighborhoods where the concept of locality does not mean physical contiguity, that the cells comprising the neighborhood are scattered within the space, especially if the cells of the CA are not spatial in the Euclidean sense. But such examples stretch the concept of GC and we will avoid them here. In Figures 5.2(a), (b), and (c), the cells comprising each neighborhood are symmetrically arranged whereas in Figure 5.2(d), there is no symmetry, although the property of uniformity which means that every

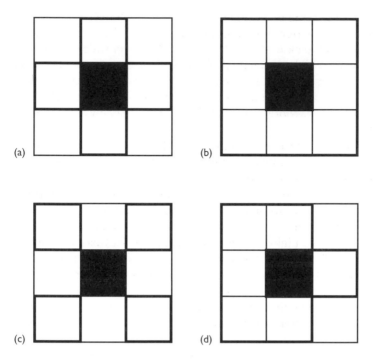

Figure 5.2 Local neighborhoods: (a) von Neumann; (b) Moore; (c) displaced von Neumann; (d) asymmetric.

cell in the system has the same form of neighborhood, imposes a meta regularity on the automata. The neighborhood in Figure 5.2(a) is called the von Neumann neighborhood, in contrast to the complete 3×3 cell space in Figure 5.2(b) which is the Moore neighborhood, and these are the most usual and the most general. The pattern in Figure 5.2(c) is a symmetrically displaced version of the von Neumann neighborhood whereas that in Figure 5.2(d) is more randomly configured, although its form must have meaning to the problem in hand. Within the complete 3×3 cellular space, there are a total of $\sum_n (9! / (9-n)! \, n!)$ combinations or forms (where the summation over n is taken from 1 to 9). This gives 511 possible neighborhoods whose cells are contiguous to one another within the 3×3 space and this admits an enormous variety of patterns that might be generated by such CA. In this context however, we will largely deal with the Moore neighborhood.

If we turn to patterns within the neighborhood which generate different transition rules, the number of possibilities is even greater. Let us assume that each cell is either on or off, the two state characterization that we will use for most of the examples in this chapter. Then for the von Neumann neighborhood in Figure 5.2(a) which consists of 5 cells, there are 2^5 or 32 different configurations of on-off cell patterns that affect the transition rule.

If the transition rule is one of switching the central neighborhood cell on or off if a certain set of patterns occurs, then there are 2^{32} possible automata that might result. In the case of the Moore neighborhood which has 9 cells, then the number of possible automata is 2^{512} which is an astronomical number, twice as large as the number of elementary particles in the universe! (Toffoli and Margolus, 1987). There are of course not two possible neighborhoods – the von Neumann and Moore – but 511 for which these computations can be envisaged.

Patterns inside neighborhoods interact with the shape of neighborhoods themselves and this complicates the concatenation of neighborhoods with patterns. But the point is that this shows that enormous variety can be generated by thinking in these terms about the kinds of patterns that might be computed using cellular automata. Our discussion is devoid, however, of any meaning to the pattern of cells but it does illustrate the possibility that any conceivable pattern might be computed using CA. Readers will have to think a little about these implications and if they find difficulty in envisaging these possibilities, first think of all the possible neighborhoods one might construct based on a 3×3 grid – begin to write them out to convince yourself that there are 511 possibilities or 512 if the neighborhood with no cells is included. Then think about the number of different on-off configurations within the 3×3 space which might trigger some action or transition. Start to write these out too. You will give up quite soon but at least this will demonstrate the enormous array of possible structures that can be built from simple bits using the cellular automata approach.

There are three distinct sets of initial conditions which are associated with different types of growth model and different characterizations of states. The simplest condition involves an automata which begins to operate on a single cell which is switched on in a two-state (on/off, developed/non-developed) system. The best example is the one we have already examined in Figure 5.1 which is the growth of a pattern from a single seed, often placed at the center of the cellular space. The second condition is a straightforward generalization of this to more than one seed. These two types of condition might be thought of as invoking growth processes within a single city or system of cities, for example. The third condition involves a system that is already complete in that every cell is already in one state or the other, and that the emphasis is on changing states, not upon further growth. Of course, this is often a matter of interpretation for the idea of developed and non-developed states might be seen in a similar way where the focus is on change rather than growth. We will explore all three of these starting points in the examples below. As regards boundary conditions, our examples involve 100×100 cellular spaces where the edges of the square grid are joined in toroid fashion, a usual structure for computer screens, where the space wraps round in the horizontal (east-west) and vertical (north-south) directions. We will stop several of our examples before the wrap around occurs, thus imposing arbitrary stopping rules as boundary conditions but in this context, our boundary conditions will always be of the simplest kind.

We will also demonstrate transition rules of both a deterministic and stochastic (random) form. Our first examples will be deterministic: whenever a transition of state in a cell takes place, this will be an exact function of the number and/or pattern of occupied cells in its neighborhood. We then vary this rule in that we assign probabilities of transition to any change in state where these probabilities might vary with the number of cells occupied in the neighborhood, for example. We will mainly deal with two-state systems although in one case we will illustrate a three-state system. Finally, we will describe several applications to cities and related ecologies, before concluding with suggestions for further reading.

5.4 Basic cellular automata: growth from single seeds

Let us begin our illustrative examples by returning to the first one shown in Figure 5.1 in its 1D and 2D forms. In one sense, these automata are almost trivial in that they grow under the most unrestrictive neighborhood rule to fill the entire space available. Their rate of growth is regular, with space and time synchronized through the general equation $N(r) = (2r + 1)^D$ or $N(t) = (2t + 1)^D$ where D is the dimension of the system. Most CA generated from a single seed will utilize transitions which do not lead to their entire space being filled in such a regular way but it would seem intuitively attractive to be able to measure their space-filling by the same equations. Interesting patterns will fill more than a 1D line across the space and less than the entire 2D space and thus we might expect the dimension D to lie between 1 and 2 for such automata. Such fractional dimensions imply that such automata are fractal as we will show, and although we do not have the time to digress into the fractal dynamics of CA, there are very strong connections between the patterns generated by CA and fractal geometry which we will exploit in a casual way (Batty and Longley, 1994).

Our first examples deal with transition rules that generate a change in cell state (from off to on) when only a fixed number of cells are active/on in the neighborhood. In Figure 5.3, we show the possibilities. Figure 5.3(a) shows the pattern generated when a cell changes state from off to on when *one* and *only one* cell is active in the Moore neighborhood. Figure 5.3(b) shows the pattern when *one* or *two* cells are active, and Figure 5.3(c) shows the pattern when *one* and *only one* cell is active in the von Neumann neighborhood. If there are more than two cells active in the Moore neighborhood and more than one active in the von Neumann neighborhoods, the automata generate space that is filled entirely (as in Figure 5.1 for the Moore neighborhood). It is quite clear that the patterns generated in Figure 5.3 are fractal or fractal-like; they are self-similar in that a basic motif, which in turn is a function of the neighborhood rules, is repeated at different scales, and as the associated

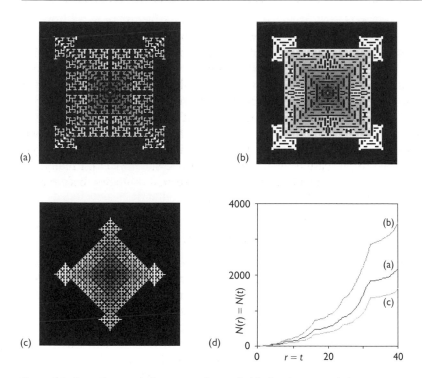

Figure 5.3 Fractal growth from a single seed: (a) development if there is one
developed cell in Moore neighborhood; (b) development if there are one
or two developed cells in Moore neighborhood; (c) development if there is
one developed cell in von Neumann neighborhood; (d) growth paths for
(a), (b), (c).

space-filling graphs show, they fill more than the 1D space and less than the
2D space. In fact because of the way they are generated, the dimension D in
$N(r) = (2r + 1)^D$ and $N(t) = (2t + 1)^D$ must be determined statistically but in
each case, these dimensions are close to but less than 2, i.e. $D \approx 1.947$ for
Figure 5.3(a), $D \approx 1.945$ for Figure 5.3(b), and $D \approx 1.997$ for Figure 5.3(c).

In this chapter, we will illustrate all our examples using the analogy between
CA and the way cities are developed, although readers are encouraged to
think of other examples pertaining to their own fields of interest and expertise.
The patterns generated in Figure 5.3 are highly structured in that the assumptions embodied within their transition rules are very restrictive. However,
these are reminiscent of a wide array of idealized city forms, such as in the
various writings of Renaissance scholars concerning the optimum layout and
size of cities. In fact, these kinds of CA provide excellent analogs for generating those highly stylized residential districts in cities where rules concerning
vehicular-pedestrian segregation are used as in the neighborhoods of English
and American New Towns which originated from the pioneering housing

layout in Radburn, New Jersey. They are also suitable for showing how the rules of town layout are exercised in imperial towns such as Roman castra and colonial towns of the New World, e.g. Savannah, Georgia (Batty, 1997).

Most cities do not grow with such restrictive conditions on development; completely deterministic, regular patterns are the exception rather than the rule. To make such automata probabilistic, it is necessary to specify that a change in state will occur with a certain probability if a particular condition or set of conditions in the neighborhood is (are) met. Thus the transition rule only operates with a certain probability. For example, in the case of the complete space-filling automata in Figure 5.1, a cell which has within its Moore neighborhood an already developed cell, is developed but only with a given probability ρ which effectively means that it is developed only $\rho \times 100\%$ of the time the transition rule is met. This is usually achieved by considering the probability to be a threshold, above which the cell is not developed, below which it is. For example, if the probability of development were $\rho = 0.2$, then a random number between 0 and 9999, say, is drawn. If the number is 2000 or greater, the cell is not developed; if less, it is developed. It is usually argued that probabilities reflect uncertainty about the decision in question, or variety in preferences which is captured through noise in the transition rule.

We show these kinds of automata generated by the probabilistic rule in Figure 5.4. In Figure 5.4(a), we show the pattern where the probability threshold is fixed at 0.2 with the boundary condition at $r = 40$ which is reached after 66 time iterations. If we were to tighten this threshold, then all that would occur is that the same kind of pattern would be generated faster as the probabilities are independent on each cycle of the automata, i.e. if a cell has a 20% chance of being developed and it is not, it still has a 20% chance of being developed the next time, and so on. Although the patterns are different in terms of the 'time' each cell is developed even when the probability thresholds are different, the space gradually fills up and in the limit, the entire space is filled as in Figure 5.1 with $D = 2$. If sparser structures with $1 < D < 2$ are to be generated, then it is necessary to make the sequence of probabilities dependent; for example, if the threshold is $\rho = 0.2$ the first time and the cell is not selected, then the threshold might become $\rho \times \rho = 0.2 \times 0.2 = 0.04$ the second time and so on. In these randomized automata, $N(r)$ and $N(t)$ are no longer synchronized or even lagged. In Figure 5.4(b), we show the pattern generated to $r = 40$, for a probability threshold of 0.5 where the neighborhood rule is the same as that used to generate the fractal pattern shown in Figure 5.3(a).

So far, we have shown how very different patterns might be simulated by altering transition rules based on the two standard Moore and von Neumann neighborhoods and by introducing probabilistic thresholds into the exercise of transition rules. However, we can also change the nature of the neighborhood by making certain cells in the Moore neighborhood illegal for

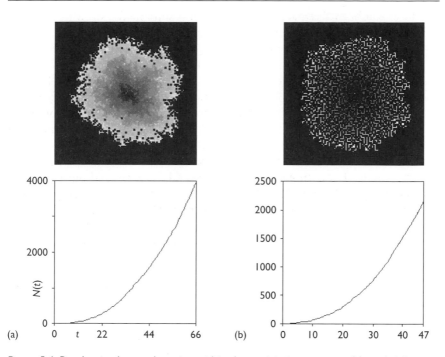

Figure 5.4 Randomized growth: noise within deterministic automata: (a) probability threshold ρ = 0.2 with development in the Moore neighborhood as in Figure 5.1; (b) probability threshold ρ = 0.5 with development in the Moore neighborhood as in Figure 5.3(a).

development. This implies a more general principle of placing a mask over the cellular space to restrict certain areas, thus artificially changing the nature of the entire space. Here however we will use such masks solely in the 3 × 3 Moore neighborhood to show how different configurations of cells can lead to different patterns. This does not destroy the uniformity assumption in that all the neighborhoods and transitions are still the same. Moreover, note that the von Neumann neighborhood is a subset of the Moore neighborhood (as shown previously in Figure 5.2(a)) in that it is formed by making the diagonal cells to the center cell illegal for development. In general, this making of cells illegal is equivalent to altering the transition rules so that certain cells are made ineligible for activating a change in state but this is also a more graphic way of illustrating how CA can produce very different forms.

In Figure 5.5, we show four typically different neighborhood masks which influence the growth from a central seed site. If you refer to Figure 5.3(c) which shows the diagonal grid of growth from the von Neumann neighborhood, it is tempting to ask how a similar horizontal grid of alternating on-off cells might be generated. This is possible by simply displacing the von Neumann neighborhood where the mask in Figure 5.5(a) shows how this is

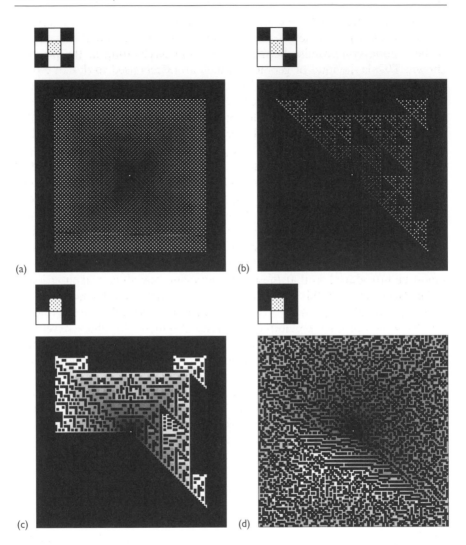

Figure 5.5 Masked neighborhoods: restrictions on development: (a) displaced von Neumann; (b) Sierpinski's gasket; (c) superblock layout; (d) superblock layout with $\rho = 0.5$ and no limit on r.

accomplished. Note that the black cells in the mask show those that are legal, the white those that are illegal, and the stippled square is the origin. If this mask is reduced to only the diagonal half of the 3×3 grid as in Figure 5.5(b), the resultant pattern which is grown to the edge of the screen is a class of fractal known as the Sierpinski gasket (Batty and Longley, 1994). It has a fractal dimension $D \approx 1.585$ which is also confirmed from estimation of D in $N(r) = (2r + 1)^D$.

In Figure 5.5(c), we have configured the legal neighborhood as a figure 7 shaped block which, if viewed as a superblock of housing, might be assumed to be a good compromise between access and daylighting in the position shown. This is the kind of geometry which architects used to design estates of superblocks in the 1950s and 1960s and its replication is reminiscent of the kinds of residential layout seen in municipal housing. In fact, it would be a relatively simple matter to add some meta-transition rules which enable green space and appropriate road access to be incorporated, thus providing reasonable simulations of the kind of geometry that actually exists. We can of course apply all the automata we have introduced to these modified neighborhoods, including the 505 neighborhoods which we have not shown in the examples so far. In Figure 5.5(d), for example, we show how we can apply a probabilistic threshold to the neighborhood of Figure 5.5(c) but this time letting the automata wrap around the screen. This produces a more realistic pattern which begins to suggest how these transition rules might be embodied as shape grammars and how physical constraints on development might be introduced so that these automata can reproduce real situations.

The last example in this section introduces an entirely new theme. So far we have only considered growth – new development in which cells can be switched on. Now we consider replacement or redevelopment, in which developed cells which reach a certain 'age' are first turned off – emptied of development or demolished, thus entering the development process once again. This is much more akin to how development actually takes place. Buildings age, are renewed or demolished, and new buildings are then built on the same sites. For any CA, we have a complete development history, and thus we know the age or time at which the cell was first or last developed. What we can now do is introduce an age limit parameter which when reached by any cell, causes that cell to become vacant, and thus eligible for redevelopment or new development.

In most of the simple examples so far, space has been synchronized with time, and development (growth) has proceeded in waves or bands at increasing distances from the central seed site. If an age threshold is introduced in any of the examples so far, then waves of development and redevelopment can be seen pulsing out from the central seed site as the automata are run indefinitely without any boundary conditions. This is clearer in the deterministic models but it is also apparent in the probabilistic where the thresholds are greater than $\rho = 0.5$. In Figure 5.6(a), we show how two waves pulsate through the structure where the age threshold is set at 15 years. In Figure 5.6(b), we show the long-term age trajectory of the system in terms of the average age of cells. Here the age profile increases at an increasing rate until the first wave of redevelopment kicks in and then it builds up again until the second wave. The relative drop in average age associated with these pulses is gradually ironed out as the waves of redevelopment become the dominant feature of the development. These waves are in fact *emergent phenomena* in that they are a result of delayed local actions. In a sense, the growth around

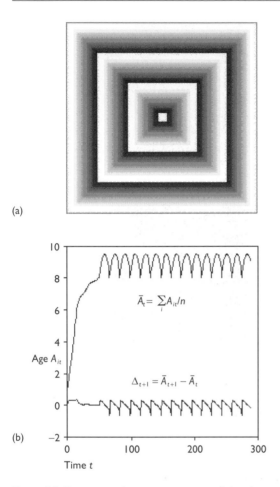

(a)

(b)

$$\bar{A}_t = \sum_i A_{it}/n$$

Age A_{it}

$$\Delta_{t+1} = \bar{A}_{t+1} - \bar{A}_t$$

Time t

Figure 5.6 Emergent phenomena: waves of development and redevelopment: (a) spatial waves of development and redevelopment; (b) waves reflected in average age of development.

the central seed site is one of increasing waves which result from local actions acting immediately on the geometry, and the waves of redevelopment are essentially these same waves which are delayed through the age threshold. We refer to waves of change in real cities but, as in these examples, although they depend upon the interactions associated with the local dynamics, they cannot be predicted from just a knowledge of these local rules. In this sense, they are formed in the same way as a wave of laughter or excitement ripples through an audience, which depends on how one's neighbors are reacting, not upon any macro-property of the system-at-large. It is thus necessary to observe the system in motion before such emergent phenomena can be explained and understood.

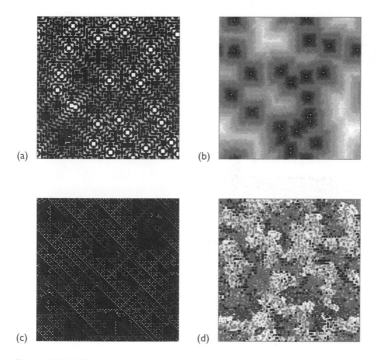

(a)

(b)

(c)

(d)

Figure 5.7 Different varieties of growth from several seeds: (a) restricted growth in the Moore neighborhood as in Figure 5.3(a); (b) complete growth in the Moore neighborhood with $\rho = 0.2$; (c) Sierpinkski's gasket following Figure 5.5(b) using many seeds; (d) superblocks from many seeds following Figure 5.5(c).

5.5 Many seeds, complete development, and self-organization

The extension of basic CA to many seeds is straightforward. The immediate interest now is in the kinds of morphologies that are generated when the different patterns around each of the seeds begin to overlap. If the automata are based on the simplest morphology in Figure 5.1, and if the different seeds generate patterns that are not synchronized spatially, then interesting overlapping patterns in the form of waves can occur. Figure 5.7(a) shows the clear overlaps in patterns that emerge when several seeds are planted and automata grown using the restrictive transition rule of Figure 5.3(a) where changes of state occur if one and only one cell exists in the neighborhood. We also show three other patterns in Figure 5.7: in Figure 5.7(b) patterns based on random growth; in Figure 5.7(c) patterns based on the Sierpinski mask (from Figure 5.5(b)); and in Figure 5.7(d), patterns based on the mask in Figure 5.5(c) with renewal/replacement at an age threshold of 15.

It is difficult to demonstrate that anything really useful comes from simply adding patterns together in this manner although there can be endless forms from which evocative and fascinating overlapping pattern waves emerge, and who is to say that some of these might not contain insightful inquiries into real systems? But unlike single seed automata where global patterns do emerge from local transitions, there is nothing other than overlap to give coherence and integrity to the many-seed patterns that are produced by such automata. However the many-seeds approach is useful for generating automata where the location of the multiple seeds is impossible to predict independently, and even for cases where CA are used for design in the manner of shape grammars (Batty, 1997). Finally, we need to note the space-filling properties of such automata. If we have regularly spaced seeds and we operate the simplest transition rule which generates the entirely filled form of $N(r) = (2r + 1)^2$, then the space is filled as $\tilde{N}(r) \approx n \, N(r)$ where n is the number of seeds. Of course when the overlaps occur, the space becomes entirely filled and this marks the boundary condition. All the other automata we have introduced can be generalized in the same way but with caveats imposed by boundary conditions, regular spacing, and space-time synchronization as we have already noted. We can thus generalize $N(r) = (2r + 1)^D$ to $\tilde{N}(r) \approx \emptyset \, N(r)$ where $\emptyset$ is a constant varying with the number of seeds but also accounting for various kinds of noise introduced where complex overlaps occur and where space and time fall out of synch.

We can now move to a somewhat different and possibly more general variety of CA which begins with an already developed system. In such cases, there need to be two distinct states other than the trivial developed/non-developed cases of the previous automata. Consider a completely developed situation is which each cell is populated by two exclusive types of household – with highbrow or lowbrow tastes in music, let us say. Imagine that each resident group 'prefers' to have at least the *same number* or *more* of their 'own kind' in their Moore neighborhood. This preference is not unusual, it is not segregation *per se* and it might even be relaxed in situations where the preference is for say at least 30%, not 50%, of their own kind to be in their neighborhood. In terms of the number of cells in the 3×3 neighborhood, this means that the resident in the central cell would prefer to have 4, 5, 6, 7 or 8 neighbors of the same kind. The transition rule embodying this preference is thus: if there are less than 4 cells of type i in the Moore neighborhood around a cell of type i, then that cell changes to state j, where i and j are the two types of cell.

This is a very different rule from any we have used so far. It involves a change in state which depends on the *nature* as well as the *number* of cells in the neighborhood, not simply the *number* which was the case in the previous growth models. It thus introduces *competition* into the automata. Imagine that the cellular space is arranged so that every other cell in the grid is of a different type or state. Highbrow residents are evenly mixed with lowbrow

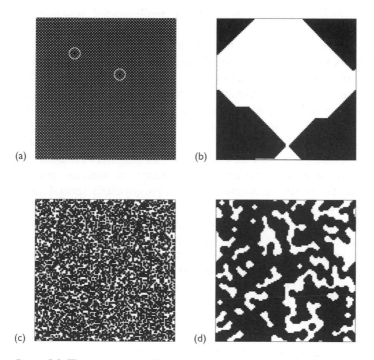

Figure 5.8 The emergence of segrated clusters: (a) regular distribution with two cell changes setting off self-organization; (b) self-organization of (a) into two large dissimilar clusters; (c) random distribution of cells with 50% black and 50% white; (d) self-organization of (c) by cell change when there are more dissimilar adjacent cells.

in a regular checkerboard fashion so that each resident has exactly four neighbors of his or her own kind. The system is balanced such that no changes of state will take place – everyone is satisfied – but the balance is precarious. If we shake the system, shuffle things a bit, the overall proportions are the same – 50% highbrow, 50% lowbrow – but what then happens is that some residents are dissatisfied with the tastes of their neighbors and they begin to change their tastes accordingly. In fact, there are two ways of making the change, either the resident might move to another cell to try to increase their number of like neighbors or they might actually change their tastes to reflect their neighbors. We will adopt the latter course in that we assume that if a resident with taste i finds him or herself surrounded by five or more neighbors with taste j, then they will assume that 'the vote' has gone against them and they will change their tastes accordingly. We show this process of self-organization in Figures 5.8(a) and 5.8(b).

Now let us see what happens when we start with a completely random configuration of tastes but in the proportion 50:50 as illustrated in Figure 5.8(c).

Applying the transition rule, then after a sufficient number of iterations (37 in this case), a new but highly segregated equilibrium emerges which is shown in Figure 5.8(d). What is remarkable about this structure is that although residents will gladly coexist with an equal number of neighbors of either type – an even mix – this does not occur. The local dynamics of the situation make any global equality impossible. You can work this through by considering how more than four cells of either kind leads to a reinforcement or positive feedback which increases segregation which in random situations is likely to be the norm. Moreover, once this begins, the local repercussions throughout the system can turn what is almost an even mix into a highly segregated pattern. Once again, this is a true property of emergence in CA for it is impossible to deduce it from a knowledge of the local dynamics. It only occurs when the local dynamics are writ-large. It was first pointed out by Thomas Schelling in 1978 as an example of how micro-motives cannot be aggregated into macro-motives, which he illustrated with respect to the emergence of highly segregated ethnic neighborhoods in US cities. A similar problem in which the residents move rather than change their tastes is worked through by Resnick (1994).

We now pose the problem of integrating this type of transition rule into those used to simulate growth and change in the urban system. First we develop a growth model for a two-income population around a single seed using the most general transition rule which changes the state of a cell to developed if at least one cell in its Moore neighborhood is developed. We impose a distance threshold of 25 units around the central seed site. If any cell is < 25 units from the seed, its development is set for low income, if ≥ 25, then the development is for high income. We fix the spatial boundary condition at which development stops at $r = 40$ units of distance. When we run the model, we generate a circular development which changes in income type once the threshold is passed and which is bounded at $r = 40$. We show this in Figure 5.9(a) where we also show the total number of cells of the two different income types, generated through space and time, in Figure 5.9(c). If we apply our preference transition rule – that if there are less than four neighbors of like kind, the cell in question changes its state to the other kind – as the growth simulation proceeds, when the first distance threshold is reached, there are never enough cells of the new kind on the edge to outweigh the existing kind and thus the development is always of the low income variety. What happens here is that on the edge of the development at the threshold, each neighborhood is mainly composed of low income and of course undeveloped cells, with very few high income. Thus we need to remove the undeveloped cells from the rule and if we now change this to ensure that at least 50% of the cells are of the same kind for the new state to be preserved, then the model will work as previously in Figure 5.9(a).

In Figures 5.9(b) and 5.9(d), we show the original growth simulation but this time with a degree of randomness imposed which means that the growth

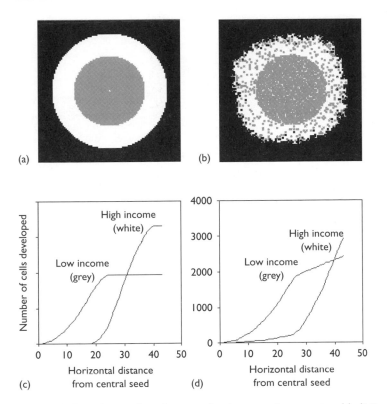

Figure 5.9 Spatial growth and segregation in a two-income city: (a) distinct location of income groups; (b) randomized location ρ = 0.2; (c) spatial segregation of income groups; (d) spatial mix of income groups.

in low and high income cells overlaps. We can of course increase the degree of overlap by reducing the probability threshold ρ, and once the threshold falls to ρ = 0.5, then the pattern is entirely random and the distance threshold has no meaning. There are now two ways in which we can apply the pre-ference transition rule. We can apply it either at the end of the process of growth, once the boundary condition has been reached, somewhat in the manner we applied it to the pattern in Figure 5.8(b), or we can apply it on every iteration as the growth proceeds. We first apply the rule to the com-pleted pattern in Figure 5.9(b) and the result is shown in Figure 5.10(a) where there is still a clear resemblance to the original growth although the randomness has disappeared. However, in Figure 5.10(b), we show what happens if we apply the preference rule to the random growth of Figure 5.9(b) as it proceeds. There is continual change and the original pattern of the central circular city arranged in two concentric rings is changed com-pletely. It is instructive to watch this as the growth proceeds. Irregularity in

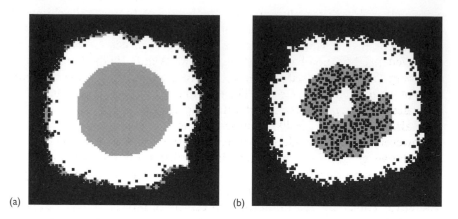

Figure 5.10 Spatial self-organization in a two-income city: (a) self-organization after the process of development; (b) self-organization at each stage of the development process.

structure is quickly removed, and small clusters which reach the threshold begin to grow at the expense of more sparsely related groups of cells. What is significant is the way the preference dynamics distorts the overall dynamics through a process of reorganization. In fact, this process might be seen as one of self-organization in that what is happening is another feature of emergence from local dynamics. Preferences are being exerted at the local scale and cells are changing their local organization to reflect this. Overall the pattern of high segregation which emerges is one of self-organization of the entire structure. In a sense, local self-organization which involves nothing other than what is happening in the local neighborhood field generates a self-organized macro-pattern which cannot be predicted from local conditions.

5.6 Applications to cities and related ecologies

The application of CA to urban systems like CA itself can be traced back to the beginning, to the first attempts to build mathematical models of urban systems which began in the 1950s. In the postwar years, social physics was in full swing and models of spatial diffusion were an important branch of these developments. Hagerstrand as early as 1950 was building diffusion models, specifically of human migration based on the excellent historical population record in Sweden, and these models although embodying action-at-a-distance through gravitation effects were close in spirit to the notion of cells being affected by changes in their immediate vicinity (Hagerstrand, 1967). These were not strict CA but cell-space models in the terminology adopted here. In the early 1960s, CA were implicit in the wave of computer models designed for land use and transportation planning. Chapin and his colleagues at North Carolina in their modelling of the land development process articulated cell-space models where changes in state were predicted as a function of a variety of factors affecting each cell, some of which embodied neighborhood effects (Chapin and Weiss, 1968). Lathrop and Hamburg (1965) proposed similar cell-space simulations for the development of western New York State. The idea that the effect of space should be neutralized in such models by adopting regular lattice structures such as grids also encouraged CA-like representations.

However, strict CA models came from another source – from theoretical quantitative geography. These were largely due to Waldo Tobler (1970, 1975, 1979) who during the 1970s worked at the University of Michigan where Arthur Burks and his Logic of Computers Group were keeping the field alive. Tobler himself first proposed cell-space models for the development of Detroit but in 1974 formally began to explore the way in which strict CA might be applied to geographical systems, culminating in his famous paper 'Cellular Geography' published in 1979. At Santa Barbara in the 1980s, Couclelis (1985, 1988, 1989), influenced by Tobler, continued these speculations, until the early 1990s when applications really began to take off as computer graphics, fractals, chaos and complexity all generated the conditions in which CA has now become an important approach in GC.

It is perhaps surprising that CA models were not explored earlier as a basis for GC in urban and related simulations, but other approaches held sway, particularly those that emphasized time rather than space *per se*. Dynamics has always held a fascination in spatial modelling. Early attempts at embodying time into operational urban models were always plagued by a simplistic view of dynamics or by difficulties of embedding appropriate dynamics within spatial representations (Forrester, 1969; Batty, 1971). By the early 1980s, however, several groups had begun to explore how developments in non-linear dynamics might be adapted to the simulation of urban change. Wilson (1981) building on catastrophe theory, Allen (1982) on adaptations of Prigogine's approach to bifurcation through positive feedback, White (1985) on notions of discontinuity in the behavior of regional systems, Dendrinos and Sonis (1990) on chaotic dynamics, and Dendrinos (1991) on predator–prey models, all set the pace. But this work downplayed the spatial element which was considered simply as a starting point. The emphasis was upon new conceptions of time. The dynamics of how spatial morphologies evolved and changed took much longer to gain the attention of researchers following this new paradigm. Nevertheless, it was from these developments in dynamics that the first explicit application of CA to real urban systems came. White and Engelen's (1993) application of CA in the light of their own work in non-linear urban dynamics was to the development of US cities such as Cincinnati where they showed how urban form could be modelled through time and how these forms were consistent with fractal geometry and urban density theory (Batty and Longley, 1994).

Strict CA, of course, does not appear immediately applicable to real systems where the definition of states, neighborhoods, and transition rules is much more general than the theory suggests. Clear statements of the theory of CA have only been produced since Wolfram's (1984, 1994) work and even now there is no definitive discussion of the ways in which strict CA might be relaxed and adapted to real systems, with the possible exception of Toffoli and Margolus's (1987) book. The key problem of adapting strict CA to generalized GC involves the issue of 'action-at-a-distance'. Numerous aspects of GC that have spatial analysis as their foundation, articulate spatial behavior as the product of 'action-at-a-distance', building on notions of gravitation, spatial autocorrelation, and network connectivity. But such notions are invariably developed from models of static systems where distance relations have clearly evolved through time. The situation for dynamic urban theory is much more confused because there is a strong argument that action-at-a-distance 'emerges' in such systems as the consequence of local action through time, as a product of the successive build-up and compounding of local effects which give rise to structures that reflect much wider interaction. There has been little or no research into such possibilities; in fact, it is the development of CA in this context which has raised the idea of these effects – so long taken for granted as exogenous – being endogenous to, and emergent from such computation. As cells evolve and change, building their overall spatial potential through local neighborhood action or decision, it is possible to conceive of all system-wide effects as being embodied in this potential. Therefore, it is only necessary to ever act behaviorally according to what this potential is within its local neighborhood for this potential takes account of wider action-at-a-distance effects. Simply by examining the position of a cell in the system, these effects are known. This is an issue that clearly requires considerable research.

The flurry of recently developed applications can be divided into those which deal with hypothetical systems in contrast to real systems, as well as those which are strict CA in contrast to cellspace (CS) models where definitions of local neighborhood and transition rules are relaxed. Other significant features involve relationships with GIS software, the extent to which multiple states are dealt with, and the starting points for such simulations based on limited numbers of seeds or already developed landscapes. There are now some 20 or more applications of CA to cities. Most deal with hypothetical examples which emphasize some aspect of urban structure or dynamics such as the segregation of land uses, the diffusion or migration of resident populations, and so on. Applications to real cities are still quite rare. The simulation of the growth dynamics of San Francisco and the Bay Area by Clarke et al. (1997) which is based on their earlier wildfire simulation using CA (Clarke et al., 1994) and Wu and Webster's (1998) simulation of the growth of Guandong Province are perhaps the most realistic large-scale applications which show how the process of urbanization can be modelled over long and short time scales. The residential simulation models in Wellington County, Ontario (Deadman *et al.*, 1993) and in East Amherst, New York (Batty and Xie, 1994) attempt to model the long-term dynamics of suburban growth. At a slightly coarser spatial level, White and Engelen's (1993) simulation of the growth of key land uses – residential, commercial and so on – in four US cities, Cincinatti, Milwaukee, Houston and Atlanta, show how different density patterns emerge, while Langlois and Phipps' (1995) simulation of Ottawa illustrates how land uses restructure themselves to form self-organizing but segregated patterns of the kind illustrated earlier in Figures 5.8 and 5.9. Batty (1997) also presents a coarse simulation of Buffalo, New York using the *Starlogo* software into which physical constraints on development are set.

Several CA models have been developed to illustrate certain theoretical aspects of self-organization which draw their inspiration from real cities. Portugali *et al.*'s (1994, 1997) *City* models allude to segregation patterns in Israeli cities while Couclelis's (1985) adaptation of CA to the development process is drawn from real examples in Los Angeles. There are, however, many models which do not formally appeal to CA although they mirror certain aspects of local interaction. For example, multi-agent models such as those developed by Sanders *et al.* (1997) fall into this class while models of urbanization which draw on various cell-based representations of urban growth such as de Cola's (1997) Washington–Baltimore simulation and Meaille and Wald's (1990) simulation of the Toulon region in Southern France could be regarded as CA models. What makes all these applications non-operational for planning purposes is the scale at which they are represented. Many divide complicated regions into only 100×100 cells. Even 1000×1000 scale representations are rarely sufficient to capture the kind of detail being modelled but at this point parallel computation is probably necessary and this restricts applications. But by far the most limiting feature of all CA and CS models is their difficulty in handling explicit spatial interaction between cells. Let us elaborate once again.

Action-at-a-distance is the central organizing principle behind spatial modelling. Consequently most applications of CA break with the rule that truly local neighborhoods condition growth and change, preferring to embody some action-at-a-distance within larger neighborhoods within which the effect of individual cells might embody some distance decay. These cell space models are writ large in empirical applications, for example in those developed by Batty and Xie (1994), Xie (1994) and White and Engelen (1993). Strict CA appears to have been invoked in the Bay Area model (Clarke *et al.*, 1997) although its description like many others leaves these important details unclear. Moreover in several applications, the scale of space – the size of the cells – and the extent to which these contain all interaction within is rarely explicit. Couclelis (1985, 1988, 1989) has emphasized many times in her papers, at best, CA must be primarily seen as a metaphor for urban growth and change: '. . . systems of the kind considered here have such volatile behavior that models representing them may be useful as 'metaphors' or conceptual organizing schemata, than as quantitative planning models' (Couclelis, 1989, p. 142).

There are, however, ways in which CA might provide much more sophisticated representations of urban systems than anything so far. When systems grow in space, they invariably diffuse and the repeated action of local rules can be so structured as to represent the cumulative buildup of interaction potential. This might be no more than when a cell is occupied. It interacts locally with all other occupied cells in its neighborhood from then on and thus cells which are occupied first always have the greatest cumulative 'potential' interaction through time. Potential might thus vary directly with the age of a cell – when it was first occupied – and in this way centrality can be reinforced as action-at-a-distance 'emerges' from such repeated local actions across space. This is the principle exploited by Batty (1998) in his exploration of how edge of city effects – bifurcations – might be introduced into urban development patterns. Another direction in developing these spatial aspects has been by Couclelis (1997) in her introduction of explicit interaction effects and the notion of proximal space into CA models. This framework has been operationalized by Takeyama and Couclelis (1997) who use geo-algebra as a basis for generalized GC modelling which relies upon cell-space representation but enables both implicit and explicit, local and global interactions to be consistently modelled inside geographic information systems.

The various applications to hypothetical urban systems which dominate the field all stress different aspects of the theory of CA. There are several strong links to GIS based on the notion that CA and CS models use a form of representation which is the same as raster-based/pixel-based GIS. Wagner (1997) shows how standard CA models can be represented and modelled within the raster-based system Idrisi while Itami (1988, 1994) who was the first to exploit this coincidence illustrates how CA has been developed within the Map Analysis Package (MAP). This parallelism between CA and GIS has been noted by many authors, particularly White and Engelen (1997) in their recent island models such as that for St. Lucia, and by Wu and Webster (1998) who have used Arc/Info to map CA output. Links to traditional urban and regional models of the spatial interaction kind have been presented in formal terms by Batty and Xie (1997) and more directly by Sembolini (1997). Formal representations of spatial competition have been embodied in CA by Benati (1997) while the development of polymorphous-polycentric cities has been explored by Batty (1998) and Wu (1997). Ways of estimating CA models to fit space-time series based on neural nets have been developed by Campos (1991). Different possible urban forms have been explored by Batty and Xie (1997) and by Cecchini (1996) who demonstrate how street as well as areal patterns can be coordinated. Indeed, Cecchini and Viola (1990,

1992) emphasize how all possible 'urban' worlds might be generated, refer-ring to these automata as *Ficties* (FICTious CitIES) which represent applica-tions of strict CA to hypothetical land use systems operating under different transition regimes.

One of the most significant features of CA is their ability to self-organize activities into specific patterns such as clusters which imply segregation. The kind of ghetto-ization modelled by Schelling (1978) referred to earlier has been picked up by Portugali *et al.* (1994; 1997) in their *City* models, by Phipps (1989) and Phipps and Langlois (1997) in their ecological and urban models, and by Benati (1997) who illustrates how the movement of traders or firms in a 2D space to optimize their markets is a process of self-organization. Finally, although several applications deal with simple two-state systems such as urban/non-urban, an increasing number have begun to explore the much trickier question of competition between land uses in the traditional manner postulated in urban economic theory in the von Thunen model, for example (Phipps and Langlois, 1997; Wu, 1997).

5.7 Conclusions

CA clearly brings many unique aspects to GC modelling, as much to define the limits of what is modellable and what is computable as to define new methods of GC (Couclelis, 1985). Emphasis on the principles of locality and transition in neighborhoods, on the question of action-at-a-distance, strikes at the very heart of the issue of the way in which potential and density are created temporally in spatial systems and the way new foci of potential, such as edge cities, emerge through positive feedback and bifurcations in the development process. Scale and space are still confused within such model-ling for there has been little research into appropriate levels at which urban systems might be constituted as automata. Applications show scales of cell and neighborhood which range across as many as three orders of magni-tude, while the competition and interaction between different states in cells often representing different uses of land or buildings in an urban context, largely remain to be worked out. CA impose a degree of regularity on the world which must be modified whenever applications are developed but, as yet, the effect of irregular cell shapes, disconnected neighborhoods, the rep-resentation of streets and linear features as cells and their interaction with areal features all define the research frontier.

Relationships to other approaches to modelling at the macro-level involv-ing traditional zone-based interaction models are being developed, but at finer scales, there is little research as yet on how CA connect to shape grammars for example (March and Stiny, 1985), and to micro-simulations in which space occupies a more partial role as in applications to social systems (Epstein and Axtell, 1996). Technically CA are limited by the way they are processed and special parallel architecture machines are required for problems which contain the detail necessary for acceptable applications, something more than the many simple pedagogical demonstrations that already exist. In fact, CA directly throws up discussion of the role of simultaneity in processes of change in dynamic systems. The CA approach provides immediate confrontation with the order in which events take place in systems simulation and thus on questions of the appropriate size of the spatial and temporal unit of representation. Links to wider and deeper theories of dynamics such as self-organized criticality (Bak *et al.*, 1989) and bifurcation theory (Allen, 1997) must be explored if CA is to extend our insights into the functioning of cities and other ecologies. All these directions provide obvious and important themes for further research.

5.8 Further reading

Although no one has yet written a general expository text on CA which gradually builds up basic ideas and translates these into generic applications – this is something we have tried to do here for geographic systems, albeit very briefly. Toffoli and Margolus's (1987) book *Cellular Automata Machines: A New Environment for Modeling* is by far the best exposition to date. Notwithstanding their use of the FORTH language to represent their applications (because their text assumes the reader has access to their customized parallel PC hardware extension *CAM*), this book is still the clearest development of CA ideas to date. Of course, you no longer need their special hardware as PC speeds are more than adequate. Wolfram's (1986, 1994) collections of papers also deal with basic concepts but these are less applied. There are however many popular computer book summaries, for example that by Gaylord and Nishidate (1996). A particularly useful and broader rendition of the CA style of modelling is contained in Resnick's (1994) wonderful little volume *Termites, Turtles and Traffic Jams: Explorations in Massively Parallel Micro-Worlds*.

Agent-based simulations which build on CA have been developed in the field of artificial life, many papers of which have been published as part of the Santa Fe Series in Complexity, namely the edited volumes *Artificial Life I, II, III*, and *IV* (Langton, 1989; Langton *et al.*, 1991; Langton, 1994; Brooks and Maes, 1994). Emmeche's (1994) book *The Garden in the Machine: The Emerging Science of Artificial Life* provides a gentle introduction to this field. Finally the book by Epstein and Axtell (1996) *Growing Artificial Societies: Social Science from the Bottom Up* provides a recent focus with strong implications for human geography.

In urban modelling, the original papers by Tobler (1970, 1975, 1979) are worth studying as are those by Couclelis (1985, 1988, 1989). The original application to Cincinnati and other US cities by White and Engelen (1993) is important. Good edited collections of papers on the application of CA to urban systems now exist, namely Besussi and Cechinni's (1996) edited volume *Artificial Worlds and Urban Studies*; and the special issue of *Environment and Planning B, Volume 23* (issue 2 March, 1997). There are review papers by Itami (1988, 1994) who was first in developing CA within GIS from a landscape planning perspective, and the pedagogic guide by Batty (1997) linking CA to urban form provides a simple primer for urban planners. Links to other GC methods especially those associated with the new science of complexity theory such as chaos and fractals can be found in the books by Batty and Longley (1994), Turcotte (1992), Nijkamp and Reggiani (1992), Dendrinos and Sonis (1990), and Allen (1997).

There are several useful web sites which provide excellent leads to CA software, particularly through concepts dealing with artificial life (A-Life). Brian Hill's A-Life Software homepage *http://www.ccnet.com/~bhill/ elsewhere.html* contains many examples and pointers to pages where you can download software, in this case for Macintosh machines. The Santa Fe homepage on A-Life at *http://alife.santafe.edu/alife/* is another good starting point and the *Starlogo* group's homepage *http://starlogo.www.media.mit.edu/ people/starlogo/* at MIT's Media Lab is worth looking at.

References

Albin, P. (1975) *The Analysis of Complex Socio-Economic Systems*, Lexington, MA: Lexington Books.

Allen, P. M. (1982) 'Evolution, modelling and design in a complex world', *Environment and Planning B*, 9, pp. 95–111.

Allen, P. M. (1997) *Cities and Regions as Self-organizing Systems: Models of Complexity*, New York: Gordon and Breach.

Axelrod, R. (1984) *The Evolution of Cooperation*, New York: Basic Books.

Bak, P., Chen, K. and Creutz, M. (1989) 'Self-organized criticality in the "Game of Life"', *Nature*, 342, pp. 780–782.

Batty, M. (1971) 'Modelling cities as dynamic systems', *Nature*, 231, pp. 425–428.

Batty, M. (1997) 'Cellular automata and urban form: a primer', *Journal of the American Planning Association*, 63, 2, pp. 266–274.

Batty, M. (1998) 'Urban evolution on the desktop: simulation using extended cellular automata', *Environment and Planning A*, 30, pp. 1943–1967.

Batty, M. and Longley, P. A. (1994) *Fractal Cities: A Geometry of Form and Function*, London: Academic Press.

Batty, M. and Xie, Y. (1994) 'From cells to cities', *Environment and Planning B*, 21, pp. s31–s48.

Batty, M. and Xie, Y. (1997) 'Possible urban automata', *Environment and Planning B*, 24, pp. 175–192.

Benati, S. (1997) 'A cellular automaton model for the simulation of competitive location', *Environment and Planning B*, 24, pp. 205–218.

Besussi, E. and Cecchini, A. (eds) (1996) *Artificial Worlds and Urban Studies*, Conveni. n.1, Venezia, Italia: DAEST.

Brooks, R. A. and Maes, P. (eds) (1994) *Artificial Life IV*, Cambridge, MA: MIT Press.

Burks, A. W. (ed.) (1970) *Essays on Cellular Automata*, Urbana-Champaign, IL: University of Illinois Press.

Campos, M. de M. (1991) 'Cellular geography and neural networks: learning socio-spatial dynamics from examples', Department of Regional Science, University of Pennsylvania, Philadelphia, PA, unpublished paper.

Cecchini, A. (1996) 'Urban modelling by means of cellular automata: a generalized urban automata with help on-line (AUGH)', *Environment and Planning B*, 23, pp. 721–732.

Cecchini, A. and Viola, F. (1990) 'Eine stadtbausimulation', *Wissenschaftliche Zeitschrift der Hochschule fur Architektur und Bauwesen*, 36, pp. 159–162.

Cecchini, A. and Viola, F. (1992) 'Ficties – fictitious cities: a simulation for the creation of cities', Paper presented to the International Seminar on Cellular Automata for Regional Analysis, DAEST, Universitario di Architettura, Venice, Italy.

Chapin, F. S. and Weiss, S. F. (1968) 'A probabilistic model for residential growth', *Transportation Research*, 2, pp. 375–390.

Clarke, K. C., Brass, J. A. and Riggan, P. J. (1994) 'A cellular automaton model of wildlife propagation and extinction', *Photogrammetric Engineering and Remote Sensing*, 60, 11, pp. 1355–1367.

Clarke, K. C., Hoppen, S. and Gaydos, L. (1997) 'A self-modifying cellular automaton model of historical urbanization in the San Francisco Bay area', *Environment and Planning B*, 24, pp. 247–261.

Cola, L. de. (1997) 'Using space/time transformations to map urbanization in Baltimore/Washington', *AutoCarto Proceedings*, 13, pp. 200–209.

Couclelis, H. (1985) 'Cellular worlds: a framework for modeling micro-macro dynamics', *Environment and Planning A*, 17, pp. 585–596.

Couclelis, H. (1988) 'Of mice and men: what rodent populations can teach us about complex spatial dynamics', *Environment and Planning A*, 29, pp. 99–109.

Couclelis, H. (1989) 'Macrostructure and microbehavior in a metropolitan area', *Environment and Planning B*, 16, pp. 141–154.

Couclelis, H. (1997) 'From cellular automata models to urban models: new principles for model development and implementation', *Environment and Planning B*, 24, pp. 165–174.

Deadman, P., Brown, R. D. and Gimblett, P. (1993) 'Modelling rural residential settlement patterns with cellular automata', *Journal of Environmental Management*, 37, pp. 147–160.

Dendrinos, D. S. (1991) *The Dynamics of Cities: Ecological Determinism, Dualism, and Chaos*, London: Routledge.

Dendrinos, D. S. and Sonis, M. (1990) *Turbulence and Socio-Spatial Dynamics*, New York: Springer Verlag.

Emmeche, C. (1994) *The Garden in the Machine: The Emerging Science of Artificial Life*, Princeton, NJ: Princeton University Press.

Epstein, J. M. and Axtell, R. (1996) *Growing Artificial Societies: Social Science from the Bottom Up*, Cambridge, MA: Brookings/MIT Press.

Forrester, J. W. (1969) *Urban Dynamics*, Cambridge, MA: MIT Press.

Gardner, M. (1970) 'The fantastic combinations of John Conway's new solitary game of "Life"', *Scientific American*, 223, pp. 120–123.

Gaylord, R. J. and Nishidate, K. (1996) *Modeling Nature: Cellular Automata Simulations with Mathematica*, Santa Clara, CA: Springer/Telos.

Hagerstrand, T. (1967) *Innovation Diffusion as a Spatial Process*, Chicago, IL: University of Chicago Press.

Holland, J. (1975 [1992]) *Adaptation in Natural and Artificial Systems*, Ann Arbor, MI: University of Michigan Press; reprinted in 1992 by MIT Press, Cambridge, MA.

Itami, R. M. (1988) 'Cellular worlds: models for dynamic conceptions of landscapes', *Landscape Architecture*, 78 (July/August), pp. 52–57.

Itami, R. M. (1994) 'Simulating spatial dynamics: cellular automata theory', *Landscape and Urban Policy*, 30, pp. 27–47.

Langlois, A. and Phipps, M. (1995) *Cellular Automata, Parallelism, and Urban Simulation*, Final Report on the Activities of the SUAC Project, Department of Geography, University of Ottawa, Ottawa, Ontario, Canada.

Langton, C. G. (ed.) (1989) *Artificial Life*, Redwood City, CA: Addison-Wesley.

Langton, C. G. (ed.) (1994) *Artificial Life III*, Redwood City, CA: Addison-Wesley.

Langton, C. G., Taylor, C., Farmer, D. and Rasmussen, S. (eds) (1991) *Artificial Life II*, Redwood City, CA: Addison-Wesley.

Lathrop, G. T. and Hamburg, J. R. (1965) 'An opportunity-accessibility model for allocating regional growth', *Journal of the American Institute of Planners*, 31, pp. 95–103.

Macrae, N. (1992) *John von Neumann*, New York: Pantheon Press.

March, L. and Stiny, G. (1985) 'Spatial systems in architecture and design: some history and logic', *Environment and Planning B*, 12, pp. 31–53.

Meaille, R. and Wald, L. (1990) 'Using geographical information systems and satellite imagery within a numerical simulation of regional urban growth', *International Journal of Geographical Information Systems*, 4, pp. 445–456.

Nijkamp, P. and Reggiani, A. (1992) *Interaction, Evolution, and Chaos in Space*, Berlin: Springer Verlag.

Phipps, M. (1989) 'Dynamical behavior of cellular automata under the constraint of neighborhood coherence', *Geographical Analysis*, 21, pp. 197–215.

Phipps, M. and Langlois, A. (1997) 'Spatial dynamics, cellular automata, and parallel processing computers', *Environment and Planning B*, 24, pp. 193–204.

Portugali, J., Benenson, I. and Omer, I. (1994) 'Sociospatial residential dynamics, stability and instability within the self-organizing city', *Geographical Analysis*, 26, pp. 321–340.

Portugali, J., Benenson, I. and Omer, I. (1997) 'Spatial cognitive dissonance and sociospatial emergence in a self-organizing city', *Environment and Planning B*, 24, pp. 263–285.

Poundstone, W. (1985) *The Recursive Universe*, New York: William Morrow.

Resnick, M. (1994) *Termites, Turtles and Traffic Jams: Explorations in Massively Parallel Micro-Worlds*, Cambridge, MA: MIT Press.

Sanders, L., Pumain, D., Mathian, H., Guerin-Pace, F. and Bura, S. (1997) 'SIMPOP: a multiagent system for the study of urbanism', *Environment and Planning B*, 24, pp. 287–305.

Schelling, T. S. (1978) *Micromotives and Macrobehavior*, New York: W. W. Norton Ltd.

Sembolini, F. (1997) 'An urban and regional model based on cellular automata', *Environment and Planning B*, 24, pp. 589–612.

Takeyama, M. and Couclelis, H. (1997) 'Map dynamics: integrating cellular automata and GIS through geo-algebra', *International Journal of Geographical Information Science*, 11, pp. 73–91.

Tobler, W. R. (1970) 'A computer movie simulating population growth in the Detroit region', *Economic Geography*, 42, pp. 234–240.

Tobler, W. R. (1975) 'Linear operators applied to areal data', in Davis, J. C. and McCullaugh, M. J. (eds) *Display and Analysis of Spatial Data*, New York: Wiley, pp. 14–37.

Tobler, W. R. (1979) 'Cellular geography', in Gale, S. and Olsson, G. (eds) *Philosophy in Geography*, Dordrecht, The Netherlands: D. Reidel, pp. 279–386.

Toffoli, T. and Margolus, N. (1987) *Cellular Automata Machines: A New Environment for Modeling*, Cambridge, MA: MIT Press.

Turcotte, D. (1992) *Fractals and Chaos in Geology and Geophysics*, New York: Cambridge University Press.

Ulam, S. M. (1962) 'On some mathematical problems connected with patterns of growth of figures', *Proceedings of Symposia in Applied Mathematics*, 14, pp. 215–224.

Ulam, S. M. (1976) *Adventures of a Mathematician*, New York: Charles Scribner.

Wagner, D. F. (1997) 'Cellular automata and geographic information systems', *Environment and Planning B*, 24, pp. 219–234.

White, R. W. (1985) 'Transitions to chaos with increasing system complexity: the case of regional industrial systems', *Environment and Planning A*, 17, pp. 387–396.

White, R. W. and Engelen, G. (1993) 'Cellular automata and fractal urban form: a cellular modelling approach to the evolution of urban land use patterns', *Environment and Planning A*, 25, pp. 1175–1193.

White, R. W. and Engelen, G. (1997) 'Cellular automaton as the basis of integrated dynamic regional modelling', *Environment and Planning B*, 24, pp. 235–246.

Wilson, A. G. (1981) *Catastrophe Theory and Bifurcation in Urban and Regional Modelling*, Berkeley, CA: University of California Press.

Wolfram, S. (1984) 'Cellular automata: a model of complexity', *Nature*, 31, pp. 419–424.

Wolfram, S. (1986) *Theory and Applications of Cellular Automata*, Singapore: World Scientific.

Wolfram, S. (1994) *Cellular Automata and Complexity: Collected Papers*, Reading, MA: AddisonWesley.

Wu, F. (1997) 'An experiment on generic polycentricity of urban growth in a cellular automata city', Department of City and Regional Planning, Cardiff University of Wales, Cardiff, UK, unpublished paper.

Wu, F. and Webster, C. J. (1998) 'Simulation of land development through the integration of cellular automata and multicriteria evaluation', *Environment and Planning B*, 25, pp. 103–126.

Xie, Y. (1994) *Analytical Models and Algorithms for Cellular Urban Dynamics*, unpublished PhD dissertation, State University of New York at Buffalo, Buffalo, NY.

Chapter 6

Geospatial expert systems

Tony Moore

6.1 Introduction

The division of computational science that has come to be known as expert
systems (ES) has its origins in the broader discipline of artificial intelligence
(AI), where it still resides. Put very simply, the broad aim of artificial intelligence is to simulate human reasoning (Laurini and Thompson, 1992). Expert
systems are the most mature products to emerge from this field (Raggad, 1996),
dating back to the mid-1960s. Since that time, when researchers at Stanford
University developed a program that used chemical expert knowledge to
automatically deduce molecular structure (Durkin, 1996), a plethora of definitions for the emergent technology have been put forward. The following
gives an indication of how the use of expert systems has expanded to encompass nearly every scientific discipline in that time (Cress and Diesler, 1990).

> 'Expert systems are computer systems that advise on or help solve real-
> world problems requiring an expert's interpretation and solve real-world
> problems using a computer model of expert human reasoning reaching
> the same conclusion the human expert would reach if faced with a
> comparable problem.'
>
> (Weiss and Kulikowski, 1984)

In the literature, expert systems are also known as knowledge-based systems
(Skidmore *et al.*, 1996), reflecting the physical computer manifestation of
what the expert knows rather than what is actually known by the expert:
'. . . developed for representing "knowledge" about some domain and for
supporting procedures for deriving inferences about the domain from some
knowledge base.' (Smith and Jiang, 1991). The knowledge base can also be
called a logistical base and comprises rules governed by the inference engine
(an integral part of an expert system), which is a set of procedures for
undertaking some kind of reasoning (Laurini and Thompson, 1992).

In addition, Robinson and Frank (1987) have said that expert systems
should: interact with humans in natural language; function despite some

errors in the data and uncertain judgmental rules; contemplate multiple competing hypotheses simultaneously; and explain their reasons for requesting additional information when needed. These expert system elements and processes will be returned to and elaborated on in more detail in the course of the chapter.

In a survey conducted by Durkin (1996) it was estimated that about 12 500 expert systems have been developed. Couple this with the assertion that expert systems have received a great deal of attention in the professional literature, computing literature and government agencies (Robinson *et al.*, 1986) and one begins to picture just how extensively expert systems have been embraced. Cheng *et al.* (1995) even go so far as to say that expert system research has emerged as an identifiable field for scientific discourse, i.e. it is seen as a distinct field of study in itself.

Despite the large number of developments, only an estimated 10% of medium to large expert systems actually succeed (Keyes, 1989). Raggad (1996) gives a similar figure for medium-to-large expert system success, despite acknowledging the growing interest in expert systems. There may be any number of reasons why expert systems fail; here are a few of them as quoted by Oravec and Travis (1992): the 'Tower of Babel' syndrome; lack of feasibility prototyping (Raggad (1996) expands on this as the failure of developers to analyse carefully user needs and the context within which the system will be used); weak inference engines; hard, slow and tedious required knowledge formulation; and inadequate knowledge (i.e. uncertain, incomplete, controversial, spatial and temporal, hypothetical, self-knowledge, meta-knowledge etc.). Raggad (1996) adds that problems arise because the expert is treated as the end user, which rarely happens.

On balance, Durkin (1996) has remarked that progress since the inception of expert systems has not lived up to the initial successes and resultant hopes. This said, expert systems have come far, and still have enormous potential. Some signs of this optimism can be observed when regarding the status of expert systems in geography. Fischer (1994) has noted that artificial intelligence has received an 'explosion' of interest in the last few years. Furthermore, Fischer asserted that there was no longer any question that expert systems (and neural networks) would be integral in building the next generation of intelligent geographical information systems (GIS). The reason why there is plenty of scope for use of expert systems in this subject area is that GIS without intelligence have a limited chance to effectively solve spatial decision support problems in a complex or imprecise environment. This air of optimism has been present since the early days of expert system application to geography.

> '. . . benefits to be accrued from taking the AI approach far outweigh any possible disadvantages.'
>
> (Fisher *et al.*, 1988)

In general, this chapter reviews and assesses the application of expert systems to the geospatial disciplines. It aims to be a picture of the status of expert systems in geography. Firstly, the history of expert systems is outlined, before exploring the differences between expert systems and conventional systems. Then aspects relating to the physical form of the expert system are outlined and their coupling to GIS explained. Next there is a review of the current status of expert systems related to geography. Topical issues such as knowledge acquisition are examined, before a final consideration of the practical aspects of building expert systems. This is followed by an examination of further opportunities for expert systems in the geospatial sciences. Finally, selected examples illustrating the elements, processes (building, coupling), tasks (knowledge representation) and structure (object orientation) of the expert system are summarized. There is also a further reading section, structured by application.

6.2 Historical review

6.2.1 History and origins of artificial intelligence and expert systems

Experiments in artificial intelligence began in the late 1950s, but initially concentrated on games playing and solving puzzles. A subsequent shift in emphasis, with the knowledge stored being the subject, resulted in more useful and powerful applications being developed (Dantzler and Scheerer, 1993). Out of this change of approach the birth of expert systems came about. The first expert system can be traced back to the mid-1960s at Stanford University. A group of researchers there were developing a computer program with a chemical application that could deduce the structure of complex molecules from mass spectrograms at a performance level rivalling that of human experts. It was called DENDRAL. Knowledge from an expert chemist was encoded and used as the driving force of the program.

During the 1960s and 1970s, expert systems were developed by researchers looking for ways to better represent knowledge. The number of such developments was small, but their contribution was valuable. The noted expert systems MYCIN and PROSPECTOR were built in this phase. Based upon these successes, more money was put into the technology in the 1980s, leading to growth. This was helped when there was a shift in emphasis from overstretching the technology (by purporting to develop the definitive expert system that could solve problems even the experts could not) to developing expert systems for narrow domains and mundane tasks in the mid-1980s (Dantzler and Scheerer, 1993; Fischer, 1994; Durkin, 1996).

In terms of hardware and software, the 1970s heralded expert system development on powerful workstations with declarative languages such as Prolog and Lisp. Because of this exclusivity, only a select few scientists were involved. Since, there has been a move towards PCs in the 1980s, with prot-agonists mainly building on existing expert system shells. This accessibility meant that expert system development was more widespread. For the record, the vast majority of programs have been developed on a PC with the aid of a shell (Durkin, 1996).

6.2.2 Differences between expert systems and conventional systems

Before highlighting the differences between expert systems and conventional systems, it should be noted that when one builds an expert system one goes through a number of stages that closely resemble classical systems analysis. These include identification, conceptualization, prototyping, creating user interfaces, testing and redefinition, and knowledge-base maintenance (Bobrow *et al.*, 1986).

It has been said that expert systems and conventional systems differ in four critical aspects (Table 6.1). Also, it should be said that approaches to building conventional and expert systems are different in terms of their relative conceptual models. There is a waterfall approach to the develop-ment of conventional systems, as opposed to a spiral model for expert system development. In this spiral model, successive prototypes are developed and tested cyclically (Boehm, 1988).

Synonymous with the differences exhibited above are the differences between data and knowledge. Consider that, with data, some automated process collects material, while with knowledge, expertise collects the mater-ial (Smith and Smith, 1977). While data tries to describe reality at the level of instances, knowledge works with abstractions and entity types to do the same (Wiederhold, 1986). At the system level, the knowledge base by necessity involves richer semantics for interpretation than does a database, which is more concerned with efficient storage and retrieval (Brodie and Mylopoulos, 1986). Generally, the knowledge base is represented in terms of

Table 6.1 The differences between expert systems and conventional systems (from Dantzler and Scheerer, 1993)

	Conventional systems	Expert systems
Goal	Conventionally it is to implement algorithms	In an expert system it is to capture and distribute expertise
Focus	Conventionally on data	With an expert system it is on knowledge
Approach	Algorithmic data processing approach	Heuristic reasoning approach
Output	Calculated result	Generate one or more decisions/analyses

rules (Smith and Jiang, 1991) and can support recursive queries (Naqvi, 1986). Data does not have these characteristics. Finally, Smith and Jiang (1991) have noted the inability of relational databases to effectively handle deductive and incomplete information. Alternatives include the object-oriented approach or logic-based approaches, which are covered later on. These approaches can store and manipulate deductive rules of reasoning and data, and can answer queries based on logical derivation coupled with some means of handling incomplete data.

6.2.3 Building expert systems

Once a problem has been defined, the first step in developing a knowledge base is the construction of a conceptual model of the problem domain (Hayes-Roth *et al.*, 1983). Conceptual modelling is an analysis of knowledge acquired from human experts (Chan and Johnston, 1996). Historically, the construction of expert systems has mostly been concerned with logic-based approaches in terms of a declarative language with rules, an example of which is PROLOG (Smith and Jiang, 1991). More will be said about this in the section on knowledge representation.

More recently, knowledge-based techniques have typically taken the form of expert system shells (Fischer, 1994). The expert system shell or 'skeleton' allows the specialist to focus on the knowledge base rather than the workings, which it already provides, e.g. EMYCIN and KAS (Knowledge Acquisition system) are the shells for MYCIN and PROSPECTOR respectively (these two are elaborated upon in the examples section), but with all domain-specific knowledge removed. Shells provide the builder with a number of tools for effective use of the inference engine. They are editing, debugging, consult-the-user and explanation (help) functions (Robinson *et al.*, 1986).

6.2.4 Elements and processes of an expert system

6.2.4.1. Elements

It has been noted that ordinary computer programs organize knowledge on two levels: data and program. Most expert systems organize knowledge on three levels: facts, rules and inferences (Robinson and Frank, 1987). These three levels correspond to two independent core parts of the expert system according to Robinson *et al.* (1986). These are a domain independent inference engine and a domain specific knowledge base (covering both facts and rules).

Expert 'rules' model behaviour of, and functions relating to, a theme. 'Facts' describe single values, such as basic information or events. Other than the core elements of the expert system, there are two other basic parts, a module for knowledge acquisition and a module for interfacing with the user (Laurini and Thompson, 1992).

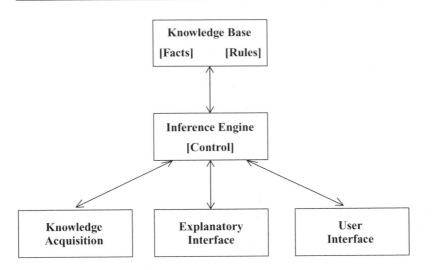

Figure 6.1 The configuration of a typical rule-based expert system (reprinted from Robinson, V. B., Frank A. U. and Blaze, M. A., 'Expert systems applied to problems in geographic information systems: introduction, review and prospects', copyright 1986, pp. 161–173, with permission from Elsevier Science.

Figure 6.1 shows the configuration of a typical rule-based expert system, displaying the interactions between the elements introduced earlier. From application to application, this arrangement may change in terms of conceptual form and nomenclature (see some of the configurations in the latter part of this chapter), though essentially the workings are the same. The principal components are described in detail later in this section.

Fisher *et al.* (1988) describe the processes within the expert system as a 'recognize-act cycle'. Firstly, the inference engine checks the situation parts of each production rule to see if the input facts and embedded facts will allow evaluation of the rule. If so then the rule is selected. After this process, the selected rules form a 'conflict set' of rules. Then the inference engine takes the first production rule in the 'conflict set' and implements it. This is called forward chaining, and is one of a few inference processes to be explored in further detail in this section. Finally, the cycle is repeated until all of the possible information has been extracted from known facts and rules. The explanation module will then tell the user how the expert system reached a conclusion.

The process of forward chaining is also known as deduction. It is used for 'What if?' scenarios. Therefore, if a condition A is true and the rule A → B can be found in the rule base, then we can deduce that B is also true. There is a reverse process to forward chaining, predictably called backward chaining or abduction (reasonable explanation). It is mostly used for discovering the reasons behind a situation. In short, if B is true and the rule A → B applies, then by abduction A is also true. There are two other less advertised processes. Induction occurs when two facts are always concomitant and it would be reasonable to assume that there is a rule expressing a relationship between them. In formal terms, if A is true and B is also true then the rule A → B applies. Finally, transitivity involves the interplay of two rules. If A → B and B → C we conclude that A → C is true (Laurini and Thompson, 1992).

6.2.4.2 Control and search

The terms forward and backward chaining are also used in connection with search strategies used to traverse the rule base, or state-space. In state-space search, operators can search in a forward direction from a given initial state to a goal state (also called data-driven search) (Robinson *et al.*, 1986). This implies that there is no knowledge of the goal in the system (Fisher, 1990). Alternatively, the search can occur in a backward direction from a given goal to initial state (also called goal-driven search) (Robinson *et al.*, 1986). This implies that there is some knowledge about the goal in the system (Fisher, 1990). The appropriateness of either method depends upon the nature of state-space and the particular problem involved (Robinson *et al.*, 1986).

Searches in state-space are conducted with the root node as the starting point, from which progress to child nodes (one of which is the goal) is the next stage. There are several types of search: depth-first search, breadth-first search and any number of heuristic ('rule-of-thumb') search methods. The latter is the most popular method of search used, as an applicable heuristic can be chosen for the specific problem addressed. As an example, two best first algorithms (the simplest of heuristic search methods) are outlined here. In 'costed search', the lowest cost child node is removed, then the children of that investigated, and so on, until the goal is reached, or there are no more child nodes to investigate. In 'branch-and-bound search', the lowest cost child node is expanded. This continues until all links are exhausted and the cheapest path to the goal chosen (Fisher, 1990).

6.2.5 Knowledge representation

According to Kartikeyan *et al.* (1995) there are three conceptual models to represent knowledge: rule-based (Wharton, 1987); frame-based (McKeown, 1987); and blackboard architecture (Hayes-Roth *et al.*, 1983). The choice of method is dictated by the nature of the problem concerned.

The rule base contains procedural knowledge and therefore can be programmed using conventional languages. There are several ways in which domain-dependent knowledge can be encoded, which incorporates searching of many paths in the knowledge-base, not all of which lead to solutions.

The following methods are activated by the control structure whenever certain conditions hold in the data. Several languages allow for this triggering by patterns, of which Prolog has historically been the most popular. Here, first-order predicate logic is used to represent knowledge in terms of formulae, e.g. Jack gave Anne a book = GIVE(Jack, Anne, book).

Production rules have been extensively used to encode knowledge. They comprise a series of IF–THEN statements, which performs an action if a certain condition is met. Alternatively, logical representation can be used, e.g. A1 & A2 & . . . & An → B. This notation means that B is true when A1, A2,. . . , An are true. The formula can also be thought of as a procedure for producing a state satisfying condition B. So, B can be a goal or an assertion or true if A1 → An was true (Robinson *et al.*, 1986).

Another group of knowledge encoding methods are semantic networks and frames, which can be traced as part of the heritage of object orientation (Smith and Jiang, 1991), more of which in the next section. Semantic nets were introduced as a means of modelling human associative memory (Quillian, 1968). They are the principal data structures used by the search procedures responsible for assembling together the locations of a multi-component object (Nilsson, 1980). In a semantic net, an object is represented as nodes in a graph, with the relations between them being labelled as arcs, e.g. in first-order predicate logic, the representation of arc W between nodes A and B is W(A, B). An advantage of using semantic nets lies in their inherent indexing property. This means that objects that are often associated, or conceptually close, can be represented near to each other in the network (Robinson *et al.*, 1986).

Frames are similar knowledge structures in which all knowledge about an object is stored together in an approximately modular fashion (Minsky, 1975). Frame-based development programs (or hybrid tools) came to rival rule-based expert system shells as object orientation took off within artificial intelligence (Durkin, 1996). Given the current popularity of object orientation, it is appropriate that it should be elaborated upon.

6.2.5.1 Object orientation

As indicated in the last section, object orientation owes much to the semantic net and frame knowledge structures. Where object oriented programming is concerned, it is easier to teach children than computer programmers set in the ways of procedural programming. The reason is that object oriented programming represents the way we perceive reality. It seemed only natural that scientists involved in artificial intelligence would eventually link up to it.

The object-oriented paradigm remedies some of the shortfalls that have been noticed in rule-based knowledge bases. In this case, the knowledge base and inference engine have been observed as being closely entwined (i.e. the action is the task of the inference engine). The knowledge base should not be so 'hard-wired' into the system, as it may need to be modified to meet specific demands. It is best kept as a separate entity from the inference engine. Alternatively, rules can be arranged as a hierarchy of objects. The knowledge base is called upon by the inference engine 'Does this rule apply?' 'This one?', etc., until a rule is found that satisfies the operative words and the derived data. This is then repeated for the next tier in the hierarchical object structure. At this stage no action is taken on the rules. Appropriate action is implemented by the inference engine once the levels in the hierarchy have been traversed (Moore *et al.*, 1996).

It is also important for an object-oriented system to be able to intelligently process some semantically imprecise spatial operators, e.g. 'close to', 'between', 'adjacent to'. To process such imprecise queries, knowledge about contexts (or user perspectives) can be introduced to the system (e.g. Subramanian and Adam, 1993; Pissinou *et al.*, 1993). There will be more about handling imprecision later in this chapter.

A further method of building object-oriented systems is the responsibility-driven approach, or client-server model (Subramanian and Adam, 1993; Lilburne *et al.*, 1996). For example, in the case of a spatial expert system shell, the ES shell could be the client and the GIS the server, or vice versa.

6.2.6 Knowledge engineering

Knowledge engineering is a term reputed to have been first coined by Ed Feigenbaum, one of the original pioneers of expert systems in the mid-1960s (Dantzler and Scheerer, 1993). It is one of the greatest challenges in building expert systems (Scott *et al.*, 1991), indeed Fisher *et al.* (1988) go as far as to say '. . . perhaps the major effort in developing an expert system'. The predominant process in knowledge engineering, knowledge acquisition, has been defined as the transfer and transformation of problem-solving expertise from some knowledge source to a computer program (Buchanan *et al.*, 1983). Sources for such problem-solving expertise include human experts, textbooks and scientific journals (Robinson *et al.*, 1986).

Knowledge engineering in general involves the codifying of human knowledge, a method by which the expert's knowledge and ways of reasoning can be understood (Laurini and Thompson, 1992). The knowledge engineer chooses a specific paradigm, within which facts and rules can be elicited. There is a parallel between this and software development but for expert systems the choice of paradigm is not obvious, dependent on the application (Robinson *et al.*, 1986). When new knowledge becomes available, it has to be confirmed as consistent with existing knowledge (Laurini and Thompson, 1992).

6.2.7. Efficiency measures

The computational resources needed for most expert system operations is appreciable, so any measure to streamline the processes and reduce the time taken for them to operate would be seen as desirable. Starting with the initial input, the decomposition of the user's problem or scenario into subproblems is an essential process for expert system operation. These subproblems can be solved individually by the system and merged again to provide a solution to the main problem (Nau, 1983).

Turning to the rule base, the use of geographically specific rules can be used to restrict operations therein. This improves the efficiency of the inference engine's search strategy on each decision unit. This is facilitated by having a geographic application list for each rule (Davis and Nanninga, 1985). This enhances the flexibility of the system, which restricts linkages to be used for a specific knowledge-based application. This is infinitely more desirable than calculating all the spatial relationships required (Lilburne et al., 1996).

6.2.8 Error modelling in expert systems

For an expert system being used in any commercial or academic environment, the user will want to know how much significance the output possesses in order to assess its validity. Therefore some measure of the quality of results is essential for the following practical reasons. Firstly, there will be assurance that any investment for development will be potentially safe. Without such a measure, the comparison of different analyses would be difficult (Burrough, 1992). The knowledge of the accuracy of any information required for decision making is important where there is a range of data types and reliabilities (Miller and Morrice, 1991). Next, future data collection and sampling strategies would benefit from such information, and finally analysis would be 'anarchical' without quality control (Burrough, 1992). It is therefore important to understand the statistical meaning of each dataset for final investigation (Moon and So, 1995).

Terms such as 'believe', 'may' and 'highly unlikely' are inherently uncertain in terms of quality to the user. Conventionally, these can be converted to probabilities and are combined, usually using Bayes' theorem (Fisher et al., 1988). However, there are other modes of combination.

Bayes' theorem is a probabilistic approach to error modelling, calculating uncertainty about the likelihood of a particular event occurring, given a piece of evidence. Within error modelling in expert systems, the Dempster–Schafer theory is increasingly being used instead. The Dempster–Schafer theory is the theory of belief functions. It has an advantage over Bayesian methods in that it can be used where evidence is lacking. It embodies the representation of ignorance in probability theory. A belief of zero applied means complete ignorance (as opposed to complete falsehood in probability theory parlance). The Dempster–Schafer theory generalizes the Bayesian approach by replacing single-point probabilities with belief intervals (Scheerer, 1993). The two methods are both accurate and effective if used correctly (Moon and So, 1995).

6.2.9 Expert systems and GIS

It has been stated that the application of expert systems to GIS has been well established. Historically, the problem domains for expert systems in GIS have been automated map design and generalization, terrain and feature extraction, geographical digital databases/user interfaces, and geographic decision support (Robinson *et al.*, 1986).

This section deals with the methods by which expert systems and GIS can be linked or coupled. It should be noted that those striving to integrate expert systems and GIS (for the benefits that they would both give each other) have not done as well as hoped due to exaggerated claims when such initiatives were first mooted (Lilburne *et al.*, 1996).

Referring to Figure 6.2, the first of the linking methods is loose coupling, where expert systems and GIS are 'loosely' integrated by communication links, a communication channel that transfers data from GIS to the expert system. This is called a 'loosely coupled standalone system'. It is also possible

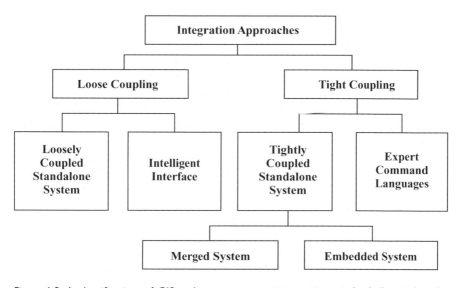

Figure 6.2 A classification of GIS and expert system integration methods (reproduced with permission, the American Society for Photogrammetry and Remote Sensing. Zhu and Healey, 'Towards intelligent spatial decision support', 2, pp. 877–886.)

to build an 'intelligent interface' to a GIS. For example, the expert knowledge about spatial modelling procedures can be incorporated via this route. However, loose coupling does not provide expert systems with the spatial data handling capabilities of GIS.

A more effective means of linking is tight coupling, which integrates expert systems and GIS with communication links in such a way that GIS appears to be an extension of expert system facilities, and vice versa. One appears as a shell around the other. A 'tightly coupled standalone system' can be either a merged system, with expert systems as a subsystem of GIS functionality, or an embedded system, where existing GIS facilities are enhanced with expert system functionality. The second type of tight coupling is 'expert command languages', where expert system reasoning is added to GIS macro or command languages (Zhu and Healey, 1992).

According to Stonebraker and Hearst (1989), two more options are available to integrate expert systems and GIS. Firstly, the database management system (DBMS) can be enhanced on a limited (e.g. data access) and rule-based (e.g. knowledge acquisition, representation techniques, reasoning) level. Alternatively, a fully integrated system, such as a truly deductive DBMS, could be built. Finally, spatial decision support systems (SDSS) are explicitly designed to support a decision research process for complex spatial problems. They provide a framework for integrating database management systems with analytical models, graphical display and tabular reporting capabilities, and relevantly the expert knowledge of decision makers (Densham, 1991). Examples of SDSS will be outlined later in the chapter.

6.3 Current status

6.3.1 The status of geography in expert system research

Today, there are many different expert system applications. The top three by some margin, as highlighted by the survey conducted by Durkin (1996), were business, manufacturing and medicine. Where is geography in all this? Fischer (1994) has noted that research in expert systems has 'lagged' in the context of GIS. This is down to a number of factors. Firstly, GIS is less mature than non-spatial disciplines in both conceptual and physical terms. Secondly, GIS models more complex phenomena than the standard model. Finally, spatial problems are highly complex or ill-structured in nature. Openshaw (1995) has also remarked on the lack of progress of artificial intelligence in geography, despite a significant increase in the rate of growth of machine intelligence quotient (MIQ) in consumer products and industrial systems. Expert systems technology suffers from a lack of formalism in geography, as in many other scientific disciplines (Robinson et al., 1986). This is put down to a lack of awareness and 'artificially induced philosophically inspired barriers' that ignore anything scientifically based.

6.3.2 Where expert systems in GIS are lacking

Expert systems are regarded as the traditional view of AI in geography (Openshaw, 1995). Fischer and Nijkamp (1992) noted a lack of analytical modelling functionality in expert systems. They also found a low level of intelligence in terms of knowledge representation and processing. This is a major hampering factor of current systems. Fischer (1994) has commented on the analytical modelling shortfall – it is not well defined or easily represented in a rules context.

Despite initial attractive designs, most expert systems do not exist in their fully realized and finalized form. This is a comment which can be backed up by the cross-section of applications contained in the examples section of this chapter. The reasons why relate to the open-ended nature of many applications, and the difficulties encountered in gaining pertinent expert knowledge – an often quoted problem.

The trouble is that knowledge acquisition is lengthy, requiring patience on the part of the expert and the knowledge engineer. It is a very poorly understood aspect of the expert system development process (Robinson et al., 1986). Knowledge acquisition is regarded as the most serious barrier to efficient expert system development though expert systems for narrow domains of knowledge are easier to develop than those that need creative or common-sense answers (Yazdani, 1984). Possible strategies to ease this difficult process could include a move to reduce dependency on experts as much as possible. Methods such as model-based reasoning, case-based reasoning and exploration-based learning have been exploited for this purpose (Fischer, 1994).

6.3.3 Should expert systems be used in geography?

There have been suggestions that expert systems are obsolete in geography (Openshaw, 1995), the main arguments being that, by design, expert systems cannot perform better than human experts. Furthermore, no human experts are good enough. Openshaw asserted that the way forward lies in developing systems at superhuman levels, through an expert system that encompasses the knowledge provided by several expert sources and goes beyond the capability of any one human. Also, the trend of expert systems being applied to mundane tasks, as observed by Durkin (1996), would leave the expert more time to work on issues that mattered.

One major academic advantage of expert systems development is that it is essential to fully specify the knowledge of any subject at a number of different levels. It puts the knowledge of one or more experts at the disposal of users and enables the efficient dissemination of that knowledge. Furthermore, preparation of the rule base provides insights into a domain and forces the protagonist to think systematically (Fisher, 1989). Finally, Miller (1994) asserted that if the application backgrounds of expert systems become more diverse then the extent to which technology and expertise can be married becomes a more valuable research domain.

6.3.4 How expert systems within GIS have changed

The field of expert systems in geography has changed, even in the last five years. Smith and Jiang (1991) stated that the application of knowledge-based techniques in GIS were largely motivated by issues relating to ease of constructing applications and ease of system use, as opposed to issues relating to computational power and efficiency. Whilst the ease of programming and using the system remains important, there has since been an appreciable change in the size, speed and economics of high performance computing (Openshaw and Abrahart, 1996), which has shifted precedence to computational power and efficiency. Indeed, regarding the remark that implementations of knowledge-based techniques have yet to be made computationally efficient (Smith and Jiang, 1991), it would seem that expert systems have much to gain from this new era of high performance computing. Smith and Jiang (1991) also predicted object-orientated expert system popularity in the short term, and logic-based popularity in the long term. Although object-orientation is currently popular, judging from the more recent expert system applications, it may be too early to say whether the logic-based approach is coming into vogue. It was noted that AI languages have fallen out of favour because of incompatibility with conventional databases and lack of trained personnel (Fischer, 1994). They have also fallen foul of trends towards widespread use of expert system shells for development purposes using personal computers (Durkin, 1996). This trend can also be observed in the more recent expert systems applications.

6.3.5 Practical aspects of expert system development

The time, money and effort expended on the development of a typical expert system is appreciable (Robinson et al., 1986), so it would be wise to take measures to reduce the chance that it will fail. To this end, Raggad (1996) described the development of an expert system consultation quality control model, an output-based evaluation process for expert system services. It is hoped that it will apply local (i.e. concerned with one end-user category) corrections to prevent expert systems failure. Rapid prototyping could also be used where there is a need to assuage doubts about the expert system. This is desirable where there is a high degree of uncertainty in the specification of the expert system, e.g. if detailed user requirements cannot be provided initially, but are likely to evolve with the system (Fedra and Jamieson, 1996).

Finally, and more generally, it should be noted that use of terms such as artificial intelligence and expert systems may have fallen out of vogue but in fact are being described in subtle terms, e.g. 'intelligent application tools'. The irony is that the AI capability is still there but under a different label (Durkin, 1996).

6.4 Further opportunities and expectations

This section takes a look at current expert systems research as a whole and from the findings identifies future opportunities for the spatial sciences. As quoted by Robinson *et al.* (1986), expert systems development is most likely to follow those already emerging. Of course, there is no harm in suggesting initiatives that have lain dormant for a while.

6.4.1 Getting round the knowledge acquisition problem

Much has been said about the problem of knowledge acquisition. A way has been suggested to overcome this lack of understanding via direct interaction between the domain expert and the program, thus bypassing the knowledge engineer. This is facilitated by having the program 'taught' by the expert by feeding it problems and seeing how it reacts, making amendments and adding knowledge as appropriate (Davis and Lenat, 1982). Alternatively, the discourse characteristic of knowledge acquisition could be expanded to encompass the conceptual modelling stage of system design. This acquisition is no longer seen as expertise transfer, but a co-operative and communicative process between the knowledge engineer and expert (Chan and Johnston, 1996).

Knowledge acquisition can be observed as a 'bottleneck' in developing knowledge-based systems. The manual approach to this suffers from experts unable to articulate their reasoning rules. On the other hand, the automated approach (which induces rules from a set of training cases) suffers from a lack of training cases. Jeng *et al.* (1996) have put forward an integrated approach that uses the strengths of both, in having human experts responsible for solving problems, and utilizing an inductive learning algorithm for reasoning and consistency checking.

The last suggestion in this section concerns an efficient knowledge-acquisition support method which is required for the improvement and maintenance of the knowledge base in durability evaluation of a ship bridge deck. A method to automatically acquire fuzzy production rules is proposed. It makes joint use of a neural network as a subsystem. The evaluation function of genetic algorithms can be provided with the weights from the neural network. In this way it is possible to acquire new knowledge where knowledge is difficult to acquire in the field (Furuta *et al.*, 1996). How well this works is a matter for further research.

6.4.2 Expert systems on the edge

This section details real-time, emergency or responsive expert systems. Typical applications include command and control in military intelligence, emergency and delivery vehicle dispatch, emergency search and rescue, medical imaging, seaport and traffic control (Williams, 1995).

Recently, rule-based expert systems have been put forward as means for implementing complex decision-intensive processes for real-time embedded systems (Chun, 1996). Though ideally suited, many real-time applications (e.g. process control situations) will require inferencing through-puts exceeding those currently provided by commercial expert system shells. Expert System Compilation and Parallelization Environment (ESCAPE) is a prototype tool being programmed for such a purpose. Commercial off the shelf (COTS) expert systems are slow and not ideal for real-time processing. However, there is a need to combine with a COTS expert system for convenience and cost-effectiveness.

ESCAPE incorporates knowledge compilation. This involves generating more complex representations of knowledge in the inference engine (Tong, 1991), as well as discovering a more natural (encapsulated procedural) representation of invoking knowledge clusters from the inference engine (Chun, 1996). This is combined with a template-based approach to building expert systems, which means identifying high-level reasoning tasks which can be generic across domains (Clancey, 1992). The approach is to build expert systems from scratch and study or identify in terms of an existing 'template', then to identify a new 'template' to explain the reasoning of the system (Chun, 1996).

Williams (1995) detailed the Tactical Military Intelligence Processing System (TMIPS), which is designed to aid a decision maker operating in rapidly changing spatially oriented environments in what is known as 'responsive' GIS. In addition, imprecise data is accommodated. Using one of a set of hypothesis selections, the system can monitor the dynamic situation in progress, to support the user wishing to have an insight about how the situation will progress in the future. The system can make predictions itself by using 'cached' expert knowledge. In the future, a knowledge acquisition tool, which assists experts to provide additional knowledge for the system, forming part of the 'cached' knowledge, is planned. Also, representation of temporal or spatio-temporal data will need to be borne in mind.

6.4.3 Expert systems for all: the Internet

The important role of the Internet and World Wide Web (WWW) in global communications and dissemination as well as being a huge source of information is commonly known and needs no introduction here. This section deals

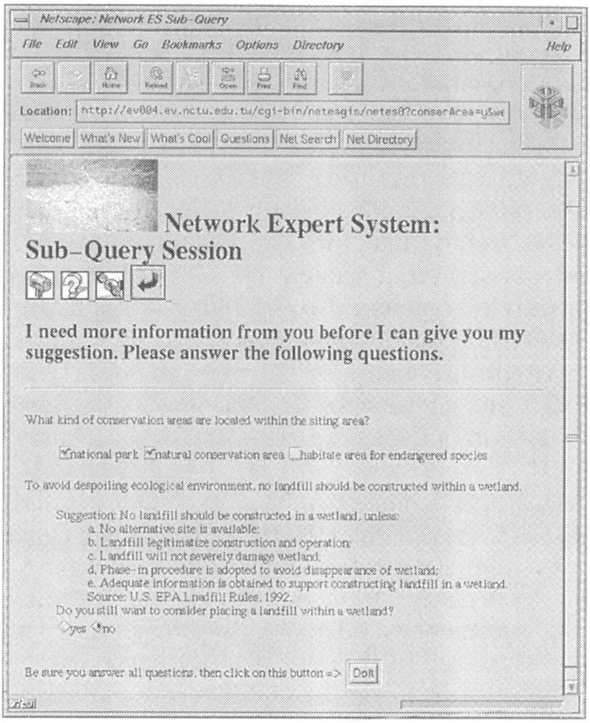

Figure 6.3 An Internet page from the landfill siting expert GIS (Kao *et al.*, 1996).

with some of the benefits to be gained from implementing expert systems on the Internet.

A prototypical network expert geographic information system for landfill siting has been proposed. It has a forward chaining knowledge base derived from the domain's literature. The actual siting analysis occurs in a GIS and is evaluated by triggered rules from the expert system. The expert system and GIS are combined to give the strengths of both. What is novel about this application is that it can be accessed from the Internet (Figure 6.3), cutting distribution and any installation or management on the part of the user (Kao *et al.*, 1996).

A different approach is that of Hardisty (1996). Hardisty described EPISys, an expert system that models hydrodynamic characteristics in the River Humber using equilibrium theory. The results were 96 hour forecast matrices issued in real time, which were found to compare favourably with real observations. Making full use of the WWW, these forecast matrices were updated twice a week for the benefit of all.

6.4.4 Laying it all bare: expert systems with vision

Another interesting suggestion is that of Koseki *et al.* (1996) who proposed an architecture for a hybrid (object-oriented) expert system development. It combines expert problem solving functions and other conventional computational functions using visual programming technology. The novelty of this application is that the knowledge is visually represented in the form of decision tables and decision trees.

In devising data models or knowledge representation schemes, a relational-linear quadtree data model has been put forward as a data structure based on space. Most of the currently used models are developed for non-spatial or image processing or computer graphics applications. In modelling for a vision knowledge base, expert vision systems firstly make use of the knowledge about a given class of scenes and the objects they contain, as well as knowledge about how the image was derived from the scene. They then apply this scene knowledge to the factual knowledge to recognize, detect or locate objects of different types (Wang, 1991).

6.4.5 Other future directions

Current technology of expert system inference engines does not go beyond the matching of symptoms to rules. In other words, there is no dynamic reorganization or substitution when certain symptoms are not available or are not included in the knowledge base. Compiler Knowledge Dictionary Expert System (CKDES) demonstrates how load, compile, sort and grouping techniques can both simplify and improve the organization or reorganization of the knowledge base. Also, it can take unorganized text knowledge, load (scan) it in and reorganize it into a workable and efficient knowledge base. During the load and compile processes a knowledge dictionary (KD) is developed. This provides the required information for dynamic reorganization (DR), which provides alternatives to allow the future inference engine to carry on with its work, where with current inference engines, the program would just stop (Harding and Redmond, 1996).

Finally, the conceptual structure of a synergistic expert system (SES) model has been described (Beeri and Spiegler, 1996), by which a decision maker can use several experts to solve a complex problem needing a consolidated solution – a form of modified blackboard approach. An arrangement of experts or expert systems are configured in series or in parallel to suit a particular problem. The object-oriented approach means that it is flexible and adaptable to decision making situations. The parallel approach is suggested as a case of group decision support systems (GDSS), hence the term synergy. One of the benefits of this set-up is that it offers a unified view of the information available at the macro SES level, together with the specifics of each expert system.

6.5 Illustrative examples

6.5.1 Early geographic expert systems

One of the most noted expert systems with an earth sciences application has been PROSPECTOR, which was developed to assist field geologists (Alty and Coombs, 1984). The original system was designed to provide three major types of advice: the assessment of sites for the existence of certain deposits; the evaluation of geological resources in a region; and the identification of the most favourable drilling sites. It should be noted that Katz (1991) has remarked that despite initial success in discovering a mineral deposit, none have since been found using PROSPECTOR.

One key feature of geological expert knowledge is that it is incomplete and uncertain. This uncertainty may rest both with the knowledge underlying problem-solving and with the evidence available to the user upon which a conclusion is to be reached. Because of this uncertainty the system needs to use a form of non-definitive reasoning, manifested in this case by the use of conditional probabilities and Bayes' theorem.

The structure of the PROSPECTOR model can be described as spaces connected by rules. A space may be some observable evidence or a hypothesis; each space has a probability value indicating how true it is. Rules have the role of specifying how a change in the probability of one space can be propagated to another. A model is built up by connecting spaces with rules in the form of a network (Robinson *et al.*, 1986).

GEOMYCIN (Davis and Nanninga, 1985) has been developed from EMYCIN, which is itself an 'empty' (i.e. devoid of context-specific rules) version of MYCIN, an expert system used for the diagnosis of infectious blood diseases. GEOMYCIN incorporates geographically equivalenced parameters, geographic data files, and rules that are geospatially specific. These capabilities have been utilized to build a realistic demonstration expert system for fire behaviour in a major Australian national park.

Another case in point involves the use of metadata as knowledge being used in content-based search. This has been used to create a knowledge-based GIS (KBGIS). KBGIS-II handles complex spatial objects by dynamic optimization. The KBGIS-II conceptual design is based on fulfilling five requirements. Firstly, to handle large, multilayered, heterogeneous databases of spatially-indexed data, which was achieved. In addition to this, the ability to query such databases about the existence, location and properties of a wide range of spatial objects was planned. Finally, such a system was designed to be interactive, have flexibility and have a learning capability (Smith *et al.*, 1987). An application of KBGIS-II concerned with the design and implementation of a declarative GIS query processor was detailed by Menon and Smith (1989).

6.5.2 Building expert systems

Building a new expert system is often a costly and time-consuming business. Possibly a better approach is to develop expert system building tools for domain experts, which support precise and imprecise rule-based knowledge. Leung and Leung (1993) describe an attempt to construct a 'fuzzy-logic-based intelligent expert system shell for knowledge-based GIS' (FLESS) to construct rule-based expert systems. Internally, there were four subsystems:

1   A knowledge acquisition subsystem consisting of management modules for objects, facts, fuzzy terms, rules and system properties.
2   A fuzzy knowledge base for storing the above.
3   A consultation driver consisting of: an inference engine (which extracts knowledge from the fuzzy knowledge base and makes inferences via rules and facts); a linguistic approximation routine (which maps fuzzy subsets onto linguistic expressions for any conditions drawn by FLESS); and a review management module.
4. A maps display module (displays the inference and analysis results in map form).

(Leung and Leung, 1993)

6.5.3 Knowledge representation

Use of the Hayes-Roth *et al.* (1983) blackboard model is another interesting approach that is illustrated by reference to an example involving travel counselling. The blackboard model is used to incorporate modules capable of collectively undertaking geographical planning tasks, such as an associated object-oriented knowledge base and a route planner mimicking human planning behaviour. The blackboard paradigm is employed to model human planning.

Many specialists make co-operative decisions which lead to the making of a tentative plan. This is all arranged on the 'blackboard', which is a global data structure for the retention of problem solving information. There is a hierarchical arrangement of information on the blackboard. In the case of the geographical planner program (Geoplanner), there are three levels, each requiring a planner – conceptually, from top to bottom; task level, concept level and route level. It was found to be powerful enough to address complex issues, such as geographical planning, though flexible enough to permit additional planning techniques (Leroux and Li, 1990).

6.5.4 Expert system elements and processes

As an example of expert system elements and processes, this section describes a remote sensing application that demonstrates search techniques and knowledge representation in a domain. The automated analysis of

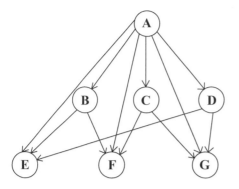

Figure 6.4 An example knowledge representation schematic (from Kartikeyan *et al.*, 1995). (Copyright 1995 IEEE.)

the human expert's interaction in remotely sensed image classification is still in its 'adolescence'. Kartikeyan *et al.* (1995) detail a simple model for spectral knowledge representation. They also outline a method of quantifying knowledge through an evidential approach as well as an automatic knowledge extraction technique for training samples. These methods were used to facilitate land cover analysis on two datasets. The inference engine offered hypotheses (which can be true or false) to test. For instance, observe this knowledge representation schematic (Figure 6.4).

There are three base land cover classes: water (E), vegetation (F) and non-vegetation (G). There are also superclasses such as B, which is a subset of water and vegetation. Based on this test, a given pixel will have an appropriate value or set of values. An iterative process is implemented that ends when one class is decided upon, or no new state is reached after an iteration. All the while, hypotheses are used to decide the next state. These hypotheses are tested through a rule-based approach. There are three possible results of the test: no rules pass; all passed rules correspond to the same hypothesis; and a set of passed rules corresponds to more than one hypothesis. The results were compared with contemporary digital techniques. It was found that commission errors were avoided, and non-spectral and collateral knowledge could be incorporated. The accuracy derived using only spectral knowledge was comparable with standard digital methods. Further investigations may include the extraction and representation of non-spectral knowledge, and also the use of geographic or other ancillary information to minimize efforts in ground truth collection.

A second example is the work of Cress and Diesler (1990), which illustrates the need for a more efficient and knowledgeable production of geological engineering maps (GEM). GEMs should portray objective information in order to best evaluate the engineering involved in regional planning. In this example, GEM production is automated by using a KBGIS approach. It utilizes a second expert system which is used to convert the KBGIS into FORTRAN. FORTRAN runs more efficiently than the Goldworks KBGIS environment, and therefore it is used as the GEM production system. Conversion between the two was facilitated by a conversion expert system (CES). There is a six-stage process for GEM production, from choosing parameters, through map acquisition, map analysis, map output, GEM acceptance, to map field checking. In the development of the KBGIS, knowledge in a GIS environment was considered beneficial for the automation of the production system.

The expert system shell used was example-based which permitted the creation of the knowledge base, incorporating various parameters, the result and a confidence factor (or weight). Weights ranged from high, as with a map constructed from detailed field-tested data, to low, signifying large areas interpolated from sparse data. From this expert system shell, a rule is induced and displayed in a decision tree format. This is optimized, including only those parameters necessary for the creation of unique classes, and is followed by the development of an appropriate classification scheme. Initially, arbitrary classification boundaries were set. Therefore, if unique classes were formed then a final decision tree was made and subsequently converted into IF–THEN structured rules. If unique classes were not formed then the heuristic was repeated. The rules are used to deduce new facts from the existing fact base. This is controlled by the inference engine by way of three techniques: forward chaining, backward chaining and goal-directed forward chaining. It was found that the KBGIS could be used to produce an accurate GEM even though the execution time required would be large. Furthermore, easy development and modification of the rule base was regarded as essential. Finally, it was found that the development of a KBGIS for GEM production was a major improvement over existing production methods.

6.5.5 Object-orientation

The COAstal Management Expert System (COAMES) has been offered as an object-oriented expert system, consisting of a user interface, an object-oriented data model, an object-oriented knowledge base (incorporating both the expert's factual knowledge and the process knowledge embodied in models) and most importantly an inference engine. In the conceptual design, the system avoids the knowledge base being hard-wired into the system (as is often the case) by arranging rules as a hierarchy of objects (Moore *et al.*, 1996). Development of the spatial expert shell (SES) involved the combining of a GIS and an enhanced expert system shell (Smart Elements, which consists of Nexpert Object combined with a GUI developers kit). A client-server set-up was initiated (with Smart Elements the client and Arc/Info the server). SES consists of a group of spatial classes with predefined state and behaviour, e.g. GIS elements – display, vector, raster – are all grouped under a top level class called gisObject (Lilburne *et al.*, 1996).

6.5.6 Efficiency measures

Tanic (1986) detailed URBYS, a tool that was intended to assist urban planners and local decision makers in the form of an expert system computer-based implementation of their own urban rules. The system analyses an urban area and advises on what action should be taken. A measure of efficiency is the use of meta-knowledge rules, by which a specific group of tools can be grouped for a specific kind of urban analysis. The rule interpreter works by either forward chaining or backward chaining. URBYS itself consists of an urban database, an expert system and an interface. It helps in urban planning by adopting the following approaches. Firstly, methods should be natural and close to that of the expert. Secondly, the system's knowledge should be easily accessible and changeable enough not to affect the system's integrity. Finally, the database should account for empirical observations.

6.5.7 Error modelling

Skidmore *et al.* (1996) have outlined the Land Classification and Mapping Expert System (LCMES) (Figure 6.5). The objectives of the study were to construct an intuitive user interface for commercial GIS to make it open to all, to rigorously test the accuracy of expert system output by comparing statistical output with conventional output, and to evaluate if expert system output was of an accuracy that would be considered operational. The methodology incorporates Bayes' theorem, in which knowledge about the likelihood of a hypothesis occurring, given a piece of evidence, is represented as a conditional probability. Two methods exist for linking evidence with the hypothesis: forward chaining (inference works forward from the data or evidence to the hypothesis), and backward chaining (inference flows from the hypothesis to the data).

The successful integration of a Bayesian expert system with a commercially available GIS for mapping forest soils has been facilitated. In this application there were five target soil landscape classes utilizing a digital terrain model, a vegetation map and the soil scientist's knowledge incorporated in the process. It was found that the map drawn by the expert system was as accurate as the map drawn by the soil scientist, statistically, with a 95% confidence interval. Having said this, there were disparities in the visual attributes of the resultant maps.

For some applications, the methodology adopted by conventional remote sensing classification techniques is insufficient or not accurate enough. Kontoes *et al.* (1993) have explored the incorporation of geographical context information from a GIS and how it can be used to remedy these disparities to some degree. In this case, soil maps and buffered road networks have been used as additional data layers to classify SPOT images for estimates of crop acreage. Also, a knowledge base containing both image context rules

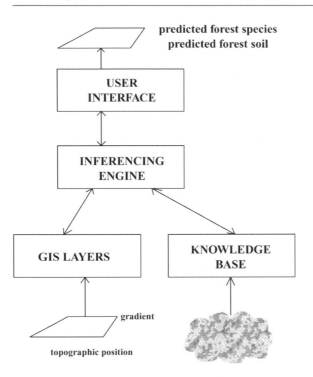

Figure 6.5 Configuration of the Land Classification and Mapping Expert System (LCMES) (reproduced with permission, the American Society for Photogrammetry and Remote Sensing. Skidmore *et al.*, 'An operational GIS expert system for mapping forest soils', 62, pp. 501–511).

and geographical context rules was used. The Dempster–Schafer model was used to combine probabilities from the rule base and image classifier. The Dempster–Schafer model 'weighs up' class evidence from the image classifier then the rules. As a result, the accuracy of classification was found to improve by 13%.

Dempster–Shafer theory is useful in that it can be used for groups of classes as well as single classes, suiting the hierarchical nature of land cover classification. Rules are tested on the ancillary information attributed to a pixel; if this is satisfied then the rule is 'fired', generating a piece of evidence, a 'support value' for a particular class or subclass, which is put onto a 'blackboard' (Hayes-Roth *et al.*, 1983) or temporary storage area. In the end, the class or superclass of maximum belief is allocated to each pixel.

6.5.8 Coupling expert systems and GIS

This section concentrates on the issues arising from the coupling of expert systems and GIS. Loh *et al.* (1994) developed INFORMS-TX (Integrated Forest Resource Management System for National Forests in Texas). It is the result of the integration of a rule-based expert system with a commercial GIS through the use of a relational database management system (CLIPS [an expert system shell developed by NASA], Arc/Info and Oracle). Heuristic rule-bases are selected and applied to pertinent data layers, based on an initial query in the GIS. Any results from the rule base are then stored in the database. Resource managers and specialists are not necessarily conversant with computer technology, so a consistent, friendly and unified user interface was deemed essential for data sharing and a common framework for problem solving. The database also stored metadata, which described the data's availability and location. This allowed for management and use of different rule bases and models to do different tasks. The rule-base design consisted of heuristic knowledge rules (derived from a number of knowledge engineering sessions with experts), computational rules (IF–THEN rules with numerical weights ranging from −1 to +1) and I/O protocol rules (all data tasks are performed by these rules). The expert system module itself does not maintain a separate database. Through the interface between the rule base manager and database manager it communicates with the RDBMS.

According to Varghese and O'Connor (1995) (see also Evans *et al.*, 1993) an expert geographic information system is a tool that integrates the functions of an expert system (Nexpert Object) and a geographic information system (Arc/Info). Two ways were suggested to enable this in a route planning context. Firstly, by allowing one to have the control of the other, the transfer of data can be facilitated. In this case, the expert system shell's C interface and Arc/Info macro language were used to build the interface between the two. In this way, dual control is also enabled, so that Arc/Info controls Nexpert if the emphasis is on intensive spatial analysis, and Nexpert controls Arc/Info if the converse applies. Secondly, coupling can be effected by establishing a data link between the two, for example a common format. Despite certain software-related limitations, this was generally a successful attempt in automating tedious and repetitive route planning tasks.

As an example of use of expert systems within a decision support environment, WaterWare (Fedra and Jamieson, 1996) has been put forward as a decision support system for river-basin planning. It has been designed to integrate the capabilities of GIS, database management systems, modelling techniques, optimization procedures and most relevantly, expert systems (in the context of handling some of the more complex queries in a problem-specific manner). Furthermore, it is a completely open, modular system with different degrees and mechanisms of coupling at various levels of integration, presenting the user with a common logical structure for hands-on analysis and information retrieval.

6.6 Summary

This chapter has provided a broad overview of the expert systems field and, in particular, its application to geography, i.e. geoexpert or spatial expert systems. The elements, processes, tasks involved, efficiency and quality measures and topical issues have been detailed and exemplified. The evidence suggests that, though there has been an appreciable amount of research in this area (for a more comprehensive picture, follow up some of the references in the next section), progress has been slow compared to other disciplines, mostly due to the complex nature of geospatial problems (Fischer, 1994). There are also general concerns such as the knowledge acquisition bottleneck to take into account (Robinson *et al.*, 1986; Fisher *et al.*, 1988; Scott *et al.*, 1991). However, there is much to be optimistic about, both generally and within geography. Based on the extent to which expert systems have been adopted in a multidisciplinary context, they still have great potential (Durkin, 1996). Without such embodiment of intelligence, geography or GIS has little chance of solving complex spatial decision support problems (Fischer, 1994). Future research may lie in developing ways of easing knowledge acquisition, real-time expert systems, use of the internet and visual expert systems.

Acknowledgements

My thanks to Kevin Morris (Plymouth Marine Laboratory) and Grahame Blackwell (University of Plymouth) for making helpful suggestions and proof reading the chapter.

Further reading

General
Frank (1984) – LOBSTER – logic-programming paradigm.
Raal, Burns and Davids (1995) – decision support tool to make GIS available to non-GIS literate people.
Robinson, Frank and Karimi (1987) – survey of 20 expert systems.
Zhu and Band (1994) – multisource data integration.

Environmental
Dantzler and Scheerer (1993); Scheerer (1993); Scheerer and Dantzler (1994) – tactical oceanography – information acquisition, interpretation and management.
Folse, Mueller and Whittaker (1990) – simulating mountain lion movement.
Pearson, Wadge and Wislocki (1992) – Landslide hazard assessment.
Robinson, Frank and Blaze (1986) – ORBI – DSS for resource planning and classification system for environmental data.

Ecological
Davis *et al.* (1991) – decision support system for coastal vegetation.
Lam (1993) – combining ecological modelling, GIS and expert systems: a case study of regional fish species richness model.
Loh and Hsieh (1995) – expert systems to spatially model secondary succession on a savanna woodland site.

Miller (1994) – coupling knowledge-based systems and GIS, model of vegetation change.
Miller and Morrice (1991); Miller (1994) – predicting changes in upland vegetation of Scotland using expert systems and GIS.

Hydrological
Merchant (1994) – DRASTIC model for groundwater capability.
Smith, Zhan and Gao (1990) – extracting channel networks from noisy DEM data.
Tim (1996) – hydrological/water quality expert systems.

Soil mapping
Skidmore *et al.* (1991) – use of expert systems and ancillary data to map forest soils.
Zhu *et al.* (1996) – infer and represent information on the spatial distribution of soil.

Socio-economic
Barath and Futo (1984) – geographic decision support systems (socio-economic).
Heikkila (1990) – Modelling fiscal impacts using expert GIS: theory and strategy.
Sarasua and Jia (1995) – integration of a GIS and KBES for pavement management.

Engineering
Evans, Djokic and Maidment (1993) – investigation of expert systems and GIS in civil engineering.
Spring and Hummer (1995) – use of engineering knowledge regarding accident causation to identify hazardous locations.

Land use
Chandra and Goran (1986) – GEODEX – evaluating site suitability for specific land use activities.
Goldberg, Alvo and Karani (1984) – FES – Forestry Expert System – landcover change.
Mackay, Robinson and Band (1992) – KBLIMS (Knowledge Based Land Information Manager and Simulator).
Wei, Jianbang and Tianhe (1992) – land use suitability.

Cartography
Freeman and Ahn (1984) – AUTONAP – cartographic name placement.
Robinson and Jackson (1985) – MAP-AID for map design.
Yue *et al.* (1991) – a statistical cartographic expert system for China.

Remote sensing
Goodenough *et al.* (1995a) – an intelligent system (SEIDAM – System of Experts for Intelligent Data Management) for calibrating AVIRIS spectrometer data.
Goodenough *et al.* (1995b) – Methodology for creating sequence of intelligent expert systems (SEIDAM).
Morris (1991) – extraction of 3D structural parameters from remotely sensed imagery and DEMs.
Srinivasan and Richards (1993) – analysis of mixed data types for photo-interpretation.

References

Alty, J. L. and Coombs, M. J. (1984) *Expert Systems – Concepts and Examples*, Manchester: NCC Publications.

Barath, E. and Futo, I. (1984) 'A regional planning system based on artificial intelligence concepts', *Pap. Reg. Sci. Ass.*, 55, pp. 135–146.

Beeri, Y. and Spiegler, I. (1996) 'Synergistic expert systems', *Decision Support Systems*, 17, 2, pp. 73–82.

Bobrow, D. G., Mittal, S. and Stefik, M. J. (1986) 'Expert systems: perils and promise', *Comm ACM*, 29, pp. 880–894.

Boehm, B. W. (1988) 'A spiral model of software development and enhancement', *Computer*, 21, 5, pp. 61–72.

Brodie, M. L. and Mylopoulos, J. (1986) 'Knowledge bases versus databases', in Brodie, M. L. and Mylopoulos, J. (eds), *On Knowledge Base Management Systems: Integrating Artificial Intelligence and Database Technologies*, New York: Springer Verlag, pp. 83–86.

Buchanan, B. G., Barstow, D., Bechtel, R., Bennett, J., Clancey, W., Kulikowski, C., Mitchell, T. and Waterman, D. A. (1983) 'Constructing an expert system', in Hayes-Roth, F., Waterman, D. A. and Lenat, D. B. (eds) *Building Expert Systems*, Reading: Addison-Wesley, pp. 127–168.

Burrough, P. A. (1992) 'Development of intelligent geographical information systems', *International Journal of Geographical Information Systems*, 6, 1, pp. 1–11.

Chan, C. W. and Johnston, M. (1996) 'Knowledge modeling for constructing an expert system to support reforestation decisions', *Knowledge-Based Systems*, 9, 1, pp. 41–59.

Chandra, N. and Goran, W. (1986) 'Steps toward a knowledge-based geographical data analysis system', in Opitz, B. K. (ed.) *Geographic Information Systems in Government*, Hampton, VA: A. Deepak.

Cheng, C. H., Holsapple, C. W. and Lee, A. (1995) 'Citation-based journal rankings for expert system research', *Expert Systems*, 12, 4, pp. 313–322.

Chun, R. K. (1996) 'Software integration of real-time expert systems', *Control Engineering Practice*, 4, 1, pp. 83–88.

Clancey, W. (1992) 'Model construction operators', *Artificial Intelligence*, 53, pp. 1–115.

Cress, J. J. and Diesler, R. R. P. (1990) 'Development and implementation of a knowledge-based GIS geological engineering map production system', *Photogrammetric Engineering and Remote Sensing*, 56, 11, pp. 1529–1535.

Dantzler, H. L. and Scheerer, D. J. (1993) 'An expert system for describing and predicting the coastal ocean environment', *Johns Hopkins APL Technical Digest*, 14, 2, pp. 181–192.

Davis, J. R. and Nanninga, P. M. (1985) 'GEOMYCIN: Towards a geographic expert system for resource management', *Journal of Environmental Management*, 21, pp. 377–390.

Davis, J. R., Nanninga, P. M., Biggins, J. and Laut, P. (1991) 'Prototype decision support system for analysing impact of catchment policies', *J. Water Res. Planning Mgmt.*, 117, 4, pp. 399–414.

Davis, R. and Lenat, D. B. (1982) *Knowledge-Based Systems in Artificial Intelligence*, New York: McGraw-Hill.

Densham, P. (1991) 'Spatial Decision Support Systems', in Maguire, D., Goodchild, M. and Rhind, D. (eds) *Geographical Information Systems – Volume 1: Principles*, London: Longman, pp. 403–412.

Durkin, R. (1996) 'Expert systems: a view of the field', *IEEE Expert*, 11, 2, pp. 56–63.

Evans, T. A., Djokic, D. and Maidment, D. R. (1993) 'Development and application of expert geographic information systems', *Journal of Computing in Civil Engineering*, 7, 3, pp. 339–353.

Fedra, K. and Jamieson, D. G. (1996) 'The "WaterWare" decision-support system for river-basin planning. 2. Planning capability', *Journal of Hydrology*, 177, pp. 177–198.

Fischer, M. M. (1994) 'From conventional to knowledge-based geographic information systems', *Computers, Environment and Urban Systems*, 18, 4, pp. 233–242.

Fischer, M. M. and Nijkamp, P. (eds) (1992) *Geographic Information Systems, Spatial Modeling and Policy Evaluation*, Berlin: Springer.

Fisher, P. F. (1989) 'Expert system application in geography', *Area*, 21, 3, pp. 279–287.

Fisher, P. F. (1990) 'A primer of geographic search using artificial intelligence', *Computers and Geosciences*, 16, 6, pp. 753–776.

Fisher, P. F., Mackaness, W. A., Peacegood, G. and Wilkinson, G. G. (1988) 'Artificial intelligence and expert systems in geodata processing', *Progress in Physical Geography*, 12, pp. 371–388.

Folse, L. J., Mueller, H. E. and Whittaker, A. D. (1990) 'Object-oriented simulation and geographic information systems', *AI Appl. Nat. Resour. Manag.*, 4, 2, pp. 41–47.

Frank, A. U. (1984) 'Extending a network database with Prolog', in *1st Int. Workshop on Expert Database Systems*, Kiawah Island, South Carolina: Benjamin Cummings.

Freeman, H. and Ahn, J. (1984) 'AUTONAP – An expert system for automatic name placement', in *Proc. Int. Symp. on Spatial Data Handling*, Zurich, Switzerland: Zurich-Irchel, pp. 544–571.

Furuta, H., He, J. and Watanabe, E. (1996) 'Fuzzy expert system for damage assessment using genetic algorithms and neural networks', *Microcomputers in Civil Engineering*, 11, 1, pp. 37–45.

Goldberg, M., Alvo, M. and Karani, G. (1984) 'The analysis of Landsat imagery using an expert system: forestry applications', in *Proc. AUTOCARTO-6*, Ottawa: ASPRS/ACSM, 2, pp. 493–503.

Goodenough, D. G., Charlebois, D., Bhogal, P., Heyd, M., Matwin, S., Niemann, O. and Portigal, F. (1995a) 'Knowledge based spectrometer analysis and GIS for forestry', in *International Geoscience and Remote Sensing Symposium*, Piscataway, NJ: IEEE, 1, pp. 464–470.

Goodenough, D. G., Bhogal, P., Charlebois, D., Matwin, S. and Niemann, O. (1995b) 'Intelligent data fusion for environmental monitoring', in *International Geoscience and Remote Sensing Symposium*, Piscataway, NJ: IEEE, 3, pp. 2157–2160.

Harding, W. T. and Redmond, R. T. (1996) 'Compilers and knowledge dictionaries for expert systems: inference engines of the future', *Expert Systems with Applications*, 10, 1, pp. 91–98.

Hardisty, J. (1996) 'EPISys: elements of an expert system', in Report of the 2nd LOIS GIS Workshop, personal communication.

Hayes-Roth, F., Waterman, D. A. and Lenat, D. B. (1983) 'An overview of expert systems', in Hayes-Roth, F., Waterman, D. A. and Lenat, D. B. (eds) *Building Expert Systems*, Reading: Addison-Wesley, pp. 3–39.

Heikkila, E. (1990) 'Modeling fiscal impacts using expert GIS theory and strategy', *Computers, Environment and Urban Systems*, 14, 1, pp. 25–35.

Jeng, B., Liang, T.-P. and Hong, M. (1996) 'Interactive induction of expert knowledge', *Expert Systems with Applications*, 10, 3–4, pp. 393–401.

Kao, J. J., Chen, W. Y., Lin, H. Y. and Guo, S. J. (1996) 'Network expert geographic information system for landfill siting', *Journal of Computing in Civil Engineering*, 10, 4, pp. 307–317.

Kartikeyan, B., Majumder, K. L. and Dasgupta, A. R. (1995) 'Expert system for land cover classification', *IEEE Transactions on Geoscience and Remote Sensing*, 33, 1, pp. 58–66.

Katz, S. S. (1991) 'Emulating the Prospector expert system with a raster GIS', *Computers and Geosciences*, 17, 7, pp. 1033–1050.

Keyes, J. (1989) 'Why expert systems fail', *AI Expert*, 4, 11, pp. 50–53.

Kontoes, C., Wilkinson, G. G., Burrill, A., Goffredo, S. and Megier, J. (1993) 'An experimental system for the integration of GIS data in knowledge-based image analysis for remote sensing of agriculture', *International Journal of Geographical Information Systems*, 7, 3, pp. 247–262.

Koseki, Y., Tanaka, M., Maeda, Y. and Koike, Y. (1996) 'Visual programming environment for hybrid expert systems', *Expert Systems with Applications*, 10, 3–4, pp. 481–486.

Lam, D. C. L. (1993) 'Combining ecological modeling, GIS and expert systems: a case study of regional fish species richness model', in Goodchild, M. F., Parks, B. O. and Steyaert, L. T. (eds) *Environmental Modelling with GIS*, Oxford: Oxford University Press, pp. 270–275.

Laurini, R. and Thompson, D. (1992) *Fundamentals of Spatial Information Systems*, London: Academic Press.

Leroux, P. and Li, K. F. (1990) 'An AI-assisted geographical planner', in *1st International Conference on Expert Planning Systems*, London: IEE, pp. 64–69.

Leung, Y. and Leung, K. S. (1993) 'An intelligent expert system for knowledge-based geographical information systems. 1. The tools', *International Journal of Geographical Information Systems*, 7, 3, pp. 189–213.

Lilburne, L., Benwell, G. and Buick, R. (1996) 'GIS, expert systems and interoperability', in *Proceedings 1st International Conference on GeoComputation*, Leeds: University of Leeds, 2, pp. 527–541.

Loh, D. K. and Hseih, Y-T. C. (1995) 'Incorporating rule-based reasoning in the spatial modeling of succession in a savanna landscape', *AI Applications*, 9, 1, pp. 29–40.

Loh, D. K., Hseih, Y-T. C., Choo, Y. K. and Holtfrerich, D. R. (1994) 'Integration of a rule-based expert system with GIS through a relational database management system for forest resource management', *Computers and Electronics in Agriculture*, 11, pp. 215–228.

McKeown, D. M. (1987) 'The role of artificial intelligence in the integration of remotely sensed data with geographic information systems', *IEEE Transactions on Geoscience and Remote Sensing*, GE-25, pp. 330–348.

Mackay, D. S., Robinson, V. B. and Band, L. E. (1992) 'Development of an integrated knowledge-based system for managing spatio-temporal ecological simulations', in *GIS/LIS Proceedings*, Bethesda, MD: ASPRS, 2, pp. 494–503.

Menon, S. and Smith, T. R. (1989) 'A non-procedural spatial query processor for GIS', *Photogrammetric Engineering and Remote Sensing*, 55, 11, pp. 1593–1600.

Merchant, J. W. (1994) 'GIS-based groundwater pollution hazard assessment – A critical review of the DRASTIC model', *Photogrammetric Engineering and Remote Sensing*, 60, 9, pp. 1117–1128.

Miller, D. R. (1994) 'Knowledge-based systems coupled with geographic information systems', in Nievergelt, J., Roos, T., Schek, H-J. and Widmayer, P. (eds) *IGIS '94: Geographic Information Systems International Workshop on Advanced Research in Geographic Information Systems Proceedings*, Berlin: Springer Verlag, pp. 143–154.

Miller, D. R. and Morrice, J. (1991) 'An expert system and GIS approach to predicting changes in the upland vegetation of Scotland', in *GIS/LIS '91 Proceedings*, Bethesda, MD: ASPRS, 1, pp. 11–20.

Minsky, M. (1975) 'A framework for representing knowledge', in Winston, P. H. (ed.) *The Psychology of Computer Vision*, New York: McGraw-Hill, pp. 211–277.

Moon, W. M. and So, C. S. (1995) 'Information representation and integration of multiple sets of spatial geoscience data', in *International Geoscience and Remote Sensing Symposium*, Piscataway, NJ: IEEE, 3, pp. 2141–2144.

Moore, A. B., Morris, K. P. and Blackwell, G. K. (1996) 'COAMES – Towards a coastal management expert system', in *Proceedings 1st International Conference on GeoComputation*, Leeds: University of Leeds, 2, pp. 629–646.

Morris, K. P. (1991) 'Using knowledge-base rules to map the three-dimensional nature of geological features', *Photogrammetric Engineering and Remote Sensing*, 57, 9, pp. 1209–1216.

Naqvi, S. (1986) 'Discussion', in Brodie, M. L. and Mylopoulos, J. (eds) *On Knowledge Base Management Systems: Integrating Artificial Intelligence and Database Technologies*, New York: Springer Verlag, p. 93.

Nau, D. S. (1983) 'Expert computer systems', *Computer*, 16, pp. 63–85.

Nilsson, N. J. (1980) *Artificial Intelligence*, Palo Alto, California: Tioga Press.

Openshaw, S. (1995) 'Artificial intelligence and geography', *AISB Quarterly*, 90, pp. 18–21.

Openshaw, S. and Abrahart, R. J. (1996) 'Geocomputation', in *Proceedings 1st International Conference on GeoComputation*, Leeds: University of Leeds, 2, pp. 665–666.

Oravec, J. A. and Travis, L. (1992) 'If we could do it over we'd . . . Learning from less-than successful expert systems projects', *Journal of Systems Software*, 19, pp. 141–146.

Pearson, E. J., Wadge, G. and Wislocki, A. P. (1992) 'An integrated expert system/GIS approach to modeling and mapping natural hazards', in Harts, J., Ottens, H. F. L. and Scholten, H. J. (eds) *EGIS '92: Proceedings Third European Conference on Geographical Information Systems*, Utrecht: EGIS Foundation, 1, pp. 762–771.

Pissinou, N., Makki, K. and Park, E. K. (1993) 'Towards the design and development of a new architecture for geographic information systems', in *Proceedings of 2nd International Conference in Knowledge Management*, New York: ACM, pp. 565–573.

Quillian, M. R. (1968) 'Semantic memory', in Minsky, M. (ed.) *Semantic Information Processing*, Cambridge: MIT Press.

Raal, P. A., Burns, M. E. R. and Davids, H. (1995) 'Beyond GIS: decision support for coastal development, a South African example', in *Proceedings COAST GIS '95: International Symposium on GIS and Computer Mapping for Coastal Zone Management*, Cork: University College Cork, pp. 271–282.

Raggad, B. J. (1996) 'Expert system quality control', *Information Processing and Management*, 32, 2, pp. 171–183.

Robinson, G. and Jackson, M. (1985) 'Expert systems in map design', in *Proc. AUTOCARTO-7*, Washington, DC: ASPRS/ACSM, pp. 430–439.

Robinson, V. B. and Frank, A. U. (1987) 'Expert systems for geographic information systems', *Photogrammetric Engineering and Remote Sensing*, 53, 10, pp. 1435–1441.

Robinson, V. B., Frank, A. U. and Blaze, M. A. (1986) 'Expert systems applied to problems in geographic information systems: introduction, review and prospects', *Computers, Environment and Urban Systems*, 11, 4, pp. 161–173.

Robinson, V. B., Frank, A. U. and Karimi, H. A. (1987) 'Expert systems for geographic information systems in natural resource management', *AI Applications in Natural Resource Management*, 1, pp. 47–57.

Sarasua, W. A. and Jia, X. (1995) 'Framework for integrating GIS-T with KBES: a pavement management example', *Transportation Research Record*, 1497, pp. 153–163.

Scheerer, D. J. (1993) 'Reasoning under uncertainty for a coastal ocean expert system', *Johns Hopkins APL Technical Digest*, 14, 3, pp. 267–280.

Scheerer, D. J. and Dantzler, H. L. (1994) 'Expert system tools for describing and predicting the coastal ocean environment', in *IEEE Oceans Conference Record*, Piscataway, NJ: IEEE, 2, pp. 11–16.

Scott, A. C., Clayton, J. E. and Gibson, E. L. (1991) *A Practical Guide to Knowledge Acquisition*, Reading, MA: Addison-Wesley.

Skidmore, A. K., Ryan, P. J., Dawes, W., Short, D. and O'Loughlin, E. (1991) 'Use of an expert system to map forest soils from a geographical information system', *International Journal of Geographical Information Systems*, 5, 4, pp. 431–445.

Skidmore, A. K., Watford, F., Luckananuny, P. and Ryan, P. J. (1996) 'An operational GIS expert system for mapping forest soils', *Photogrammetric Engineering and Remote Sensing*, 62, 5, pp. 501–511.

Smith, J. M. and Smith, D. C. P. (1977) 'Database abstractions, aggregation and generalization', in *Association for Computing Machinery Transactions on Database Systems*, 2, 2, pp. 105–133.

Smith, T. R. and Jiang, J. (1991) 'Knowledge-based approaches in GIS', in Maguire, D., Goodchild, M. and Rhind, D. (eds) *Geographical Information Systems – Volume 1: Principles*, London: Longman, pp. 413–425.

Smith, T. R., Peuquet, D., Menon, S. and Agarwal, P. (1987) 'KBGIS-II: A knowledge-based geographical information system', *International Journal of Geographical Information Systems*, 1, 2, pp. 149–172.

Smith, T. R., Zhan, C. and Gao, P. (1990) 'A knowledge-based, two-step procedure for extracting channel networks from noisy DEM data', *Computers and Geosciences*, 16, 6, pp. 777–786.

Spring, G. S. and Hummer, J. (1995) 'Identification of hazardous highway locations using knowledge-based GIS: a case study', *Transportation Record*, No. 1497.

Srinivasan, A. and Richards, J. A. (1993) 'Analysis of spatial data using knowledge-based methods', *International Journal of Geographical Information Systems*, 7, 6, pp. 479–500.

Stonebraker, M. and Hearst, M. (1989) 'Future trends in expert database systems', in Kerschberg, L. (ed.) *Expert Database Systems*, Redwood City: Benjamin Cummings, pp. 3–20.

Subramanian, R. and Adam, N. R. (1993) 'Design and implementation of an expert object-oriented geographic information system', in *Proceedings of 2nd International Conference in Knowledge Management*, New York: ACM, pp. 537–546.

Tanic, E. (1986) 'Urban planning and artificial intelligence: the URBYS system', *Comput. Environ. Urban Systems*, 10, 3–4, pp. 135–154.

Tim, U. S. (1996) 'Emerging technologies for hydrologic and water-quality modeling research', *Transactions of the ASAE*, 39, 2, pp. 465–476.

Tong, C. (1991) 'The nature and significance of knowledge compilation', *IEEE Expert*, 6, 2, pp. 88–93.

Varghese, K. and O'Connor, J. T. (1995) 'Routing large vehicles on industrial construction sites', *Journal of Construction Engineering and Management*, 121, 1, pp. 1–12.

Wang, F. (1991) 'Integrating GIS and remote sensing image analysis systems by unifying knowledge representation schemes', *IEEE Transactions on Geoscience and Remote Sensing*, 29, 4, pp. 656–664.

Wei, Z., Jianbang, H. and Tianhe, C. (1992) 'Application of expert systems to land resources research', *Computers, Environment and Urban Systems*, 16, 4, pp. 321–327.

Weiss, S. M. and Kulikowski, C. A. (1984) *A Practical Guide to Designing Expert Systems*, Totawa, NJ: Rowman and Allenheld.

Wharton, S. W. (1987) 'A spectral knowledge-based approach for urban land-cover discrimination', *IEEE Transactions on Geoscience and Remote Sensing*, GE-25, pp. 272–282.

Wiederhold, G. (1986) 'Knowledge versus data', in Brodie, M. L. and Mylopoulos, J. (eds) *On Knowledge Base Management Systems: Integrating Artificial Intelligence and Database Technologies*, New York: Springer Verlag, pp. 77–82.

Williams, G. J. (1995) 'Templates for spatial reasoning in responsive geographical information systems', *International Journal of Geographical Information Systems*, 9, 2, pp. 117–131.

Yazdani, M. (1984) 'Knowledge engineering in Prolog', in Forsyth, R. (ed.) *Expert Systems: Principles and Case Studies*, London: Chapman and Hall, pp. 191–211.

Yue, L., Zhang, L., Xiaogang, C. and Yingming, Z. (1991) 'The establishment and application of the geographic mapping database by city/county unit in China', *International Journal of Geographical Information Systems*, 5, 1, pp. 73–84.

Zhu, A-X. and Band, L. E. (1994) 'Knowledge-based approach to data integration for soil mapping', *Canadian Journal of Remote Sensing*, 20, 4, pp. 408–418.

Zhu, A-X., Band, L. E., Dutton, B. and Nimlos, T. J. (1996) 'Automated soil inference under fuzzy-logic', *Ecological Modelling*, 90, 2, pp. 123–145.

Zhu, X. and Healey, R. (1992) 'Towards intelligent spatial decision support: integrating geographic information systems and expert systems', in *GIS/LIS Proceedings*, Bethesda, MD: ASPRS, 2, pp. 877–886.

Chapter 7

Fuzzy modelling

Peter Fisher

7.1 Introduction

In the classic text, *Explanation in Geography*, Harvey (1969) briefly identifies one of the fundamental problems of deterministic description and analysis of geographical phenomena, the identification of the geographical individual (individuation). He states that the individual results from either properties or position of that individual. In the description and analysis of geographical phenomena and location, however, we widely use vague terms, as adjectives, nouns, and verbs, which make it hard or impossible to clearly define the geographical individual in any meaningful way which does not involve an arbitrary cut off.

For example, if we describe or wish to analyse the major cities in Europe (Figure 7.1), both the terms major and Europe are arbitrary. They rely on the drawing of a line (a boundary) in geographical space or in one or more attribute properties so that a specific group of cities are defined. What about a city which is just outside the geographical area? Or one which is just smaller in any parameter being used to define the major cities? Why should they be excluded from the analysis? One response might be that they should be included, and so the arbitrary boundaries are redrawn, and the subject of the question then moves to another city. Redrawing the boundaries might have no end, until all land areas and all settlements are included (arguably until the total population on the Eurasian continent, if not the surface of the whole globe, is included). Within a political entity (which Europe is not), the boundary might be considered absolute, a clear division of space, but the definition of major cities is still problematic. Major is a vague adjective, describing the primacy of the cities to be examined, and Europe is a vague noun, a geographic region without unequivocal limits.

In this chapter we will first examine the nature of vagueness, although space here precludes a comprehensive exposition. Discussion will then move on to examine classic set theory to show what can be judged to be wrong with it in terms of vagueness, and then we will examine fuzzy sets as an alternative. The most important topic of fuzzy set theory will follow, namely

Figure 7.1 The 'major' cities of Europe defined within the mapping options of Excel '95™.

the definition of fuzzy memberships. Next fuzzy logic is examined, and discussion of extension of fuzzy set theory, and of dissenting opinions, follows. In conclusion, some other types of soft set theories are indicated, which may prove useful to GeoComputation (GC) in the future.

7.2 Sorites paradox

The process of argument used in the preceding section is well known in philosophy and is called the Sorites paradox (Sainsbury, 1995; Williamson, 1994). In its original formulation the argument went like this:

> If one grain of sand is placed on a surface is there a heap?
> If a second grain is placed with the first, is there a heap?
> If a third grain is placed with the first two, is there a heap?
> If a ten millionth grain is placed with the 9 999 999 grains, is there a heap?

In the first three instances, where $n = 1$ to 3, the answer is clearly 'No'. If the last question, when $n = 10\ 000\ 000$, were asked on its own the answer would almost certainly be 'Yes', but if the process of questioning progresses from the first to the last question, with an increment of one grain each time, the logical answer is still 'No'; it is never the case that a non-heap is turned into a heap by the addition of a single grain. That answer, however, is clearly wrong, and that is why it is a paradox. The person answering will undoubtedly feel increasingly uncomfortable as the process advances, but they will not, logically, be able to answer 'Yes, there is now a heap when there was not before' at any point. The heap to non-heap argument by the removal of grains works the same way; there is always a heap.

There is no attempt here to say that 10 000 000 sand grains do not constitute a heap of sand. Rather the purpose is to question the logic and the argument, and to acknowledge that if we have a number of people in a room, and a particular group of sand grains in a pile, they may well not agree on whether there is a heap of sand. Heap is a vague concept. The same process of argument characterizes the definition of very many geographical phenomena. Concepts such as the heap, which are Sorites susceptible, are widely acknowledged as being vague concepts (Russell, 1923; Williamson, 1994). The exact causes of vagueness, need not be discussed here, but are comprehensively reviewed by Williamson (1994).

The style of argument (little-by-little) which characterizes the Sorites paradox is supposed to have been used first by the Greek philosopher Eubulides of Miletus (Barnes, 1982; Burnyeat, 1982; Williamson, 1994). A similar argument occurs between Jehovah and Abraham in the Bible (Genesis, 18, 23–33) resulting in a temporary stay of execution for Sodom on the off-chance that there might be ten righteous citizens in the city. Aspects of the problem of vagueness, and of the possibility of vague objects, have for long been illustrated with geographical examples, such as 'Where is Snowdon?' (Sainsbury, 1989), and 'How long is the Thames?' (Williamson, 1994), although not all commentators agree that there are vague objects (Evans, 1978).

Fuzzy set theory is an attempt, gaining very widespread currency (Kosko, 1993; McNeill and Freiberger, 1993), to address the problem posed by vagueness, as demonstrated by the Sorites paradox, to traditional sets and logics. Fuzzy set theory as such was first suggested by Zadeh (1965) in a seminal paper. It actually built on the work of Kaplan and Schott (1951) and Black (1937), although no reference was made to these works to avoid the philosophical problems of vagueness and the pitfalls inherent in entering that argument (Kosko, 1993). But Zadeh did not have his own way, and the validity of fuzzy set theory ran, and still runs, into considerable opposition (Haack, 1974; Zadeh, 1980; Kosko, 1990; Laviolette and Seaman, 1994).

The adoption of fuzzy set theory by geographers could be a case study in the diffusion of concepts. The first paper suggesting that there might be something worth the attention of geographers in the area of fuzzy set theory seems to have been by Gale (1972), a behavioural geographer. A subsequent paper by Pipkin (1978) also explored its application to that area, while

Ponsard (1977) examined settlement hierarchies and Leung (1979) was more interested in land use planning and decision making.

Robinson and Strahler (1984) were among the first to suggest that fuzzy sets may have far more to offer modern geographical methods and computer models, in suggesting that it presented a logical basis for the storage and analysis of satellite imagery and GIS. Following these papers, the late 1980s and 1990s has literally led to hundreds of examples of the application of fuzzy sets to many areas of geographical endeavour. A recent examination of an on-line bibliographical catalogue for the geographical sciences (GeoBooks) yielded nearly 250 entries using the word fuzzy sets or fuzzy logic.

7.3 Classic sets and logic

Classic set theory is based in the writing of Aristotle, as formalized by logicians and mathematicians. Key characters in the development and our present understanding of this set theory were Boole, Venn and Cantor, and the sets which result are referred to as Boolean and Cantor sets, and are often illustrated in Venn diagrams (Figure 7.2). Figure 7.2(a) is the standard representation of a pair of sets, A and B. They are shown within a rectangle, the universe, and with a hard boundary. The boundary can be in single or multiple attributes, or in space or in time. Boolean sets can become the basis of logical operations which involve the combination of the sets. The union (Figure 7.2(b)) is the region occupied by either A or B, the intersect (Figure 7.2(c)) is the region occupied by both A and B, and the inverse (Figure 7.2(d)) is the region occupied by neither A nor B. Whether or not an object belongs to a set can also be portrayed as a line graph (Figure 7.3), effectively a cross-section through the Venn diagram, where belonging is shown by code 1 and non-belonging by code 0, a binary coding. This is indicated as a value of either 0 or 1, or $\{0,1\}$. Union and intersection can also be shown (Figure 7.3(b), (c) respectively).

If the belonging of an individual to a set is based on the possession of a threshold value of a particular property, and the diagnostic possession of the property is the subject of error (owing to observation, measurement, etc.), then there is a probability of the property being observed or measured correctly, and so whether or not an individual is a member of a set can be assigned a probability ($p(x)_A$; the probability of x being in A) as can its probability of belonging to B ($p(x)_B$). Then the probability of that object being in the intersect of sets A and B is given in the classic equation (7.1), the union is given in equation (7.2), and the inverse in equation (7.3).

$$p(x)_{A \cap B} = p(x)_A \bullet p(x)_B \tag{7.1}$$

$$p(x)_{A \cup B} = p(x)_A + p(x)_B - p(x)_{A \cap B} \tag{7.2}$$

$$p(x)_{A'} = 1 - p(x)_A \tag{7.3}$$

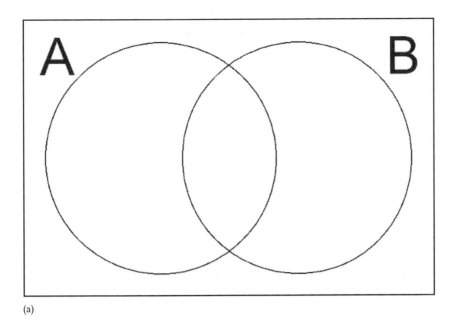

(a)

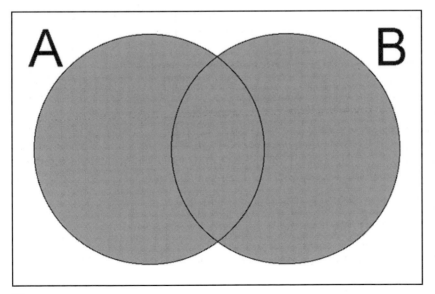

(b)

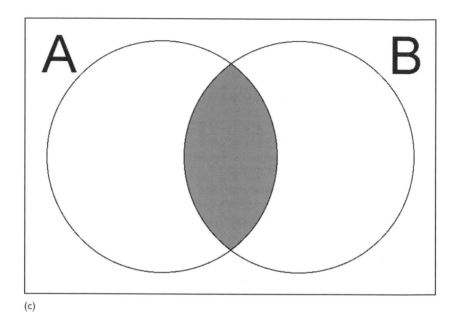

(c)

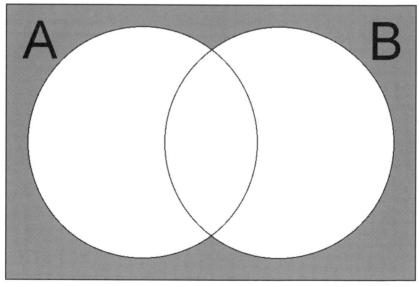

(d)

Figure 7.2 Venn diagrams of two sets: (a) shows A and B, and the rectangular area indicating the universe the sets occupy; in (b) the shaded area shows the union of A or B; in (c) the shaded area shows the intersection of A and B; and in (d) the inverse of the union of A or B is shown.

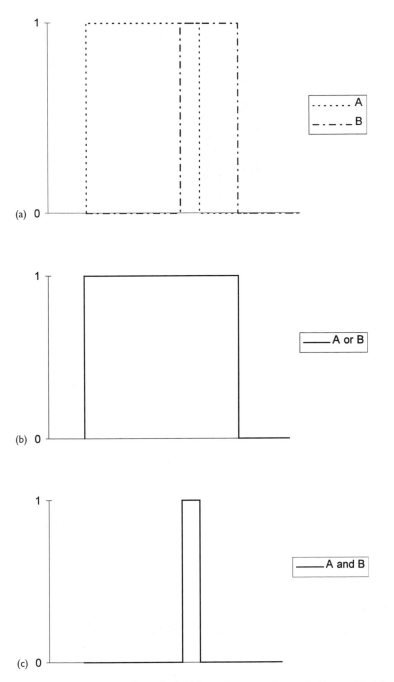

Figure 7.3 Line graphs through the Venn diagrams shown in Figure 7.2: (a) shows A and B; (b) shows the union of A or B; and (c) shows the intersection of A and B.

Examples of Cantor or Boolean sets are the set of people allowed to practise medicine as a doctor, or the set of people with degrees. Geographical Boolean sets are the census division of, at least, western countries, where a specific region is distinguished, each household is assigned to one and only one region, and so people are counted by that region; there is a simple one to many relationship between regions and people. In time, an obvious Boolean set is the year. Large crowds of people celebrate the change from one year to the next, and many time series analyses divide information into arbitrary, and hard, divisions of time.

The Boolean set is the basis of most conventional set and statistical analysis. For example, we are always advised that in preparing a statistical test of an hypothesis (such as that the means of two groups of values are equal using the t test), we must set up a null hypothesis (that they are not equal) and an alternative hypothesis (that they are equal). We then determine the value of the test statistic (t), and compare it with the threshold value which comes from tables or a calculation of the distribution by a computer. The value calculated from the observations is then compared with the threshold value, and a decision is made whether to reject the null hypothesis. In a test with a particular level of confidence, and with a particular dataset, only one outcome is valid, no matter how close the calculated test statistic is to the threshold. More advanced interpretations admit to a continuous probability of the null hypothesis being correct, or else investigators conveniently select another value of the confidence interval so that the test does support their preconceptions. In short, although hypothesis testing is a matter of clear-cut Boolean decisions, few investigators adopt a slavish adherence to rejection of the null hypothesis, admitting some vagueness as to the meaning of the threshold.

7.4 Fuzzy set theory

Zadeh (1965) first put forward fuzzy set theory as a direct response to the short-comings of Boolean or Cantor sets. Zadeh himself used the example of the tall person. The set of tall people is something which most humans have an idea about, but which no two humans agree on, and whether a particular person is in one individual's conception of the set of tall people might change with time. Furthermore, there are people about whom someone may not be prepared to commit to the idea that they are either tall or not tall. Rather, if questioned, a person might say they were nearly tall, or pretty tall, using a linguistic hedge, or qualifier to express a degree to which the person in question is within their concept of tall people.

A similar concept is the idea of baldness (Burrough *et al.*, 1992) where people are variously prepared to concede that someone with no hair at all is definitely bald, and a person with perhaps a hundred hairs is bald, but people with more and more hair are less and less bald, until you find someone with a small thinning of hair on the top of their head who might be described as balding, predicting a future belonging to the set of bald people, but currently only a minor affinity with the set.

In the case of both the bald and the tall person we are essentially admitting to a degree of belonging to the set of people with that characteristic. Furthermore, any person may to some extent be tall and to some extent be bald. The extent may be so small that it is effectively 0, but if we define the sets tall and not tall, many people will belong to both sets to degrees. It is this degree of belonging which Zadeh suggests is described by fuzzy set theory. Essentially the concepts of bald and tall are Sorites susceptible, and one way to decide whether an individual, an attribute or a location is appropriate to description by fuzzy set theory is to pose a typical Sorites argument.

In geography we can distinguish a large number of concepts which map comfortably into the concept of a degree of belonging, but do not fit so well with a concept of a hard Boolean set, but the degree of belonging may be a property of either the geographical individual or a location irrespective of the individual. A number of examples are reviewed in the remainder of this section.

For the purposes of government and taxation, the administrative extent of the city tends to be well defined. In either attribute definition or in spatial extent, the real extent of a city is, however, poorly defined (Batty and Longley, 1994). To many people living in any city, its administrative extent is not well known (unless perhaps they happen to live near a boundary, or are politically aware). To business and population distribution, the administrative extent is rarely an abrupt barrier to development and land use, although changes in tax rate and development incentives, for example, can make it so.

As with the city, Burrough (1989) and McBratney and De Gruijter (1992), among others, have argued that the spatial extent of a soil is poorly defined, by either location or attribute. Soils change continuously in space, forming gradations, and only rarely are those gradations abrupt. Thus the concept of a soil map as a set of well defined regions separated by distinct boundaries does not match the real-world occurrence of soil characteristics. Furthermore, the soil frequently changes at a high spatial frequency, and so is poorly captured by the Boolean map. Lagacherie et al. (1996) and Wang and Hall (1996), in rather different contexts, extend this conception by showing how uncertain soil boundaries can be added to a standard soil map. Soil mapping is aimed primarily at improved land management, and Burrough (1989), Burrough et al. (1992), Davidson et al. (1994), and Wang et al. (1990) have all argued that land evaluation which is based on soil mapping is better and more reliably treated by fuzzy sets, than by Boolean sets. They have argued that the concepts of land evaluation are more successfully implemented within fuzzy sets and have variously demonstrated that better results from the process of land evaluation are achieved with fuzzy sets based analysis.

Dale (1988), Moraczewski (1993a, b) and Roberts (1986, 1989) have all argued that the analysis of vegetation communities is better treated through fuzzy set theory than through Boolean concepts and, indeed, Dale (1988) and Roberts (1986) argue that the traditional method of ordination used in the analysis of phytosociology is a tacit admission that the vegetation at any particular location possesses partial characteristics of some central concept of the community, and ordination analysis (through principal components or factor analysis) is one method of describing the strength of any location belonging to the concepts. The same argument with respect to ordination has been made by soil researchers (Powell *et al.*, 1991). Furthermore, the method admits that one location may have similarities in its properties to more than one concept.

7.5 Fuzzy memberships

At the heart of fuzzy set theory is the concept of a fuzzy membership. A Boolean set is defined by the binary coding {0,1}, whereby an object which is in the set has code 1, and an object which is not has code 0. A fuzzy set membership, on the other hand, is defined by any real number in the interval [0,1]. If the value is closer to 0 then the object is less like the concept being described, if closer to 1 then it is more like. A person 2 m high might have a value 0.9 in the set of tall people whereas one 1.8 m high might have the value 0.2. The fuzzy membership is commonly described in formulae by μ, and, particularly, an object x has fuzzy membership of the set A, $\mu(x)_A$. The difference between the Boolean set and the fuzzy set memberships is shown in Figure 7.4. The binary Boolean set has memberships 1 and 0, while the fuzzy set has continuous membership values between 0 and 1, grading over different values of some indicator variable plotted on the x axis.

Two major approaches to defining the fuzzy membership of an object have been recognized by Robinson (1988) and Burrough (1989): the *semantic import model*, and the *similarity relation model*. An *experimental model* also needs to be acknowledged.

7.5.1 Semantic import model

The most common form of fuzzy membership definition is the semantic import model. Here some form of expert knowledge is used to assign a membership on the basis of the measurement of some property, d. There are two main different ways it can be done. The first method involves defining a set of paired values (2, 3 or 4 pairs) called fuzzy numbers which specify the critical points of the membership function defining the limits of the transition from memberships of $\mu = 0$ to $\mu = 1$. The second method is to specify a formula which relates the changes in the value of d to continuous variation in μ.

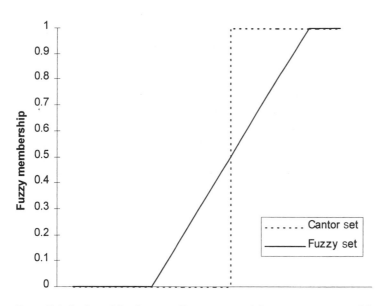

Figure 7.4 A plot of Boolean or Contor set and fuzzy set in terms of fuzzy memberships.

The simplest form of the fuzzy number method is to include two fuzzy numbers of the property ($\{d_1,0\}$, $\{d_2,1\}$) which specifies the one transition function shown in Figure 7.5(a) where $d_1 < d_2$, and d_1 and d_2 are the values of d associated with memberships $\mu = 0$ to $\mu = 1$ respectively. The triplet ($\{d_1,0\}$, $\{d_2,1\}$, $\{d_3,0\}$) are fuzzy numbers which define a triangular membership function where $d_1 < d_2 < d_3$ (Figure 7.5(b)). Alternatively four fuzzy numbers describe a trapezoid membership function ($\{d_1,0\}$, $\{d_2,1\}$, $\{d_3,1\}$, $\{d_4,0\}$ where $d_1 < d_2 < d_3 < d_4$; Figure 7.5(c)). The functions thus defined give the full range of the necessary memberships. One criticism of this approach is that it is too simplistic. It is, however, rapid to compute and especially rapid to compute derivatives of logical operations (see below), but there is no sensitivity in the form of the membership function, which may be necessary or desirable in some applications.

One way to extend the method is to specify a larger number of fuzzy numbers giving membership values associated with a number of different values of d, on both the rise (and fall) of the membership function (Figure 7.5(d)). Here, for example, the value of d is specified for every 0.1 increment in the value of μ as it rises from 0 to 1 ($\{d_1,0\}$, $\{d_2,0.1\}$, $\{d_3,0.2\}$, $\{d_4,0.3\}$, $\{d_5,0.4\}$, $\{d_6,0.5\}$, $\{d_7,0.6\}$, $\{d_8,0.7\}$, $\{d_9,0.8\}$, $\{d_{10},0.9\}$, $\{d_{11},1\}$). Altman (1994) has applied fuzzy numbers in the definition of fuzzy geographical regions such as locations near a tower and related geographical queries such as is a tower near and east of a patch of woodland (itself poorly defined).

The continuous membership function is defined through a formula such as

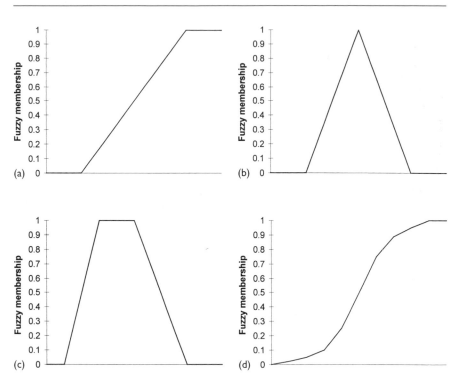

Figure 7.5 Alternative definitions of fuzzy memberships by specifying the membership values associated with critical values of an indicator variable (the x axis): in (a) two critical values are given; in (b) three values; in (c) four; and in (d) eleven.

that specified in equation (7.4). Such equations are used in many applications of fuzzy set theory. Equation (7.4) itself defines the function illustrated in Figure 7.6(a) which moves from a value of $\mu = 1$ asymptotically towards a value of $\mu = 0$. It can be reformulated to either mirror this relationship, where $\mu = 0$ asymptotically towards a value of $\mu = 1$, or using further piecewise specification, using half of the graph for those values where $\mu > 0.5$ and the mirror image for $\mu < 0.5$, causing the memberships to range from 0 to 1 (Figure 7.6(b)). This formulaic nature of the membership function has been used in many different forms in the geographical literature.

$$\mu(x_{ij}) = \begin{cases} 1 & \text{for } d \le b_1 \\ \dfrac{1}{\left[\left(1 + \left(\dfrac{d - b_1}{b_2}\right)^2\right)\right]} & \text{for } d > b_1 \end{cases} \tag{7.4}$$

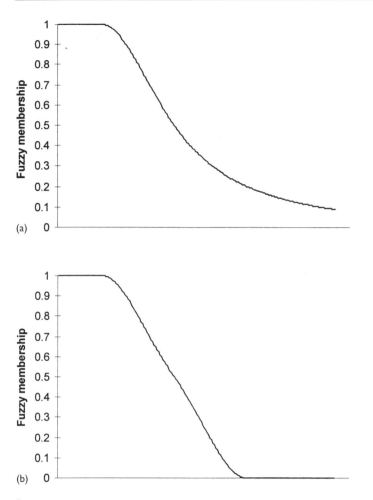

Figure 7.6 Two of the host of possible fuzzy memberships: (a) uses equation (7.4), while (b) employs the upper part of equation 7.4, with its mirror image to force the curve to meet the *x* axis.

where $\mu(x_{ij})$ is the fuzzy membership of an object in a set; d is the value of the property used to define the membership function; b_1 is a threshold value of the property at the limit of any object definitely being a member of the set; and b_2 is the value of the property (over the value of b_1) corresponding to fuzzy membership = 0.5, sometimes called the cross-over point.

A number of different applications of fuzzy set theory to soils information and land evaluation have used such formulaic approaches to define fuzzy set membership. Burrough (1989) uses a formula similar to equation (7.4) to derive a fuzzy membership for the concept of sandy soils from measurements of the proportion of a soil sample which is sand sized. Sandy is a fuzzy concept which is not clearly defined. Cation exchange capacity is an easily measured soil parameter, and a good indicator of suitability of soil for wheat cultivation, and Davidson *et al.* (1994) use the formulaic approach to semantic import to derive a fuzzy membership of the concept of soil suitable for wheat.

Fisher (1994) employed the formulaic approach to membership definition when he argued that the normal GIS-based determination of the area visible from a point in the landscape is only one version of the viewshed problem, and that a fuzzy membership function for the visible area based on the distance from the viewing point to any target location is more useful for describing a personal experience of landscape. Here a version of equation (7.4) is one form of the membership function discussed, and where d is the distance between a viewpoint and a target. The equation is used to evaluate the degree to which an object at the target is visible. The parameters b_1 and b_2 are selected in this instance to reflect the characteristics of the individual viewing the landscape (eyesight, etc.), and the target contrast with the background (colour). Alternative versions of the formula are proposed to account for solar glare, and atmospheric fog and haze.

Whether the semantic import model is used via the fuzzy number or the formulae approaches, the important point is the association of critical values of d with particular memberships. This is primarily based on expert opinion, and is fundamentally a feature of the requirements and design of the application in hand.

7.5.2 Similarity relation model

While the semantic import model of fuzzy memberships requires specification of the membership function, the similarity relation model is a pattern recognition approach which searches the data for membership values. A number of different fuzzy classification strategies have been proposed (Bezdek, 1981), although two approaches are perhaps best known the fuzzy c means (FCM) algorithm (Bezdek *et al.*, 1984) and, more recently, fuzzy neural networks (Wilkinson *et al.*, 1995). Both methods take as input a multivariate dataset of p variables by q cases. They optimize the identification of a predetermined number of groups, c, and for each of the q cases the degree to which any case resembles the properties of each of the c groups is reported as a fuzzy membership value in the range [0,1]. Thus the membership value is a function of the classification process and of the input data. Two primary variants are possible for methods which can either be supervised (where information on prototypical groups is entered) or unsupervised (where a search is executed to determine the optimal division into some number of pre-determined classes). One outcome of the process is the special case of membership values {0,1} indicating the Boolean set solution, although more normally some degree of fuzziness is induced in the outcome.

Fuzzy c means have been applied to a large number of different situations in geographical computing. Foremost among these have been studies of the application to soils and soil mapping (Odeh *et al.*, 1990; Powell *et al.*, 1991; McBratney and De Gruijter, 1992). All are based in a similar philosophical position: firstly, soil types are poorly defined in their attributes, secondly, they are poorly defined in their spatial extent, and, thirdly, the zone of intergrade between soil types in space or in variables may actually be more interesting and more important than the core areas which do fall clearly into particular classes. McBratney and Moore (1985) have also examined climate classifications of Australia and China with the FCM classification. It should be noted that Leung (1987) has used the formulaic approach to map the climates of Taiwan.

A number of researchers have also examined the application of the FCM algorithm to remotely sensed images. Robinson and Thongs (1986) show that the reservoir in Central Park, New York City, can be clearly detected, and the pixels which coincide with the shoreline are assigned partial memberships in the land and water classes detected. Fisher and Pathirana (1990) have shown that a land cover classification of a suburban area in Ohio with the FCM algorithm yields a relatively reliable classification of the proportions of each pixel which are occupied by different land covers. More recently researchers analysing satellite imagery have employed neural networks to extract fuzzy set memberships (Civico, 1993; Foody, 1995, 1996; Lees, 1996; Wilkinson *et al.*, 1995). Here the weights associated with the layer-linkages within the neural network are recovered, and interpreted as fuzzy memberships.

Other methods have been used in various different applications to effectively compute fuzzy set memberships, but they are specific to the application area. As noted above, Dale (1988) and Roberts (1986) working in plant ecology and phytosociology have pointed out that methods such as ordination and canonical correlates generate scatter plots of the degree to which observation points match a particular interpreted variable, the community. The hope is that the scatter plot will show clear clustering of data points, but more commonly it shows a spread throughout the data space. This is clearly interpretable as a recognition of the equivalence of the community with a fuzzy set and the component score, or correlate as a surrogate for a membership. Researchers looking at the application of fuzzy set theory to remote sensing have also examined a number of different approaches to membership extraction, including, perhaps most confusingly, probability. Thus Foody *et al.* (1992) show that the probabilities derived from a maximum likelihood classifier are similar to the proportions of different plant communities on the ground along a vegetation continuum, an ecotone, while Foody (1992) used a FCM approach to the same problem. Similarly, where pure pixel end members can be identified, mixture modelling can be used to derive membership grades (Foody and Cox, 1994). Bastin (in press) has executed an experiment comparing the effectiveness of maximum likelihood, mixture modelling, and FCM for membership extraction. She identified the fuzzy classifier as the most appropriate, not only giving the best level of agreement with finer resolution proportions, but also producing a well dispersed set of fuzzy membership values.

7.5.3 Experimental model

Finally, an approach based on experimental analysis should be recognized. In this approach, there is neither machine-based search nor acknowledgement by an expert of the relevance of fuzzy sets. Rather the membership function is empirically derived by experiment, usually with human subjects. If one, or a group of subjects, have a Boolean concept in mind and are asked to decide when the Boolean concept is reached, the set of responses can be used to derive the fuzzy membership function. If temperature control is the issue to be addressed, then taking a group of people, and asking them when they become cold and need the heating to come on is a way of establishing the fuzzy set of the environment being *cold*. Alternatively, one person may repeatedly be asked the same question at different times and in different contexts. When the results for one, or multiple respondents, are related to the measured temperature then a heating control system can be built and the controls verified (Kosko, 1993). Such environmental control is a typical industrial application of fuzzy sets.

The experimental approach has been used in geographical applications. Thus Edwards and Lowell (1996) have investigated the boundaries between forest stands as interpreted from aerial photographs. They argue that the spatial delineation of forest stands is inherently poorly defined and so approximated by fuzzy sets, and they used interpretations from a number of different interpreters to establish the map of fuzzy boundaries.

Robinson (1990) developed the Spatial Relations Acquisition Station (SRAS), a program to derive a membership function for spatial relationships such as near, close or far. It works by asking a user whether particular locations are near or far from a target or focus feature. The system initially presents fuzzy memberships which are based on a formula, and varies the form of the function according to the answers of the user. The user can only respond that one location is near to or not near to the focal location. Fisher and Orf (1991) used SRAS to investigate the meaning of near among students on a university campus. They found very varied opinions as to the meaning of the term, as might be expected, but they could not identify any causal influences among personal characteristics (car users, male/female, etc.).

7.6 Fuzzy logic

With memberships of fuzzy sets defined, just as with Boolean sets, it is possible to execute logical set operations directly comparable with the union, intersect and inverse of Boolean sets. These were first proposed by Zadeh (1965) in his original development of fuzzy set theory.

Fuzzy union is simply a maximum operation (equation 7.5), and can be illustrated in Figure 7.7(b). Taking the maximum value of μ for every value of d, is exactly the same as the set theoretic union operation for Boolean sets (Figure 7.3(b)). On the other hand, equation (7.6) defines the intersection of the two fuzzy sets, and takes the minimum value of μ for every value of d (Figure 7.7(c)). It is also directly comparable to the Boolean intersect (Figure 7.3(c)). Finally, the negation operation is also a generalization of the special case of the Boolean operation (equation 7.7).

$$\mu_{(A \cup B)} = \max (\mu_{(A)}, \mu_{(B)}) \tag{7.5}$$

$$\mu_{(A \cap B)} = \min (\mu_{(A)}, \mu_{(B)}) \tag{7.6}$$

$$\mu_{(A')} = 1 - \mu_{(A)} \tag{7.7}$$

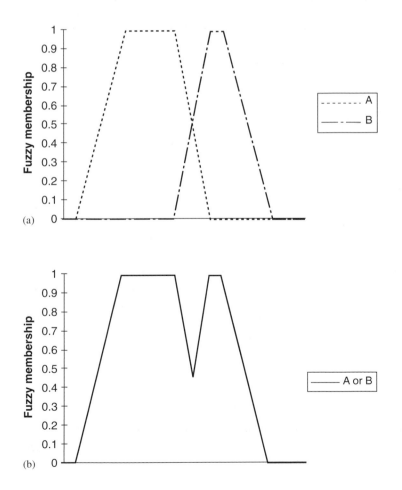

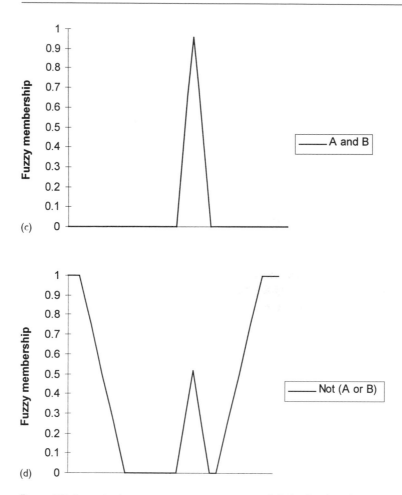

Figure 7.7 Fuzzy logic operators on two sets parallel the Boolean logic operators. Operating on (a) two sets, A and B, we can define (b) the union, (c) the intersect, and (d) the inverse, here, of the union.

The three principal fuzzy operators are a generality of the set combinations in Boolean sets. In fact, there are a large number of different proposed fuzzy union and intersect operators. Indeed, Leung (1988, p. 29) gives three further pairs of fuzzy union and intersect operators which all specialize to the correct versions of Boolean set operators, as well as a number of other operators which do not, and goes on to state that one of the advantages of fuzzy logic is that there is no unique definition of the laws of set combination.

Equations (7.5) and (7.6) and other equations which share the property of specializing to the Boolean equivalent are known as the hard union and intersect operators. There are also soft versions of the fuzzy union and intersect operators (equations (7.8) and (7.9) respectively), and these bear a remarkable, and intended, similarity to the *probability* of an object being a member of the Boolean sets (equations (7.2) and (7.1) respectively). These have the advantage that they do not maximize and minimize the memberships, and so are convenient operators for some situations.

$$\mu(x)_{A \cup B} = \mu(x)_A + \mu(x)_B - \mu(x)_{A \cap B} \tag{7.8}$$

$$\mu(x)_{A \cap B} = \mu(x)_A \bullet \mu(x)_B \tag{7.9}$$

Fuzzy set operators are extensive and readers wishing to find more are referred to texts such as those by Klir and Yuan (1995), and Kruse *et al.* (1994), or by Leung (1988) more specifically for subjects in geography and planning.

Several researchers have argued that the application of fuzzy logic to geographical problems has several advantages. Thus Burrough (1989) shows that, from observations of the amount of sand in different soil horizons, it is possible to derive fuzzy set memberships of any horizon being sandy, and then of the membership of the soil being sandy in any layer, or of being sandy throughout. Burrough *et al.* (1992) argue that a fuzzy logic approach to soil evaluation for land use provides a more rigorous approach with an outcome which better reflects the reality of the situation. Specifically, the Boolean set derived from the final fuzzy combination includes areas which meet any criteria of contiguity and suitability, but would be excluded by a Boolean analysis from the start.

7.7 Fuzzy regression

Within fuzzy set theory, there are a large number of methods being presented and the tool-kit is continually expanding. These methods seem rarely to have been explored in the geographical literature. The purpose of this section is therefore to do little more than point out some of the methods which exist and to suggest that a fuller examination by geographers interested in computing may be pertinent. Regression is a method of analysis well known to geographers, and fundamental to many quantitative analyses of geographical data; indeed, many methods are based in the general linear model and so derived from regression (Johnston, 1978). That model, however, is based on Boolean conceptualizations of the data, and a possible precision in the measurement which may be confused by error. In the instance of geographical data, it could be argued that few concepts of data are truely Boolean, and that fuzzy regression would be a significantly more appropriate tool for analysis, motivated as it is by the recognition that some data concepts are poorly defined (Klir and Yuan, 1995, p. 454). Arguably, standard methods of regression analysis have been rejected by geographers in recent years because of the mis-conceptualization of the data implicit in a

Boolean data model. This is perhaps epitomized by the use of Boolean and binary answers to opinion surveys. Indeed, Bouchon (1981; Akdag and Bouchon, 1988) has explored the novel area of fuzzy questionnaires, but they have not gained wide acceptance, although the measures used in many questionnaire surveys are inherently fuzzy in both question and response.

For long, geographical computing has been concerned with the location of facilities and the planning of land use. Recently, the analytical hierarchy process (Saaty, 1980) has been advocated and received a considerable amount of attention as an approach, embedded as it is in fuzzy set theory, for wide area land use planning, particularly with raster GIS (Banai, 1993; Eastman *et al.*, 1995; Sui, 1992). In this method, a hierachy of factors which influence land use planning is set up. The factors are grouped into similar related themes, and then a hierarchy is formed with those coming together at different levels. For example, soils and water influences might be grouped separately at the lowest level, but fused as a natural resource influence at a higher level (Banai, 1993). At each level in the hierarchy, the importance of factors is decided based on a pairwise comparison of the factors, and relative scorings of those factors. From the ensuing matrix of scores, eigenvalues are extracted to derive weightings for GIS themes which are then used in a weighted overlay of the map data. Eastman *et al.* (1995) outline the complete procedure as it is implemented in the Idrisi GIS software, including the allocation of land to particular uses based on a simple heuristic optimization procedure.

7.8 Fuzzy sets, possibility and probability

Fuzzy set memberships can be interpreted as a possibility distribution, as opposed to a probability distribution (Zadeh, 1978). Correctly this is done when the fuzzy membership is used in the analysis of a belief function within evidence theory (Klir and Yuan, 1995). Commonly, membership values, however they are derived (through semantic import, or similarity relation models or as a result of logical or mathematical operations), are interpreted as the possibility of the set value being true.

Zadeh (1968) has pointed out that there is also a probability aspect to fuzzy sets. One criticism of fuzzy sets is that they are actually Boolean. There are two aspects of this. If a particular threshold value of the membership is taken as a cut off, and the set is hardened (also known as crispened) on the basis of that threshold then the set which results is a Boolean set. Indeed, the fuzzy membership function is no more than the additive sequence of multiple Boolean sets. Furthermore, the membership function itself (no matter the method used) is a precise definition from an infinite number of possible definitions. Therefore the fuzzy set membership itself is not definite. It is subject to error and therefore has a definable probability of being correctly identified.

This has not been widely addressed in geographical research, but Lees (1996), for example, shows that neural networks can be used to derive not just the fuzzy memberships of vegetation types in a satellite image, but also the probability of the values being correct. In discussing the fuzzy viewshed, Fisher (1994) points out that if the selection of the formula of the membership function in a particular instance is a matter of personal characteristics or atmospheric conditions, then both a person with those characteristics and the weather conditions have a definable probability of returning, which would give an overall probability of the fuzzy set being correctly identified. Similarly, reported probabilities can be fuzzy, thus the probability of rain reported in a weather forecast is not a precise statement of probability but is a fuzzy probability (about 30% chance of rain), and not well represented by any crisp set concept (McNeill and Freiberger, 1993, p. 69). In short, within fuzzy set theory, there are higher-order concepts of uncertainty, which is one of the most important aspects of vagueness discussed by Williamson (1994).

7.9 Dissenting voices

Proponents of fuzzy set theory have not had the discussion all their own way. Indeed, the launch of *IEEE Transactions in Fuzzy Systems* caused the compilation of a special issue on whether or not fuzzy sets are either necessary or appropriate as a method of representation of uncertainty. The special issue is centred around a paper by Laviolette and Seaman (1994), who make a variety of points. Foremost, they claim that the need for a fuzzy sets approach is due to a misrepresentation of the set and that all questions should address membership of a crisp set. This criticism is parallel to the logical positivist view and is an epistemic statement, founded in a belief about how science should be conducted, which may not be extendible to the science of the natural world or the behavioural world we occupy. This criticism was earlier addressed by Zadeh (1980), and completely fails to address the concept of objects being vague.

Unfortunately this opinion colours their later arguments, which, for example, state that the mathematics of fuzzy sets are unnecessarily complicated. Kosko (1990) has gone furthest in arguing that, essentially, probability is a special case of fuzzy; it is within a superset defined by fuzzy sets, although this argument is dismissed by Laviolette and Seaman (1994) as irrelevant. Perhaps the most telling case made by those authors is that fuzzy logic is unnecessary since the results can be completely replicated by Bayesian probability. They ignore the point that one of the appeals of fuzzy set theory is actually the ease and intuition of computation of simple set operations. People feel comfortable with these simple operations, more so than with the unfamiliar and apparently complex formulae of a Bayesian analysis.

In philosophy and logic circles, a similar discussion has been held, although from a different point of view. The approach of the logical positivist school suggests that if something cannot be defined with precision it is not an appropriate subject for investigation either by logicians or by scientists (Ayer, 1936). Vague concepts have indeed been referred to as nonsense by that school of philosophy, and, although that was intended to be derogatory, it gave rise to *The Logic of Nonsense* (Halldén, 1949), a key text in the development of vagueness as a valid concern in philosophy. Haack (1974) has also strongly attacked fuzzy set theory and vagueness as unnecessary and inappropriate, titling her book *Deviant Logics*. The association between the philosophical concern with vagueness and fuzzy set theory should not be overdone, however, since the latter has been the subject of considerable criticism from some vague theorists (Williamson, 1994) in both the application of fuzzy logic and the conceptualizaiton of the membership function as a Boolean methodology to address non-Boolean concepts. This is undoubtedly a valid criticism, but the computability of the epistemological approach adopted by Williamson (1994), for example, has not been examined to date (although it partly coincides with the experimental approach to membership definition discussed above). Fuzzy sets may not be a satisfactory approach from a pure philosophy view, but, together with rough sets, it is the principal computational method which addresses vagueness at all.

7.10 Alternative soft set theories

A final note which needs to be added to this chapter is that fuzzy sets and fuzzy logic is just one of a number of different approaches to approximate reasoning which are grouped together under the heading of soft set theory (Pawlak, 1994). It took seven years from the publication of Zadeh's original fuzzy set paper until Gale (1972) introduced the concepts to the geographical literature, and nearly 20 years until Robinson and Strahler (1984) heralded a more widespread adoption within the context of GIS. Pawlak (1982) introduced rough sets, the first major alternative method of soft set theory, 15 years ago, but this does not appear to have been investigated by the geographical community, although it would appear to offer much. Pawlak (1994) lists four other soft set theories, including multisets, multifuzzy sets, Blizard's sets, and general sets. There is a wealth of alternative set theories to be explored, and if fuzzy set theory is anything to go by they may represent considerable advances and advantages in matching our conceptual model of the computational world, stored and analysed within computer systems, with the real world we inhabit and geographers' study.

7.11 Conclusion

Many geographical phenomena (objects, relations and processes) are most appropriately conceptualized as vague. Within current computing environments, the most widespread method to address vagueness is fuzzy sets, and it is in that context that this chapter is presented. Vagueness is fundamental to geographical phenomena and it must be addressed, where appropriate. In this chapter, I have attempted to blend the theory of fuzzy sets together with geographical examples of its use. I believe firmly that currently it represents an appropriate and worthwhile way ahead for geographical computing to answer questions people ask in a way they want to hear. I make no claim that this will remain the case into the future, but those involved in spatial theory and geographical computing need to become more conversant with general issues of soft computing and higher levels of uncertainty (probability of fuzzy sets, fuzziness of fuzzy sets and of probability). Fuzzy sets are a theoretical alternative to traditional Boolean analyses within both logical and mathematical functions, and should be exploited, or at least exhaustively explored within the broad area of GC.

Acknowledgements

In particular I must acknowledge the intellectual tutorship of John Dalrymple in the founding of my interest in the concepts which underlie fuzzy sets many years ago (although he may not recognize it), and the collegiate encouragement of Vince Robinson and Mike DeMers in the honing of the ideas within the context of geography in general and GIS in particular.

References

Akdag, H. and Bouchon, B. (1988) 'Using fuzzy set theory in the analysis of structures of information', *Fuzzy Sets and Systems*, 28, pp. 263–271.

Altman, D. (1994) 'Fuzzy set theoretic approaches for handling imprecision in spatial analysis', *International Journal of Geographical Information Systems*, 8, pp. 271–289.

Ayer, A. J. (1936) *Language, Truth and Logic*, London: Gollanz.

Banai, R. (1993) 'Fuzziness in geographical information systems: contributions from the analytic hierarchy process', *International Journal of Geographical Information Systems*, 7, pp. 315–329.

Barnes, J. (1982) 'Medicine experience and logic', in Barnes, J., Brunschwig, J., Burnyeat, M. and Schofield, M. (eds), *Science and Speculation*, Cambridge: Cambridge University Press, pp. 241–168.

Bastin, L. (1999) 'Comparison of fuzzy *c*-mean classification, linear mixture modelling and MLC probabilities as tools for unmixing coarse pixels', *International Journal of Remote Sensing*, in press.

Batty, M. and Longley, P. (1994) *Fractal Cities*, London: Academic Press.

Bezdek, J. C. (1981) *Pattern Recognition with Fuzzy Objective Function Algorithms*, New York: Plenum Press.

Bezdek, J. C., Ehrlich, R. and Full, W. (1984) 'FCM: the fuzzy *c*-means clustering algorithm', *Computers & Geosciences*, 10, pp. 191–203.

Black, M. (1937) 'Vagueness: an exercise in logical analysis', *Philosophy of Science*, 4, pp. 427–455.

Bouchon, B. (1981) 'Fuzzy questionnaires', *Fuzzy Sets and Systems*, 6, pp. 1–9.

Burnyeat, M. F. (1982) 'Gods and Heaps', in Schofield, M. and Nussbaum, M. C. (eds) *Language and Logos*, Cambridge: University Press, pp. 315–338.

Burrough, P. A. (1989) 'Fuzzy mathematical methods for soil survey and land evaluation', *Journal of Soil Science*, 40, pp. 477–492.

Burrough, P. A., MacMillan, R. A. and van Deursen, W. (1992) 'Fuzzy classification methods for determining land suitability from soil profile observations and topography', *Journal of Soil Science*, 43, pp. 193–210.

Civco, D. L. (1993) 'Artificial neural networks for land-cover classification and mapping', *International Journal of Geographical Information Systems*, 7, pp. 173–185.

Dale, M. B. (1988) 'Some fuzzy approaches to phytosociology: ideals and instances', *Folia Geobotanica et Phytotaxonomica*, 23, pp. 239–274.

Davidson, D. A., Theocharopoulos, S. P. and Bloksma, R. J. (1994) 'A land evaluation project in Greece using GIS and based on Boolean fuzzy set methodologies', *International Journal of Geographical Information Systems*, 8, pp. 369–384.

Eastman, J. R., Jin, W., Kyem, A. K. and Toledano, J. (1995) 'Raster procedures for multi-criteria/multi-objective decisions', *Photogrammetric Engineering and Remote Sensing*, 61, pp. 539–547.

Edwards, G. and Lowell, K. E. (1996) 'Modeling uncertainty in photointerpreted boundaries', *Photogrammetric Engineering and Remote Sensing*, 62, pp. 337–391.

Evans, G. (1978) 'Can there be vague objects?', *Analysis*, 38, p. 208.

Fisher, P. F. (1994) 'Probable and fuzzy models of the viewshed operation', in M. Worboys, (ed.) *Innovations in GIS 1*, London: Taylor & Francis, pp. 161–175.

Fisher, P. F. and Orf, T. (1991) 'An investigation of the meaning of near and close on a university campus', *Computers, Environment and Urban Systems*, 15, pp. 23–35.

Fisher, P. F. and Pathirana, S. (1990) 'The evaluation of fuzzy membership of land cover classes in the suburban zone', *Remote Sensing of Environment*, 34, pp. 121–132.

Foody, G. M. (1992) 'A fuzzy sets approach to the representation of vegetation continua from remotely sensed data: an example from lowland heath', *Photogrammetric Engineering and Remote Sensing*, 58, pp. 221–225.

Foody, G. M. (1995) 'Land cover classification by an artificial neural network with ancillary information', *International Journal of Geographical Information Systems*, 9, pp. 527–542.

Foody, G. M. (1996) 'Approaches to the production and evaluation of fuzzy land cover classification from remotely-sensed data', *International Journal of Remote Sensing*, 17, pp. 1317–1340.

Foody, G. M. and Cox, D. P. (1994) 'Sub-pixel land cover composition estimation using a linear mixture model and fuzzy membership functions', *International Journal of Remote Sensing*, 15, pp. 619–631.

Foody, G. M., Campbell, N. A., Trodd, N. M. and Wood, T. F. (1992) 'Derivation and applications of probabilistic measures of class membership from the maximum-likelihood classification', *Photogrammetric Engineering and Remote Sensing*, 58, pp. 1335–1341.

Gale, S. (1972) 'Inexactness fuzzy sets and the foundation of behavioral geography', *Geographical Analysis*, 4, pp. 337–349.

Haack, S. (1974) *Deviant Logic, Fuzzy Logic*, Cambridge: University Press.

Halldén, S. (1949) *The Logic of Nonsense*, Uppsala: Universitets Arsskrift.

Harvey, D. W. (1969) *Explanation in Geography*, London: Arnold.

Johnston, R. J. (1978) *Multivariate Statistical Analysis in Geography*, London: Longman.

Kaplan, A. and Schott, H. F. (1951) 'A calculus for empirical classes', *Methodos*, 3, pp. 165–188.

Klir, G. J. and Yuan, B. (1995) *Fuzzy Sets and Fuzzy Logic: Theory and Applications*, Englewood Cliff, NJ: Prentice Hall.

Kosko, B. (1990) 'Fuzziness vs probability', *International Journal of General Systems*, 17, pp. 211–240.

Kosko, B. (1993) *Fuzzy Thinking: The New Science of Fuzzy Logic*, New York: Hyperion.

Kruse, R., Gebhardt, J. and Klawonn, F. (1994) *Foundations of Fuzzy Systems*, Chichester: Wiley.

Lagacherie, P., Andrieux, P. and Bouzigues, R. (1996) 'The soil boundaries: from reality to coding in GIS', in Burrough, P. A. and Frank, A. (eds) *Spatial Conceptual Models for Geographic Objects with Undetermined Boundaries*, London: Taylor & Francis, pp. 275–286.

Laviolette, M. and Seaman, J. W. (1994) 'The efficacy of fuzzy representations of uncertainty', *IEEE Transactions on Fuzzy Systems*, 2, pp. 4–15.

Lees, B. (1996) 'Improving the spatial extension of point data by changing the data model', in *Proceeding of the Third International Conference/Workshop on Integrating GIS and Environmental Modeling*, Santa Barbara: National Center for Geographic Information and Analysis, CD-ROM.

Leung, Y. C. (1979) 'Locational choice: a fuzzy set approach', *Geographical Bulletin*, 15, pp. 28–34.

Leung, Y. C. (1987) 'On the imprecision of boundaries', *Geographical Analysis*, 19, pp. 125–151.

Leung, Y. C. (1988) *Spatial Analysis and Planning under Imprecision*, New York: Elsevier.

McBratney, A. B. and De Gruijter, J. J. (1992) 'A continuum approach to soil classification by modified fuzzy k-means with extragrades', *Journal of Soil Science*, 43, pp. 159–175.

McBratney, A. B. and Moore, A. W. (1985) 'Application of fuzzy sets to climatic classification', *Agricultural and Forest Meteorology*, 35, pp. 165–185.

McNeill, D. and Freiberger, P. (1993) *Fuzzy Logic: The Revolutionary Computer Technology that is Changing Our World*, New York: Touchstone.

Moraczewski, I. R. (1993a) 'Fuzzy logic for phytosociology 1: syntaxa as vague concepts', *Vegetatio*, 106, pp. 1–11.

Moraczewski, I. R. (1993b) 'Fuzzy logic for phytosociology 2: generalization and prediction', *Vegetatio*, 106, pp. 13–20.

Odeh, I. O. A., McBratney, A. B. and Chittleborough, D. J. (1990) 'Design and optimal sample spacings for mapping soil using fuzzy k-means and regionalized variable theory', *Geoderma*, 47, pp. 93–122.

Pawlak, Z. (1982) 'Rough sets', *International Journal of Computer and Information Sciences*, 11, pp. 341–356.

Pawlak, Z. (1994) 'Hard and soft sets', in Ziarko, W. P. (ed.) *Rough Sets, Fuzzy Sets and Knowledge Discovery*, Berlin: Springer Verlag, pp. 130–135.

Pipkin, J. S. (1978) 'Fuzzy sets and spatial choice', *Annals of the Association of American Geographers*, 68, pp. 196–204.

Ponsard, C. (1977) 'Hierarchie des places centrales et graphes psi-flous', *Environment and Planning A*, 9, pp. 1233–1252.

Powell, B., McBratney, A. B. and Macloed, D. A. (1991) 'The application of ordination and fuzzy classification techniques to field pedology and soil stratigraphy in the Lockyer Valley, Queensland', *Catena*, 18, pp. 409–420.

Roberts, D. W. (1986) 'Ordination on the basis of fuzzy set theory', *Vegetatio*, 66, pp. 123–131.

Roberts, D. W. (1989) 'Fuzzy systems vegetation theory', *Vegetatio*, 83, pp. 71–80.

Robinson, V. B. (1988) 'Some implications of fuzzy set theory applied to geographic databases', *Computers, Environment and Urban Systems*, 12, pp. 89–98.

Robinson, V. B. (1990) 'Interactive machine acquisition of a fuzzy spatial relation', *Computers & Geosciences*, 16, pp. 857–872.

Robinson, V. B. and Strahler, A. H. (1984) 'Issues in designing geographic information systems under conditions of inexactness', in *Proceedings of the 10th International Symposium on Machine Processing of Remotely Sensed Data*, Lafayette: Purdue University, pp. 198–204.

Robinson, V. B. and Thongs, D. (1986) 'Fuzzy set theory applied to the mixed pixel problem of multispectral landcover databases', in Opitz, B. (ed.) *Geographic Information Systems in Government*, Hampton: A Deerpak Publishing, pp. 871–885.

Russell, B. (1923) 'Vagueness', *Australian Journal of Philosophy*, 1, pp. 84–92.

Saaty, T. L. (1980) *The Analytical Hierarchy Process*, New York: McGraw-Hill.

Sainsbury, R. M. (1989) 'What is a vague object?', *Analysis*, 49, pp. 99–103.

Sainsbury, R. M. (1995) *Paradoxes*, 2nd edn, Cambridge: University Press.

Sui, D. Z. (1992) 'A fuzzy GIS modeling approach for urban land evaluation', *Computers, Environment and Urban Systems*, 16, pp. 101–115.

Wang, F. and Hall, G. B. (1996) 'Fuzzy representation of geographical boundaries', *International Journal of Geographical Information Systems*, 10, pp. 573–590.

Wang, F., Hall, G. B. and Subaryono (1990) 'Fuzzy information representation and processing in conventional GIS software: database design and application', *International Journal of Geographical Information Systems*, 4, pp. 261–283.

Wilkinson, G. G., Fierens, F. and Kanellopoulos, I. (1995) 'Integration of neural and statistical approaches in spatial data classification', *Geographical Systems*, 2, pp. 1–20.

Williamson, T. (1994) *Vagueness*, London: Routledge.

Zadeh, L. A. (1965) 'Fuzzy sets', *Information and Control*, 8, pp. 338–353.

Zadeh, L. A. (1968) 'Probability measures of fuzzy events', *Journal of Mathematical Analysis and Applications*, 23, pp. 421–427.

Zadeh, L. A. (1978) 'Fuzzy sets as a basis for a theory of possibility', *Fuzzy Sets and Systems*, 1, pp. 3–28.

Zadeh, L. A. (1980) 'Fuzzy sets versus probability', *Proceedings of the IEEE*, 68, p. 421.

Chapter 8

Neurocomputing – tools for geographers

Manfred M. Fischer and Robert J. Abrahart

8.1 Introduction

Neurocomputing is an emergent technology concerned with information processing systems that autonomously develop operational capabilities in adaptive response to an information environment. The principal information processing structures of interest in neurocomputing are computational neural networks, although other classes of adaptive information systems are also considered, such as genetic learning systems, fuzzy learning systems, and simulated annealing systems. Several features distinguish this approach to information processing from algorithmic and rule-based information systems (Fischer, 1994, 1995):

1. Information processing is performed in parallel. Large-scale parallelism can produce a significant increase in the speed of information processing (inherent parallelism).
2. Knowledge is not encoded in symbolic structures, but rather in patterns of numerical strength associated with the connections that exist between the processing elements of the system (connectionist type of knowledge representation) (Smolensky, 1988).
3. Neural networks offer fault-tolerant solutions. These tools are able to learn from and make decisions based on incomplete, noisy and fuzzy information.
4. Neurocomputing does not require algorithms or rule development and will often produce a significant reduction in the quantities of software that need to be developed since the same method can be used in many different application areas.

This relatively new approach to information processing offers great potential for tackling difficult problems, especially in those areas of pattern recognition and exploratory data analysis for which the algorithms and rules are not known, or where they might be known but the software to implement them would be too expensive or too time-consuming to develop. Indeed, with a neurocomputing solution, the only bespoke software that would need to be developed will in most instances be for relatively straightforward operations such as data preprocessing, data file input, data post-processing and data file output. Computer aided software engineering (CASE) tools could be used to build the appropriate routine software modules (Hecht-Nielsen, 1990).

The following sections are intended to provide a basic understanding of various tools and concepts related to the geographical application of neurocomputing. Much of this material is based on earlier work and includes several important items and considerations that are also discussed in alternative publications to which the reader is referred, for example, Fischer (1998a, b).

8.2 What is a computational neural network?

Briefly stated, a computational neural network (CNN) is a parallel distributed information structure consisting of a set of adaptive processing (computational) elements and a set of unidirectional data connections. These networks are 'neural' in the sense that they have been inspired by neuroscience. No claim is made to them being faithful models of biological or cognitive neural phenomena. In fact, the computational networks that are covered in this chapter have more in common with traditional mathematical and/or statistical models, such as non-parametric pattern classifiers, statistical regression models and clustering algorithms, than they do with neurobiological models.

The term 'computational' neural network is used to emphasize the difference between computational and artificial intelligence. Neglect of this difference might lead to confusion, misunderstanding and misuse of neural network models in geographical data analysis and modelling. Computational intelligence (CI) denotes the lowest 'level of intelligence' which stems from the fact that computers are able to process numerical (low-level) data without using knowledge in the artificial intelligence (AI) sense. An AI system, in contrast, is a CI system where added value comes from incorporating knowledge that humans possess, in the form of non-numerical information, operational rules or constraints. Neural network implementations in the form of feedforward pattern classifiers and function approximators, which are considered at a later point, are therefore CI rather than AI systems.

Increased effort is now being made to investigate the potential benefits of neural network analysis and modelling in various areas of geographical research. In particular, these computational devices would appear to offer several important advantages that could be exploited, over and above those associated with what is now becoming known as 'the traditional approach to geographical information processing'. The strongest appeal of CNNs is their suitability for machine learning (i.e. computational adaptivity). Machine learning in CNNs consists of adjusting the connection weights to improve the overall performance of a model. This is a very simple and pleasing formulation of the learning problem. Speed of computation is another key attraction. In traditional single processor Von Neumann computers, the speed of the machine is limited by the propagation delay of the transistors. However, with their intrinsic parallel distributed processing structure, computational neural networks can perform computations at a much higher rate, especially when implemented on a parallel digital computer or, ultimately, when implemented in customized hardware such as dedicated neurocomputing chips. The rapid speed at which these tools can work enables them to become ideal candidates for use in real-time applications involving, for example, pattern recognition within data rich GIS and remote sensing environments. It is also clear that the increasing availability of parallel hardware and virtual parallel machines, coupled with the spatial data explosion, will enhance the attractiveness of CNNs (or other parallel tools) for geographical data analysis and modelling. The non-linear nature of computational neural networks also enables them to perform function approximation and pattern classification operations that are well beyond the reach of optimal linear techniques. These tools therefore offer greater representational flexibilities and total freedom from linear model design. CNNs are also considered to be semi- or non-parametric devices requiring few or no assumptions to be made about the form of underlying population distributions, in strong contrast to conventional statistical models. One other important feature is the robust behaviour of CNNs when faced with incomplete, noisy and fuzzy information. Noise, in this instance, refers to the probabilistic introduction of errors into data. This is an important aspect of most real-world applications and neural networks can be especially good at handling troublesome data in a reasonable manner.

CNNs have massive potential and could be applied with much success in numerous diverse areas of geographical data analysis and environmental modelling and, in particular, to help solve the problems described below.

8.2.1 Pattern classification

The task of pattern classification is to assign an input pattern represented by a feature vector to one of several prespecified class groups. Well-known applications would include spectral pattern recognition where pixel-by-pixel information obtained from multispectral images is utilized for the classification of pixels (image resolution cells) into given *a priori* land cover categories. However, as the complexities of the data grow (e.g. more spectral bands from satellite scanners, higher levels of greyscaling, or finer pixel resolutions), together with the increasing trend for additional information from alternative sources to be incorporated (e.g. digital terrain models), then so too does our need for more powerful pattern classification tools (e.g. Fitzgerald and Lees, 1992, 1994; Fischer *et al.*, 1994; Gopal and Fischer, 1997).

8.2.2 Clustering/categorization

In clustering operations (also known as unsupervised pattern classification) there are no prespecified, or known, class group labels attached to the training data. A clustering algorithm is used to explore the data and to determine inherent similarities that exist between the various patterns that make up the data set. Each item is then identified as belonging to a cluster of similar patterns. Well-known clustering applications would include data mining, data compression, and exploratory spatial data analysis. Clustering has also been used to divide river flow data into different event types or hydrograph behaviours, where an event is taken to mean a short section of the hydrograph, and with the final products being intended for use in subsequent multi-network modelling applications (Abrahart and See, 1998). Enormous pattern classification tasks, e.g. using census-based small area statistics for consumer behaviour discrimination, have however proven to be difficult in unconstrained settings for conventional clustering algorithmic approaches, even when using powerful computers (Openshaw and Wymer, 1995).

8.2.3 Function approximation

The task of function approximation is to create a generalized model of a known or unknown function. Suppose a set of n training patterns (input–output pairs) $\{(x_1,\ y_1),\ (x_2,\ y_2),\ \ldots,\ (x_n,\ y_n)\}$ have been generated from an unknown function $\Phi(x)$ [subject to noise]. Then the task of function approximation would be to find an estimate of that unknown function. Various spatial analysis problems require function approximation. Examples would include spatial regression and spatial interaction modelling (e.g. Openshaw, 1993; Fischer and Gopal, 1994), modelling soil water retention curves (Schaap and Bouten, 1996), and mimicking multiple soil texture look-up charts (Abrahart, 1997). Spatial interpolation would be another case in point, e.g. the production of continuous surfaces from point data for subsequent use in hydrogeological applications (Rizzo and Dougherty, 1994).

8.2.4 Prediction/forecasting

In mathematical terms given a set of n samples $\{y(t_1),\ y(t_2),\ \ldots,\ y(t_n)\}$ in a time sequence $t_1,\ t_2,\ \ldots,\ t_n$ then the task of prediction or forecasting is to estimate the sample $y(t)$ at some future time (often $t_n + 1$). Time series prediction is an important task and can have a significant impact on decision making with respect to regional development and policy making. A lot of research in this area has also been concentrated on attempts at simulating the rainfall-runoff transformation process, e.g. the production of flood quantile estimates for ungauged catchments from neighbouring areas (Liong *et al.*, 1994), modelling a 5×5 cell synthetic watershed using inputs derived from a stochastic rainfall generator (Hsu *et al.*, 1995), predicting runoff for the Upper Wye Basin on a one hour time step (Abrahart and Kneale, 1997), and predicting runoff for the Leaf River basin using five-year daily data (Smith and Eli, 1995). Spatio-temporal modelling examples would include forecasting the spread of aids in Ohio (Gould, 1994) and predicting rainfall output generated from a space-time mathematical model (French *et al.*, 1992).

8.2.5 Optimization

A wide variety of problems in geographical analysis and data modelling can be posed as (non-linear) spatial optimization problems. The goal of an optimization algorithm is to find a solution that satisfies a set of constraints such that an objective function is maximized or minimized. The 'Travelling Salesman Problem' (Savage *et al.*, 1976; Papadimitriou and Steiglitz, 1977) and the question of finding optimal site locations are classic analytical examples (both being of a non-polynomial-complete nature). In these cases, the solution will often depend on a number of factors or possibilities which must all be examined and tested, with the enormous number of possible combinations often rendering such problems insoluble using conventional methods of spatial analysis (Murnion, 1996). Similar problems and constraints would also be associated with the optimization of complex computer simulation models such as those used in the prediction of groundwater levels (Rogers and Dowla, 1994).

8.3 How do computational neural networks work?

Computational neural networks are not mysterious devices despite their reputation as black box models. These modern tools are in fact just simple (usually non-linear) adaptive information processing structures. In mathematical terms, a computational neural network can be defined as a directed graph which has the following properties:

1. a state level u_i is associated with each node i;
2. a real-valued weight w_{ij} is associated with each edge ij between two nodes i and j that specifies the strength of this link;
3. a real-valued bias θ_i is associated with each node i;
4. a (usually non-linear) transfer function $\varphi_i [u_i, w_{ij}, \theta_i, (i \neq j)]$ is defined for each node i which determines the state of that node as a function of its bias, the weights on its incoming links from other nodes, and the different states of the j nodes that are connected to it via these links.

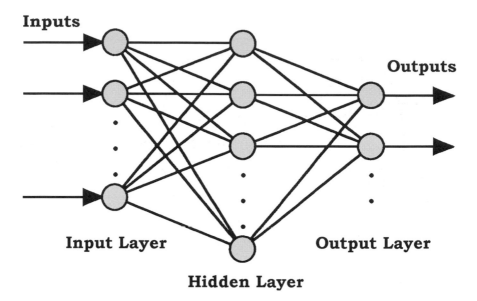

Inputs

Outputs

Input Layer   **Output Layer**

Hidden Layer

Figure 8.1 Basic configuration of a feedforward multilayered perceptron.

There are a number of standard terms in use. The 'nodes' are referred to as processing elements (PEs) or processing units. The 'edges' of the network are called connections. The connections function as unidirectional conduction paths (for signal or data flows) and transmit their information in a predetermined direction. Each PE can have numerous incoming connections. These are called 'input connections' and there is no upper limit on their number (unlimited fan-in characteristic). There is also no restriction on the number of 'output connections'. Each output connection carries an identical output signal which is the state, or activation level, of that PE. The weights are termed 'connection parameters' and it is these items that are altered during the training process and which in turn determine the overall behaviour of the CNN model.

A typical CNN architecture is shown in Figure 8.1. Circles are used to denote the processing elements, which are all linked with weighted connections, to form a network. The single output signal from each PE branches, or fans out, and identical copies of the same signal are either distributed to other PEs or leave the network altogether. The input that is presented to the network from the external world can be viewed as being a data array $x = (x_1, \ldots, x_n) \in \Re^n$ [n-dimensional Euclidean space]; and the output from the network as being another data array $y = (y_1, \ldots, y_m) \in \Re^m$ [m-dimensional Euclidean space with $m < n$]. In this notation $\Re^n$ is the n-dimensional input space of real numbers with x being an element of $\Re^n$ and $\Re^m$ is the m-dimensional output space of real numbers with y being an

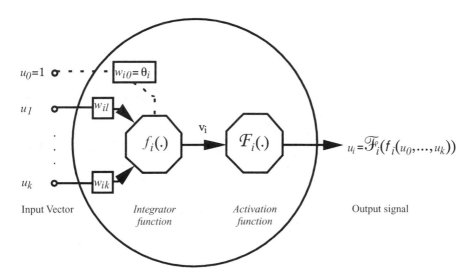

Figure 8.2 Generic processing element u_i (Fischer, 1998a).

element of $\Re^m$. The CNN, when viewed in such a manner, can therefore be thought of as being just one simple function Φ: $\Re^n \to \Re^m$.

The PEs contain internal 'transfer functions' and it is the implementation of these functions, in association with the weighted connections, that will in combination generate Φ: the so-called 'network function'. This simple generalization forms the basis for the construction and implementation of embedded CNN solutions using standard 3GL code within bespoke programs or programmed information systems. Neural solutions are in these instances treated as being standalone modules that are brought into action as and when needed using standard software procedures and associated function calls (e.g. Van den Boogaard and Kruisbrink, 1996; Abrahart, 1998).

Computational neural networks can be differentiated according to the following set of criteria:

1. their node characteristics, i.e. properties of the processing elements;
2. their network topologies, i.e. pattern of connections between the processing elements (also termed network architecture);
3. the method that is used to determine their connection weights (called learning rules, learning algorithms, machine learning, or network training).

8.4 Characteristics of the processing elements

Most notation in the field of neural networks is focused on the processing elements, their chosen method of arrangement into multiple layers, and the weighted connections that exist between them. Figure 8.2 shows the internal workings of a generic processing element. This is the basic processing element that is associated with a computational neural network and elements of this nature would be found occupying general (non-dedicated) positions within the overall structure, i.e. this PE (a) accepts inputs from other PEs and (b) sends its output signal (activation) to other PEs. Those processing elements that are not dedicated input or output units will maintain this general form and function and in so doing provide the fundamental non-linear computing capabilities that exist within each CNN. To make matters simple, where there is no confusion, we shall use the notation u_i to refer to both the processing element and the numerical activation (output) of that unit. Each element u_i computes a single (numerical) unit output or activation value. The input and output signal from the processing elements can be in the form of discrete numbers, usually taking on values $\{0, 1\}$ or $\{-1, 0, 1\}$, or it can be in the form of continuous values that will in most cases range between 0 and 1 or -1 and $+1$.

Figure 8.2 shows that the processing element, u_i, gets k input signals, $\mathbf{u} = \{u_1, \ldots, u_k\}$ which all arrive via the incoming connections that impinge on element u_i. Note that the connected elements are indexed 1 through k and that $k < i$. (where i is the total number of inputs). The corresponding connection weights associated with edge j_i between nodes j and i are w_{ij} ($j = 1, \ldots, k$). It is important to understand the manner in which the subscript of the connection weight w_{ij} is written. The first subscript refers to the PE in question and the second subscript refers to the unit from which the incoming connection originated. The reverse of this notation is also used in the neural network literature. We refer to the weights $\mathbf{w}_i = \{w_{i1}, \ldots, w_{ik}\}$ as the incoming weights for unit u_i. To simplify notation $\mathbf{W}$ is used for the vector $\mathbf{w}_i$.

Positive weights indicate reinforcement, negative weights represent inhibition, and convention dictates that for each PE there is an extra input unit u_0 whose output is always $+1$. The corresponding weight for this input w_{i0} is referred to as the bias θ_i for each unit i. The bias is otherwise treated in the same manner as any other weight and its existence accounts for the difference between k and i that was mentioned earlier. Thus, we can define the single combined $(k + 1)$ input vector

$$\mathbf{u} = [1, u_1, u_2, \ldots, u_k]^\mathrm{T} \tag{8.1}$$

where T means 'the transpose of' (in this case signifying a column not a row vector) and, correspondingly, we can define the single combined $(k + 1)$ weight (also called 'connection weight' or 'input parameter') vector

$$\mathbf{W} = [\theta, w_{i1}, \ldots, w_{ik}]^\mathrm{T} \tag{8.2}$$

where T again means 'the transpose of' (and again signifying a column not a row vector).

The basic operation that is performed within each processing element is the computation of that unit's 'activation' or 'output' signal u_i. This involves the implementation of a transfer function φ_i, which is itself composed of two mathematical functions, an integrator function f_i and an activation (or output) function $\mathscr{F}_i$ (Bezdek, 1993):

$$u_i = \varphi_i(\boldsymbol{u}) = \mathscr{F}_i(f_i(\boldsymbol{u})) \tag{8.3}$$

Typically, the same transfer function is used for all processing units within each individual layer of the computational neural network although this is not a fixed requirement.

The purpose of the integrator function f_i is to integrate the incoming activations from all other units that are connected to the processing element in question, together with the corresponding weights that have been assigned to the various incoming connections, and in so doing to transform (reduce) the incoming multiple k arguments into a single value (called 'net input' or 'activation potential' of the PE) termed v_i. In most but not all cases f_i is specified as the inner product of the vectors, $\boldsymbol{u}$ and $\boldsymbol{W}$, as follows

$$v_i = f_i(\boldsymbol{u}) = <\boldsymbol{u}, \boldsymbol{W}> = \sum_{j=0,1,\ldots,k} w_{ij} u_j \tag{8.4}$$

where $\boldsymbol{W}$ has to be predefined or learned during the training phase. In the basic case, the net input to a PE is just the weighted sum of the separate inputs from each of the k connected units plus the bias term w_{i0}. Because of the individual multiple weighting process that is used to compute v_i a degree of network tolerance for noise and missing data is automatic (Gallant, 1995). The bias term represents the offset from the origin of the k-dimensional Euclidean space $\mathfrak{R}^k$ to the hyperplane normal to $\boldsymbol{W}$ defined by f_i. In other words, bias quantifies the amount of positive or negative shift that is applied to the integrator function with respect to its zero marker in each PE. This arrangement is called a first-order processing element when f_i is an affine (linear if $w_{i0} = 0$) function of its input vector $\boldsymbol{u} = [u_1, \ldots, u_k]^T$. Higher-order processing elements will arise when more complicated functions are used for specifying f_i. For example, a second-order processing element would be realized if f_i was specified in a quadratic form, say $\boldsymbol{u}^T \boldsymbol{W} \boldsymbol{u}$, in $\boldsymbol{u}$. This might then be viewed as an alternative generalization to that which was considered in equation (8.4).

The 'activation' or 'output' function denoted by $\mathscr{F}_i(f_i(.))$ defines the output of a processing unit in terms of its total input v_i. There are various possibilities with regard to the exact specification of $\mathscr{F}_i$ but the most usual form comprises a non-linear, non-decreasing, bounded, piece-wise, differentiable function (fixed within finite asymptotic limits). With regard to computational overheads, it is also desirable that its derivative should not be difficult to calculate.

If the input data are continuous and have values that range between 0 to +1 then the logistic function is a common choice for $\mathcal{F}_i$:

$$\mathcal{F}_i(v_i) = \frac{1}{1 + \exp(-\beta v_i)} \tag{8.5}$$

where β denotes the slope parameter which has to be chosen *a priori*. In the limit, as β approaches infinity, the logistic function becomes a simple threshold function producing ones and zeros.

8.5 Network topologies

In this section we examine the pattern of the connections that exist between the processing elements, often referred to as the network architecture, and where it is possible to make a major distinction between feedforward computational neural networks, and recurrent computational neural networks.

Feedforward computational neural networks comprise those architectural configurations in which the networks do not contain directed cycles, i.e. the flow of information all goes in one direction from the input nodes (start) – via various intermediaries – to the output nodes (finish). There are no data feedback loops whatsoever. It is often convenient to organize our arrangement of nodes within each feedforward CNN into a number of distinct layers and to label each layer according to the following rule. We define an L-layer feedforward network as being a CNN wherein the processing elements are grouped into $L + 1$ layers (subsets) $L_0, L_1, \ldots, L_L$ such that if unit u in layer L_a is connected to unit u_i in layer L_b than $a < b$, i.e. the layers are numbered in ascending order commensurate with the direction of our data flow. For a strict L-layer network, we would also require that the output links from the processing elements in one layer are only connected to units in the next layer, i.e. $b = a + 1$ (as opposed to $a + 2$, $a + 3$, etc.). All units in layer L_0 are input units, all units in layers $L_1, \ldots, L_L$ are trainable processing elements, and all units in layer L_L are also output devices.

Figure 8.3 comprises a pair of architectural diagrams and is intended to illustrate various features concerning the basic layout of a multilayered feedforward CNN. Both networks have a single hidden layer and can therefore be referred to as single hidden layer feedforward networks. The shorthand notation for describing both such multilayered items would be to refer to them as 8:4:2 networks since, going from left to right, there are 8 input nodes, 4 hidden processing elements and 2 output units. The CNN in Figure 8.3(a) is said to be 'fully connected' in the sense that each and every node in the network is connected to each and every node in the adjacent forward layer. If some of the connections are missing from the network, then the network is said to be 'partially connected'. Figure 8.3(b) is an example of a partially connected CNN where the input nodes and processing elements in

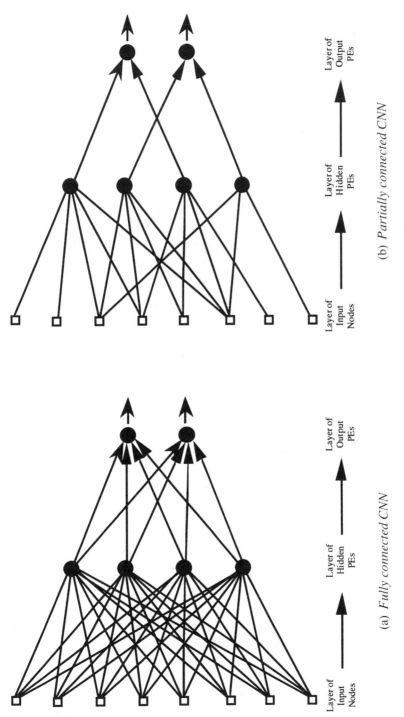

(a) *Fully connected CNN*

(b) *Partially connected CNN*

Figure 8.3 Feedforward computational neural network architectures with one hidden layer (Fischer, 1998a).

the hidden layer are connected to a partial (limited) set of processing elements in the immediate forward neighborhood. The set of localized nodes feeding an individual PE is said to constitute the 'receptive field' of the PE. Although the CNN in Figure 8.3(b) is noted to have an identical number of input nodes, hidden units and output units to that of Figure 8.3(a) the pattern of its connections nevertheless forms a specialized structure. In real world applications, when specialized structures of this nature are built into the design of a feedforward CNN, it would be to reflect prior information about the problem that has been targeted for analysis.

A recurrent (feedback) computational neural network distinguishes itself from a feedforward CNN in that it contains data processing cycles (i.e. feedback connections or data processing loops). The data are not only fed forward in the usual manner but can also be fed backwards from output units to input units, from one hidden layer back to another, or even from one node back to itself. The exact nature of each full, or partial, recurrent structure will in these instances have a profound impact on the training programme and learning capabilities of the CNN which, in turn, will affect its overall performance. Moreover, in contrast to feedforward networks, the computational processing is not defined in a unique manner according to a set of simple weighted connections because the temporal dimension must now also be considered. When the output of a PE is fed back to the same element we are also dealing with a recursive computation that has no explicit halting condition. So, at a particular instance in time, how do we tell if the fixed point of the recursive evaluation is the desired result or just one of a set of intermediate computations? To help solve this problem, it is usual to assume that each computation at each node will take a certain amount of time to process. If the arguments for a PE have been transmitted at time t then its output will be produced at time $t + 1$. A recursive process can therefore be stopped after a certain number of steps and the last computed output taken as the result of its recursive computation.

8.6 Learning in a computational neural network

In addition to the information processing characteristics associated with their individual elements, and between-network differences that arise from the use of alternative network topologies, the learning or training process forms another important distinguishing feature of computational neural networks. In the context of CNN learning, the process of training is perhaps best viewed as being a (typically) local, step-wise, steepest-gradient-based search procedure. It is operating within a multidimensional weight space and is looking for a solution (i.e. an ideal set of weights) which optimizes a prespecified objective function with or without constraints (using dedicated performance criterion to evaluate each model). Learning is performed in a progressive manner and is in most cases accomplished using an adaptive procedure referred to as the 'learning rule', 'training rule', or (machine) 'learning algorithm'.

Standard practice is to distinguish between two different types of learning situation:

1.   supervised learning problems;
2.   unsupervised learning problems.

In 'supervised learning' (also known as 'learning with a teacher' or 'associative learning'), for each example, or training pattern, there is an associated correct response (also termed 'teacher signal') which is known to the CNN output units. In 'unsupervised learning', there are no prespecified correct responses available against which the network can compare its output. Unsupervised learning is typically based on some variation of Hebbian and/ or competitive learning and in most cases involves the clustering of – or detection of similarities amongst – unlabelled patterns within a given training set. The intention in this instance is to optimize some form of comprehensive performance function or evaluation criterion defined in terms of output activities related to the processing elements within the CNN. In each application the weights and the outputs of the CNN are expected to converge to representations that capture the statistical regularities of the training data.

A wide variety of different learning algorithms is now available for solving both supervised and unsupervised learning problems, most of which have in fact been designed for specific network architectures. Most of these learning algorithms, especially those intended for supervised learning in feedforward networks, have their roots based in traditional non-linear function-minimization procedures that can be classified as being either local or global search heuristics (error minimization strategies). Learning algorithms are termed local if the computations needed to update each weight in the CNN can be performed using information that is available on a local basis to that specific weight. This requirement could, for example, be motivated by a desire to implement learning algorithms in parallel hardware. Local minimization algorithms (such as those based on gradient-descent, conjugate-gradient and quasi-Newton methods) are fast but will often converge to a local minimum (with increased chances of getting a suboptimal solution). In contrast, global minimization algorithms, such as simulated annealing and evolutionary computation, possess heuristic strategies that will enable them to escape from local minima. However, in all such cases, these algorithms are weak in either one or other of these two alternatives, i.e. good local search procedures are associated with poor global searching, and vice versa. To illustrate this point we can look at the use of gradient information. This knowledge is not just useful, but often of prime importance, in all local search procedures, yet such knowledge is not put to good use in simulated annealing or evolutionary computation. In contrast, gradient-descent algorithms, with numerous multistart possibilities are prone to encountering local minima and will thus often produce suboptimal solutions, i.e. these algorithms are weak in global search. Designing more efficient algorithms for CNN learning is thus an active research topic for neurocomputing specialists.

One critical issue for the successful application of a CNN concerns the complex relationship that exists between learning (training) and generalization. It is important to stress that the ultimate goal of network training is not to learn an exact representation of the training data, but rather to build a model of the underlying process(es) which generated that data, in order to achieve a good generalization (out-of-sample performance for the model). One simple method for optimizing the generalization performance of a neural network model is to control its effective complexity; with complexity in this case being measured in terms of the number of network parameters. This problem of finding the optimal complexity for a neural network model – although often considered crucial for a successful application – has until now been somewhat neglected in most CNN-based geographical analysis and environmental modelling applications.

In principle there are three main approaches that can be used to control over-fitting:

1. regularization techniques, i.e. adding an extra term to the error function that is designed to penalize those mappings which are not smooth;
2. pruning techniques, i.e. start with an oversized network and remove inconsequential links or nodes using automated procedures (e.g. Fischer et al., 1994; Abrahart et al., 1998);
3. cross-validation techniques to determine when to stop training, e.g. the early stopping heuristic that is demonstrated in Fischer and Gopal (1994) and Gopal and Fischer (1993).

The point of best generalization is determined from the trade-off between bias and variance associated with network output and is said to occur when the combination of bias and variance is minimized. In the case of a feedforward CNN, it is possible to reduce both bias and variance in a simultaneous manner, using a sequence of ever larger data sets, in association with a set of models that have ever greater complexities, to improve the generalization performance of the neural network solution. The generalization performance that might be achieved is however still limited according to the intrinsic noise of the data.

8.7 A classification of computational neural networks

A taxonomic classification of four important families of computational neural network models (backpropagation networks, radial basis function networks, supervised and unsupervised ART networks, and self-organizing feature maps) is presented in Figure 8.4. These particular types of computational

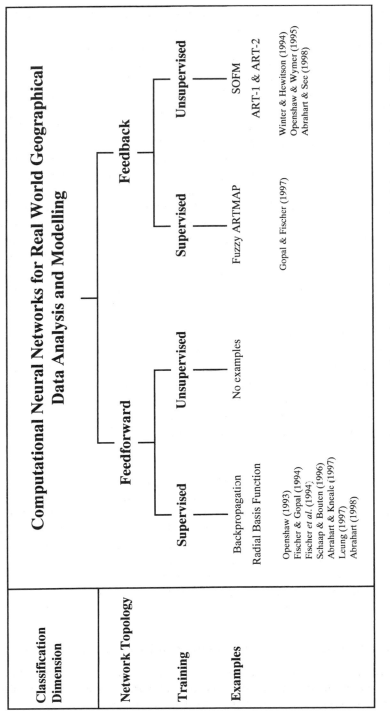

Figure 8.4 A simple fourfold taxonomic classification of computational neural networks for geographical data analysis and modelling (Fischer, 1998a).

neural network would appear to be the most attractive tools for solving real-world spatial analysis and geographical data modelling tasks. The classification has two levels: the first division is between networks with or without directed cycles; the second division is between networks that are trained with or without supervision.

8.7.1 Backpropagation CNN

Backpropagation computational neural networks have emerged as major workhorses in various areas of business and commerce, and are the most common type of neural network that has been used in geographical data analysis and modelling. These tools can be used as universal function approximators for tasks such as spatial regression, spatial interaction modelling, spatial site selection, for pattern classification in data-rich environments, or for space-time series analysis and prediction (e.g. Fischer and Gopal, 1994; Fischer et al., 1994; Abrahart and Kneale, 1997; Leung, 1997; Abrahart, 1998). In strict terms, however, backpropagation is a technique that provides an efficient computational procedure for evaluating the derivatives of the network's performance function with respect to given network parameters and corresponds to a propagation of errors backwards through the network (hence the name). This technique was first popularized by Rumelhart et al. (1986) and has since been used in countless applications. A brief introduction to some basic mathematics associated with the backpropagation training algorithm can be found in Clothiaux and Batchmann (1994) and Gopal and Fischer (1996).

In most cases backpropagation training is used with multilayered feedforward networks (also termed multilayered perceptrons) so it has become convenient to refer to this type of supervised feedforward network as a backpropagation network. Note, however, that other training options are now being incorporated as integral components within existing software packages, although the predominance of reported backpropagation applications over other training methods still shows little or no sign of changing. Each backpropagation CNN will in most cases have the following network, processing element and learning properties:

*   *Network properties:* A multilayered (typically single hidden layer) architecture;
*   *Processing element properties:* Continuous inputs and outputs; continuous non-linear sigmoid-type PE transfer functions, assuming values between 0 and 1 or −1 and +1, where evaluation of the network proceeds according to PE ordering, and with each PE computing and posting its new activation value before the next PE is examined; and output unit activations that are interpreted as the outputs for the entire CNN;

• *Learning properties:* Training involves using the backpropagation technique in combination with, typically but not necessarily, some form of gradient-descent-based learning algorithm.

Theoretical results have shown that single hidden layer feedforward networks are able to approximate arbitrary mappings arbitrarily well in the presence of noise, and that major errors will only start to arise if there are too few hidden nodes, or if the relationship being modelled is insufficiently deterministic (Hornik *et al.*, 1989). But these same authors provided little more than general guidance on how such operations could be implemented and what little guidance has been offered suggests that network training will be hard in real world applications. Moreover, there is now some debate about this all embracing statement, and it has been suggested there might well be some advantage in considering two hidden layers to provide an additional degree of representational power (Openshaw and Openshaw, 1997).

8.7.2 Radial basis function CNN

Radial basis function (RBF) networks are a special type of single hidden layer feedforward CNN in which the activation of a hidden unit is determined by the 'distance' between the 'input vector' and a 'model vector' (as opposed to computing a non-linear function of the scalar product of the input vector and a weight vector). The origins of these tools are based in exact interpolation methods that require all input vectors to be mapped in an exact manner onto their corresponding target vector. With this particular form of CNN inside each processing element in the single hidden layer, the transfer mechanism is a radial basis function. Several forms of basis have to date been considered but the most common form is still the Gaussian kernel:

$$f(x) = \exp[-(x - M)^2/\sigma^2] \tag{8.6}$$

where M and σ are two parameters representing the mean and standard deviation of the input variable x. For a particular hidden unit, i, RBF$_i$ is located at a cluster centre c_i in the associated n-dimensional input space. The cluster centre c_i is represented by the vector $[w_{1i}, \ldots, w_{ni}]$ of connection weights between the n input units and the hidden unit i. The standard deviation of this cluster defines the range for RBF$_i$. Note that RBF is monotonic, in contrast to the sigmoid function. With each argument, the transfer function in each hidden unit computes the Euclidean norm between the input vector and the centre of the kernel which is our required measure of 'vector separation' distance. The kernel function is centered at a point which is specified according to the weight vector associated with each PE and it is both the positions and widths of the individual kernels that must be learned from the training patterns. In terms of function approximation, the hidden units thus provide a set of functions that constitute a basis set for representing input patterns in the space that is spanned by the hidden units. Each output unit in the output layer will in most cases implement a simple linear summation of the radial basis functions. A number of different learning algorithms can be used with an RBF CNN. The common algorithm utilizes a hybrid learning mechanism that decouples learning at the hidden layer from that at the output layer. There are two phases. First, in the unsupervised learning phase, RBF adjustment is implemented in the hidden units using statistical clustering. This technique involves estimating kernel positions and kernel widths using for example a simple k-means based clustering algorithm. Second, in the supervised learning phase, adjustment of the second layer of connections is implemented using linear regression or gradient-descent techniques. This would involve determining the appropriate connection weights between units in the hidden and the output layers using for example a least mean squares or backpropagation algorithm. Because the output units are in most cases linear, the application of an initial non-iterative algorithm is commonplace and often sufficient. However, if need be, a supervised gradient-based algorithm can also be utilized in a further step to refine the connection parameters. A brief introduction to some basic mathematics associated with RBF networks can be found in Bishop (1995).

It is worthwhile to note that RBF networks have fast convergence properties and do not suffer from the problematic effects of local minima. However, when compared with standard backpropagation networks, the training process could indeed be orders of magnitude faster, an important disadvantage is the fact RBF networks require more training data and more hidden units to achieve the same levels of approximation.

8.7.3 ART networks

Adaptive resonance theory (ART) networks differ from the two previous types of network in that these networks are recurrent. Output from the individual processing elements is not just fed forwards from input nodes to output nodes, but it is also fed backwards, from output units to input units. ART provides the basic principles and underlying concepts that are used in these networks (Grossberg, 1976a, b). ART networks were developed as possible models of cognitive phenomena in humans and animals and thus have more biological association than did our earlier examples. ART makes use of two important items that are used in the analysis of brain behaviour: stability and plasticity. The stability–plasticity dilemma concerns the power of a system to preserve the balance between retaining previous learned patterns and learning new patterns.

In simple conceptual terms an ART network contains two main layers of processing elements: a top layer (output-concept layer F_2) and a bottom layer (input-feature layer F_1). There are two sets of weighted connections between each of the nodes in these two layers: top-down weights that represent learned patterns (expectations) and bottom-up weights that represent a scheme through which the new inputs can be accommodated. However, in more precise terms, each actual ART implementation could in fact be disassembled into three components:

1. an input processing field (F_1-layer) consisting of two parts, the input portion with input nodes and the interface portion (interconnections);
2. a layer of linear units (F_2-layer) representing prototype vectors whose outputs are acted on during competitive learning, i.e. the winner is the node with a weight vector that is closest to the input vector (closest in a Euclidean distance sense);
3. various supplemental units for implementing a reset mechanism to control the degree of matching for patterns that are to be placed in the same cluster.

The interface portion of the F_1-layer combines signals from the input portion and the F_2-layer, for use in comparing input signals to the weight vector for the cluster that has been selected as a candidate for learning. Each individual unit in the F_1-layer is connected to the F_2-layer by feedforward and feedback connections. Changes in the activations of the units and in their weights are governed by coupled differential equations.

This type of CNN is in essence a clustering tool that is used for the automatic grouping of unlabelled input vectors into several categories (clusters) such that each input is assigned a label corresponding to a unique cluster. ART networks use a simple method of representation wherein each cluster is represented using the weight vector for an individual prototype unit. Similarities drive the clustering process. Vectors that are grouped into the same cluster are similar which means that associated input patterns are close to each other in terms of input space. If an input vector is close to a prototype, then it is considered a member of the prototype's cluster, with local differences being attributed to unimportant features or to noise. When the input data and stored prototype are sufficiently similar then these two items are said to resonate (from which the name of this technique is obtained). It should be stressed that there is no set number of clusters, additional output nodes (clusters) being created as and when needed. An important item in this implementation of the stability–plasticity dilemma is the control of partial matching between new feature vectors and the number of stored (learned) patterns that the system can tolerate. Indeed, any clustering algorithm that does not have a prespecifed number of clusters, or does not in some manner limit the growth of new clusters, must have some other mechanism or parameter for controlling cluster resolution and for preventing excessive one-to-one mapping. Each ART network has a vigilance parameter (VP) that is used for this purpose and which is explained in the next section.

The ART-1 learning algorithm has two major phases. In the first phase, input patterns are presented and activation values calculated for the output neurons. This defines the winning neuron. The second phase calculates the mismatch between the input pattern and the current pattern associated with the winning neuron. If the mismatch is below a threshold (vigilance parameter) then the old pattern is updated to accommodate the new one. But if the mismatch is above the threshold then the procedure continues to look for a better existing concept-output unit or it will create a new concept-output unit. ART networks will be stable for a finite set of training examples because even with additional iterations the final clusters will not change from those produced using the original set of training examples. Thus, these tools possess incremental clustering capabilities, and can handle an infinite stream of input data. ART networks also do not require large memories for storing training data because their cluster prototype units contain implicit representation of all previous input encounters. However, ART networks are sensitive to the presentation order of the training examples, and might produce different clusterings on the same input data when the presentation order of patterns is varied. Similar effects are also present in incremental versions of traditional clustering techniques, e.g. k-means clustering is also sensitive to the initial selection of cluster centres.

Most ART networks are intended for unsupervised classification. The simplest of the ART networks are ART-1 that uses discrete data (Carpenter and Grossberg, 1987a) and ART-2 which uses continuous data (Carpenter and Grossberg, 1987b). A more recent addition to this collection is a supervised version of ART-1 called ARTMAP (Carpenter et al., 1991). There is also fuzzy ARTMAP, a generalization of ARTMAP for continuous data, that was created with the replacement of ART-1 in ARTMAP with Fuzzy ART (Carpenter et al., 1992). In this instance, fuzzy ART synthesizes fuzzy logic and adaptive resonance theory by exploiting the formal similarities between (a) the computations of fuzzy subsethood and (b) the dynamics of prototype choice, search and learning. This approach is appealing because it provides an agreeable integration of clustering with supervised learning on the one hand and fuzzy logic and adaptive resonance theory on the other. A comparison of fuzzy ARTMAP with backpropagation and maximum likelihood classification for a real world spectral pattern recognition problem is reported in Gopal and Fischer (1997).

8.7.4 Self-organizing feature map

Another important class of powerful recurrent computational neural networks are self-organizing feature maps (otherwise referred to a SOFM networks or Kohonen networks). SOFM networks are used for vector quantization and data analysis and these tools have been foremost and in the main developed by Kohonen (1982, 1988). These quantitative 'mapping tools', which are at least in part based on the structure of the mammalian brain, will in addition to the classification process also attempt to preserve important topological relationships. Although the standard implementation of these tools is for unsupervised classification and feature extraction purposes, such items can also be used as modelling tools, for example where inputs are mapped onto a response surface in an optimal manner (Openshaw and Openshaw, 1997). However, a supervised SOFM is also available, and one possible realization of an appropriate training algorithm can be found in Kasabov (1996). The underlying basis for such networks is rooted deep in vector quantization theory and their emergence as an operational geographical tool has arisen from the spatial data explosion and our associated need for large scale multidimensional data reduction capabilities. In simple conceptual terms, a SOFM consists of two layers, an input layer and an output layer, called a feature map, which represents the output vectors of the output space. The task of each SOFM is to map input vectors from the input units on to the output units or feature map (which under normal circumstances takes the form of a one- or two-dimensional array), and to perform this adaptive transformation in an ordered topological fashion, such that topological relationships between the input vectors are preserved and represented in the final product via the spatial distribution or pattern of unit activations. So, the more related two vectors are in terms of input space, the closer will be the position of the two corresponding units that represent these input patterns in the feature map. The overall idea then is to develop a topological map of input vectors such that similar input vectors would trigger both their own units and other similar units in proportion to their topological closeness. Thus a global organization of the units and associated data is expected to emerge for the training programme.

In more detail the essential characteristics of SOFM networks can be summarized as follows:

- *Network properties:* A two-layer architecture where the input layer is fully connected to the output layer (Kohonen layer) and whose units are arranged in a two-dimensional grid (map). The map units have local interaction capabilities which means that changes in the behaviour of one unit will have a direct effect on the behaviour of other units in its immediate neighbourhood.
- *Processing element properties:* Each output unit is charcterized by an n-dimensional weight vector and contains a linear processing element. Each feature map unit computes its net input on a linear basis and non-linearities come into being when the selection is made as to which unit 'fires'.
- *Learning properties:* Unsupervised learning in a network is the adaptive modification of the connection weights associated with local interacting units in response to input excitations and in accordance with a competitive learning rule (i.e. weight adjustment of the winning unit and its neighbours). The weight adjustment of the neighbouring units is instrumental in preserving the topological ordering of the input space.

SOFM networks can also be used for front end pattern classification purposes or for other important decision making processes. Abrahart and See (1998) and See and Openshaw (1998) have used SOFM networks to perform data-splitting operations prior to the implementation of more accurate multi-network river flow prediction. It is also possible to have output values from the feature map layer passed into the hidden layer of a backpropagation network on a direct feed basis.

8.8 Advantages, application domains and examples

8.8.1 Advantages

The attraction of CNN-based geographical data analysis and modelling extends far beyond the high computation rates provided by massive parallelism and the numerous advantages that are now on offer for us to exploit are perhaps best considered under the following points:

1. greater representational flexibilities and freedom from linear model design constraints;
2. built-in network capabilities (via representation, training, etc.) to incorporate rather than to ignore the special nature of spatial data;
3. greater degrees of robustness, or fault tolerance, to deal with noisy data, and missing or fuzzy information;
4. efficient operational capabilities for dealing with large, ultra-large and massive spatial data sets, together with the associated prospect of obtaining better results through being able to process finer resolution data or to perform real-time geographical analysis;
5. built-in dynamic operational capabilities for adapting connection weights commensurate with changes in the surrounding environment (dynamic learning);
6. good generalization (out-of-sample performance) capabilities that work in a specific and, in general terms, quite satisfying manner;
7. potential improvements in the quality of results associated with a reduction in the number of rigid assumptions and computational shortcuts that are otherwise introduced using conventional methodologies and techniques.

8.8.2 Application domains

CNN models, in particular hidden-layered feedforward networks, together with their wide range of different recognized learning techniques are now able to provide geographers with novel, elegant, and extremely valuable classes of mathematical tools – all based on sound theoretical concepts – for geographical data analysis and modelling. Moreover, such tools are not intended to be substitutes for traditional methods, but should instead be viewed as being non-linear extensions to conventional statistical methods such as regression models, spatial interaction models, linear discriminant functions, pattern recognition techniques, and time series prediction tools (White, 1989; Pao, 1989; Fischer and Gopal, 1994; Fischer *et al.*, 1994; Abrahart and Kneale, 1997; Abrahart, 1998). Much work has to date been done in what are now seen as being the two major domains wherein these tools are most applicable:

1. as universal function approximators in areas such as spatial regression, spatial interaction, spatial site selection and space-time series modelling;
2. as pattern recognizers and classifiers, which function as intelligent aids, and allow the user to sift through copious amounts of digital data in a fast and efficient manner, implement multidimensional data reduction based on otherwise unknown properties, and, where possible, to find patterns of interest in data-rich environments, e.g. census small area statistics and high-resolution remote sensing images.

Feedforward CNN networks, within a geographical analysis and modelling context, are often implemented for complex function approximation purposes. A simple three-stage process has therefore been proposed for the application of such tools and an illustration of this method is provided in Fischer and Gopal (1994):

1. identification of a candidate model from a range of multilayered feedforward CNN options and specific types of non-linear processing element (e.g. perceptron or radial basis function);
2. estimation of network parameters for the selected CNN model and optimization of model complexities with respect to a given training set (using regularization, network pruning or cross-validation);
3. appropriate testing and evaluation of the final CNN model in terms of its generalization capabilities (out-of-sample performance).

8.8.3 Examples

In the following paragraphs three different geographical examples are provided in order to give a general impression of what is and is not possible with regard to CNN usage.

Fischer and Gopal (1994) used a one-hidden-layer backpropagation network, with sigmoidal processing elements, to model interregional telecommunication traffic in Austria. This work involved using noisy, real-world, limited record data. An epoch-based stochastic steepest gradient descent algorithm (epoch size: 20 patterns) was used to minimize the least mean square error function and a random technique of cross-validation was used to provide an early stopping heuristic for optimization of the model. Two performance measures, average relative variance and coefficient of determination, were used to evaluate CNN performance against the traditional regression approach of a gravity-type model, the latter forming what was considered to be a statistical benchmark. The CNN solution can be viewed as a generalized non-linear spatial regression model albeit of quite specific form. This model provided superior performance to the current best practice which was doubtless in no small measure due to its more generalized functional configuration. Openshaw (1993), in a similar manner, compared CNN spatial interaction models with conventional entropy maximizing models also with good results.

Fischer and associates (Fischer et al., 1994; Fischer and Gopal, 1996; Gopal and Fischer, 1997) analysed the performance of three different neural classifiers that were used to solve a pixel-by-pixel supervised classification problem working with spectral urban land cover information from a Landsat-5 Thematic Mapper (TM) image for Vienna and its northern surroundings. They examined: a pruned one-hidden-layer perceptron, with logistic hidden units, and softmax output transformation; a two-layer radial basis network, with Gaussian hidden units, and softmax output transformation; and a fuzzy ARTMAP classifier.

The results were compared with a Gaussian maximum likelihood classification which was taken to represent best current practice. In supervised classification, individual pixels are assigned to various class groups according to the spectral properties relating to a number of prespecified training sites. The problem of discriminating between urban land cover categories is challenging because urban areas comprise a complex spatial assemblage of disparate land-cover types, including built structures, numerous vegetation types, bare soil and water bodies, each of which has different spectral reflectance characteristics. As a result, the classification of urban land cover represents a challenging problem, where current best practice tends to give poor relative performance. The classifiers were all trained on TM bands 1–5 and 7. There were 1640 training pixels and 820 testing pixels. The six-dimensional feature vectors (bands or channels) were used to discriminate between eight prespecified class groups: mixed grass and arable farmland; vineyards and areas with low vegetation cover; asphalt and concrete surfaces; woodland and public gardens with trees; low density residential and industrial areas; densely built up residential areas; water courses; and stagnant water bodies. Individual classifier performance was assessed using standard measures such as a confusion matrix, the map user's classification, the map producer's classification, and the total classification accuracies. Fuzzy ARTMAP produced an outstanding out-of-sample classification result of 99.26% on the testing data set. The error rate was less than 1/15 that of the two-hidden-layer perceptron, 1/20 that of the maximum likelihood classifier and 1/30 that of the RBF-CNN. An inspection of the classification error matrices also revealed that the fuzzy ARTMAP classifier was better at accommodating a heterogeneous class label such as 'densely built up residential areas' even with smaller numbers of training pixels. The maximum likelihood classifier experienced difficulties when dealing with impure land cover signatures. It also took a long time to process the data which would therefore create a major problem with large areas and in data-rich environments. Such problems would also be exacerbated with the use of data from powerful multichannel satellite scanners, such as the 192 channel High Resolution Infrared Imaging Spectrometer, or when working with multitemporal images, or when incorporating numerous ancillary GIS related inputs.

Abrahart and Kneale (1997) implemented a series of experiments to explore the potential use of backpropagation neural network solutions for continuous river flow forecasting on the Upper River Wye (Central Wales). Data for the Cefn Brwyn gauging station (No. 55008) were available on a one-hour time step for the period 1984–86 and comprised: recorded rainfall records (RAIN), potential evapotranspiration estimates (PET), and measured river flow values (FLOW). To predict FLOW at time t the original data were transformed to produce a fixed length moving time frame window that encompassed: annual hour-count (CLOCK), RAIN t, RAIN t-1 to t-6, PET t, PET t-1 to t-6, FLOW t-1 to t-6, and FLOW t; with subsequent conversion of CLOCK values into their sine and cosine equivalents. Each set of variables was normalized, between 0 and 1, and the complete historic record was then split into three annual data sets: 1984, 1985 and 1986. The complete data record for 1985 (normal climatic regime), in conjunction with the backpropagation algorithm, was used to train a two-hidden-layer feedforward network that had a 22:16:14:1 architecture. Network forecasting capabilities were then tested with data for 1984 (summer drought) and 1986 (normal climatic regime). The network was observed to converge in a smooth and uneventful manner with a reasonable degree of generalization being achieved after just 5000 epochs. Computed model efficiencies, calculated using a standard hydrological technique, indicated excellent results for all three periods: comprising 99.29% for the training period, with validation statistics of 90.43% for 1984, and 92.58% for 1986. Figure 8.5 contains two associated graphs relating to a 500-hour section of the training period. The top graph shows a near-exact fit between the height (peaks) and timing (rising and falling limbs) of individual storm events, with just minor differences occurring from time to time. Most noticeable is an underestimation of the highest peak and the fact that the forecasted values are often one step ahead of the actual river flow values. However, with regard to real-time forecasting, early prediction is more helpful than a delayed response.

8.9 Conclusions and outlook

Computational neural networks provide much more than just a set of novel, useful or valuable data-driven mathematical tools. Indeed, with respect to

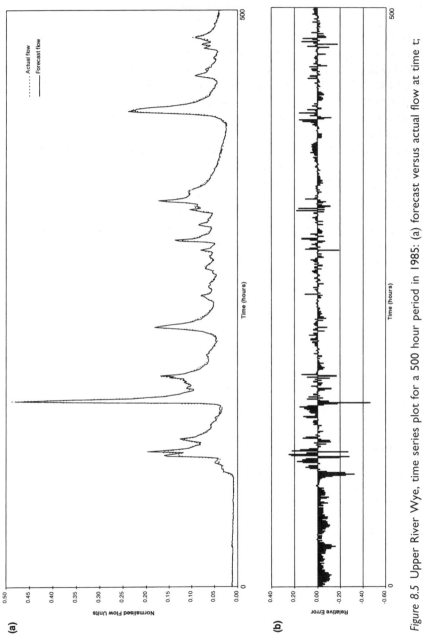

Figure 8.5 Upper River Wye, time series plot for a 500 hour period in 1985: (a) forecast versus actual flow at time *t*; (b) relative error at time *t* (Abrahart and Kneale, 1997).

geographical data analysis and modelling, these new tools provide an appropriate framework for re-engineering many well-established spatial analysis and environmental modelling techniques to meet the new large-scale data processing needs associated with GIS and GeoComputation. Application of CNN models to spatial data sets holds the potential for fundamental advances in empirical understanding across a broad spectrum of geographical related fields. To realize these advances, it is therefore important to adopt a principled rather than an *ad hoc* approach in which spatial statistics and computational neural network modelling must work together. The most important challenges in the coming years will be twofold. Firstly, to develop geographical application domain specific methodologies that are relevant to neurocomputing. Secondly, to gain deeper theoretical insights into the complex relationship that exists between learning and generalization, which is of critical importance for the success of real-world applications.

The mystique perceived by those outside the field can in part be attributed to the origins of computational neural network systems in the study of biological neural systems which, in association with the extended hype and metaphorical jargon that is rife in this area of computer science, has acted to lessen the amount of serious attention that is given to this new information processing paradigm. But – and this is important to note – numerous aspects related to the subject of computational neural networks lend themselves to rigorous mathematical analysis. This in turn provides a sound foundation on which to base an investigation into the capabilities and limitations of different CNN tools and for examining their use in real-world geographical applications. Casting such analyses in the universal language of mathematics and applying them via soundly formulated statistical frameworks would be a worthwhile positive act that could help to dispel much unwarranted mystique and avoid much potential misuse in the future.

References

Abrahart, R. J. (1997) 'First experiments in neural network mapping', in Pascoe, R. T. (ed.), *GeoComputation 97: Proceedings 2nd International Conference on GeoComputation*, Dunedin, New Zealand: University of Otago, 26–29 August 1997, pp. 375–383.

Abrahart, R. J. (1998) 'Neural networks and the problem of accumulated error: an embedded solution that offers new opportunities for modelling and testing', in Babovic, V. and Larsen, C. L. (eds) *Hydroinformatics 98: Proceedings Third International Conference on Hydroinformatics*, Copenhagen, Denmark, 24–26 August 1998, Vol. 2, Rotterdam: Balkema, pp. 725–731.

Abrahart, R. J. and Kneale, P. E. (1997) 'Exploring neural network rainfall-runoff modelling', in *Proceedings Sixth National Hydrology Symposium*, University of Salford, 15–18 September 1997, Wallingford: British Hydrological Society, pp. 9.35–9.44.

Abrahart, R. J. and See, L. (1998) 'Neural network vs. ARMA modelling: constructing benchmark case studies of river flow prediction', in *GeoComputation 98: Proceedings Third International Conference on GeoComputation*, University of Bristol, 17–19 September 1998, Manchester: GeoComputation CD-ROM.

Abrahart, R. J., See, L. and Kneale, P. E. (1998) 'New tools for neurohydrologists: using "network pruning" and "model breeding" algorithms to discover optimum inputs and architectures', in *GeoComputation 98: Proceedings Third International Conference on GeoComputation*, University of Bristol, 17–19 September 1998, CD-ROM.

Bezdek, J. C. (1993) 'Pattern recognition with fuzzy sets and neural nets', in *Tutorial Texts, International Joint Conference on Neural Networks, IJCNN'93*, pp. 169–206.

Bishop, C. M. (1995) *Neural Networks for Pattern Recognition*. Oxford: Clarendon Press.

Carpenter, G. A. and Grossberg, S. (1987a) 'A massively parallel architecture for a self-organizing neural pattern recognition machine', *Computer Vision, Graphics, and Image Processing*, 37, pp. 54 115.

Carpenter, G. A. and Grossberg, S. (1987b) 'ART 2: self-organization of stable category recognition codes for analog input patterns', *Applied Optics*, 26, 23, pp. 4919–4930.

Carpenter, G. A. and Grossberg, S. (eds) (1991) *Pattern Recognition by Self-Organizing Networks*, Cambridge, MA: MIT Press.

Carpenter, G. A., Grossberg, S. and Reynolds, J. H. (1991) 'ARTMAP: supervised real-time learning and classification of nonstationary data by a self-organizing neural network', *Neural Networks*, 4, pp. 565–588.

Carpenter, G. A., Grossberg, S., Markuzon, N., Reynolds, J. H. and Rosen, D. B. (1992) 'Fuzzy ARTMAP: a neural network architecture for incremental supervised learning of analog multidimensional maps', *IEEE Transactions on Neural Networks*, 3, 5, pp. 698–713.

Clothiaux, E. E. and Batchmann, C. M. (1994) 'Neural networks and their applications', in Hewitson, B. C. and Crane, R. G. (eds) *Neural Nets: Applications in Geography*, Dordrecht: Kluwer, pp. 11–52.

Fischer, M. M. (1994) 'Expert systems and artificial neural networks for spatial analysis and modelling: essential components for knowledge-based geographic information systems', *Geographical Systems*, 1, 3, pp. 221–235.

Fischer, M. M. (1995) 'Fundamentals in neurocomputing', in Fischer, M. M, Sikos, T. T. and Bassa, L. (eds) *Recent Developments in Spatial Information, Modelling and Processing*, Budapest: Geomarket, pp. 31–41.

Fischer, M. M. (1998a) 'Computational neural networks – a new paradigm for spatial analysis', *Environment and Planning A*, 30, pp. 1873–1892.

Fischer, M. M. (1998b) 'Spatial analysis: retrospect and prospect', in Longley, P., Goodchild, M. F., Maguire, P. and Rhind, D. W. (eds) *Geographic Information Systems: Principles, Technical Issues, Management Issues and Applications*, New York: Wiley, pp. 283–292.

Fischer, M. M. and Gopal, S. (1994) 'Artificial neural networks. A new approach to modelling interregional telecommunication flows', *Journal of Regional Science*, 34, 4, pp. 503–527.

Fischer, M. M. and Gopal, S. (1996) 'Spectral pattern recognition and fuzzy ARTMAP: design features, system dynamics and real world simulations', in *Pro-

ceedings of EUFIT'96, Fourth European Congress on Intelligent Technologies and Soft Computing, Aachen: Elite Foundation, pp. 1664–1668.

Fischer, M. M., Gopal, S., Staufer, P. and Steinnocher, K. (1994) 'Evaluation of neural pattern classifiers for a remote sensing application', Paper presented at the 34th European Congress of the Regional Science Association, Groningen, August 1994, *Geographical Systems,* 4, 2, pp. 195–223 and 235–236.

Fitzgerald, R. W. and Lees, B. G. (1992) 'The application of neural networks to the floristic classification of remote sensing and GIS data in complex terrain', in *Proceedings 17th Congress of the International Society for Photogrammetry and Remote Sensing,* Washington D. C., USA 2–14 August 1992, Vol. XXIX, Part B7, pp. 570–573.

Fitzgerald, R. W. and Lees, B. G. (1994) 'Spatial context and scale relationships in raster data for thematic mapping in natural systems', in *Proceedings SDH'94: Sixth International Symposium on Spatial Data Handling,* Edinburgh, 5–9 September 1994, pp. 462–475.

French, M. N., Krajewski, W. F. and Cuykendall, R. R. (1992) 'Rainfall forecasting in space and time using a neural network', *Journal of Hydrology,* 137, pp. 1–31.

Gallant, S. I. (1995) *Neural Network Learning and Expert Systems,* Cambridge, MA: MIT Press.

Gopal, S. and Fischer, M. M. (1993) 'Neural net based interregional telephone traffic models', in *Proceedings of the International Joint Conference on Neural Networks 2,* Nagoya: IEEE Press, pp. 2041–2044.

Gopal, S. and Fischer, M. M. (1996) 'Learning in single hidden-layer feedforward network models: backpropagation in a spatial interaction modelling context', *Geographical Analysis,* 28, 1, pp. 38–55.

Gopal, S. and Fischer, M. M. (1997) 'Fuzzy ARTMAP – a neural classifier for multispectral image classification', in Fischer, M. M. and Getis, A. (eds) *Recent Developments in Spatial Analysis. Spatial Statistics, Behavioural Modelling and Computational Intelligence,* Berlin: Springer Verlag, pp. 306–335.

Gould, P. (1994) 'Neural computing and the aids pandemic: the case of Ohio', Hewitson, B. C. and Crane, R. G. (eds) *Neural Nets: Applications in Geography,* Dordrecht: Kluwer, pp. 101–119.

Grossberg, S. (1976a) 'Adaptive pattern classification and universal recording I: parallel development and coding of neural feature detectors', *Biological Cybernetics,* 23, pp. 121–134.

Grossberg, S. (1976b) 'Adaptive pattern classification and universal recording II: feedback, expectation, olfaction and illusion', *Biological Cybernetics,* 23, pp. 187–202.

Hecht-Nielsen, R. (1990) *Neurocomputing,* Reading, MA: Addison-Wesley.

Hornik, K., Stinchcombe, M. and White, H. (1989) 'Multilayer feedforward networks are universal approximators', *Neural Networks,* 2, pp. 359–366.

Hsu, K.-L., Gupta, H. V. and Sorooshian, S. (1995) 'Artificial neural network modeling of the rainfall-runoff process', *Water Resources Research,* 31, pp. 2517–2530.

Kasabov, N. K. (1996) *Foundations of Neural Networks, Fuzzy Systems, and Knowledge Engineering,* Cambridge, MA: MIT Press.

Kohonen, T. (1982) 'Self-organized formation of topologically correct feature maps', *Biological Cybernetics,* 43, pp. 59–69.

Kohonen, T. (1988) *Self-Organization and Associative Memory,* Heidelberg: Springer Verlag.

Leung, Y. (1997) 'Feedforward neural network models for spatial pattern classification', in Fischer, M. M. and Getis, A. (eds) *Recent Developments in Spatial Analysis. Spatial Statistics, Behavioural Modelling and Computational Intelligence*, Berlin: Springer Verlag, pp. 336–359.

Liong, S. Y., Nguyen, V. T. V., Chan, W. T. and Chia, Y. S. (1994) 'Regional estimation of floods for ungaged catchments with neural networks', in Cheong, H.-F., Shankar, N. J., Chan, E.-S. and Ng, W.-J. (eds) *Developments in Hydraulic Engineering and their Impact on the Environment, Proceedings Ninth Congress of the Asian and Pacific Division of the International Association for Hydraulic Research*, Singapore, 24–26 August 1994, pp. 372–378.

Murnion, S. D. (1996) 'Spatial analysis using unsupervised neural networks', *Computers and Geosciences*, 22, 9, pp. 1027–1031.

Openshaw, S. (1993) 'Modelling spatial interaction using a neural net', in Fischer, M. M. and Nijkamp, P. (eds) *Geographic Information Systems, Spatial Modelling, and Policy Evaluation*. Berlin: Springer Verlag, pp. 147–164.

Openshaw, S. and Openshaw, C. (1997) *Artificial Intelligence in Geography*, Chichester: Wiley.

Openshaw, S. and Wymer, S. (1995) 'Classifying and regionalising census data', in Openshaw, S. (ed.) *Census Users Handbook*, Cambridge: GeoInformation International, pp. 353–361.

Pao, Y.-H. (1989) *Adaptive Pattern Recognition and Neural Networks*, Reading, MA: Addison-Wesley.

Papadimitriou, C. H. and Steiglitz, K. (1977) 'On the complexity of local search for the traveling salesman problem', *SIAM Journal of Computing*, 6, 1, pp. 76–83.

Rizzo, D. M. and Dougherty, D. E. (1994) 'Characterization of acquifer properties using artificial neural networks: neural kriging', *Water Resources Research*, 30, pp. 483–497.

Rogers, L. L. and Dowla, F. U. (1994) 'Optimization of groundwater remediation using artificial neural networks with parallel solute transport modelling', *Water Resources Research*, 30, 2, pp. 457–481.

Rumelhart, D. E., Hinton, G. E. and Williams, R. J. (1986) 'Learning internal representations by error propagations', in Rumelhart, D. E. and McClelland, J. L. (eds) *Parallel Distributed Processing: Explorations in the Microstructures of Cognition, Vol. 1*, Cambridge, MA: MIT Press, pp. 318–362.

Savage, S., Weiner, P. and Bagchi, A. (1976) 'Neighborhood search algorithms for guaranteeing optimal traveling salesman tours must be inefficient', *Journal of Computer and System Sciences*, 12, pp. 25–35.

Schaap, M. G. and Bouten, W. (1996) 'Modeling water retention curves of sandy soils using neural networks', *Water Resources Research*, 32, 10, pp. 3033–3040.

See, L. and Openshaw, S. (1998) 'Using soft computing techniques to enhance flood forecasting on the River Ouse', in Babovic, V. and Larsen, C. L. (eds) *Hydroinformatics 98: Proceedings Third International Conference on Hydroinformatics*, Copenhagen, Denmark, 24–26 August 1998, Rotterdam: Balkema, Vol. 2, pp. 819–824.

Smith, J. and Eli, R. N. (1995) 'Neural-network models of rainfall-runoff process', *Journal of Water Resources Planning and Management*, 121, pp. 499–509.

Smolensky, P. (1988) 'On the proper treatment of connectionism', *Behavioral and Brain Sciences*, 11, pp. 1–74.

Van den Boogaard, H. F. P. and Kruisbrink, A. C. H. (1996) 'Hybrid modelling by integrating neural networks and numerical models', in Müller, A. (ed.) *Hydroinformatics 96: Proceedings 2nd International Conference on Hydroinformatics*, Zurich, Switzerland, 9–13 September 1996, Rotterdam: Balkema, Vol. 2, pp. 471–477.

White, H. (1989) 'Learning in artificial neural networks: a statistical perspective', *Neural Computation*, 1, pp. 425–464.

Winter, K. and Hewitson, B. C. (1994) 'Self organizing maps – application to census data', in Hewitson, B. C. and Crane, R. G. (eds) *Neural Nets: Applications in Geography*, Dordrecht: Kluwer, pp. 71–77.

Chapter 9

Genetic programming: a new approach to spatial model building

Gary Diplock

9.1 Introducing genetic programming

The rapid speed up in computer hardware combined with the development of applicable artificial intelligence (AI) tools is opening up exciting new analysis and modelling opportunities in geography. More specifically, there is the prospect of developing new technologies for spatial data modelling to cope better with the complexity of the phenomena under investigation. Indeed, the increasing capture of spatially referenced data (largely created by the use of geographical information systems, or GIS) and the improved sophistication of data handling techniques demand that new methods be developed which can properly exploit this spatial data rich environment.

Clearly, there are certain approaches in geography that are likely to benefit from these developments, and spatial modelling is one such area. Spatial model building is, traditionally, a difficult mathematical process, especially as the systems under investigation often exhibit extremely complex, even chaotic, behaviour. Consequently, models are hard to build, and few good models exist. For example, spatial interaction models (Wilson, 1971; Fotheringham, 1983) are successfully used in a variety of applications today, but were derived in the mathematical and statistical modelling era of the 1960s and 1970s, when both spatial data and the means of using it were scarce and inadequate. A new approach to model building based on AI techniques offers the prospect of a radically different method of model development.

Evolutionary computing methods are an extremely flexible and intelligent search procedure which mimic the known mechanisms of natural evolution. Such techniques facilitate the development of less rigid model building tools, by applying significant amounts of computer power and sophisticated algorithms. There are two broad categories of evolutionary algorithms: genetic algorithms (Holland, 1975) and genetic programming (Koza, 1992). The former are more established but the latter is developing rapidly and generating a great deal of interest in a wide range of applications. The two approaches share many similar characteristics, although there are also important differences.

Genetic algorithms (GA) are widely accepted and numerous texts discuss their properties (Davis, 1991; Goldberg, 1989). They are most suited to combinatorial problems or parameter estimation. Diplock and Openshaw (1996) illustrate the use of genetic algorithms to estimate the unknown parameters of a series of spatial interaction models of increasing degrees of complexity. The genetic algorithms compared favourably with more traditional methods of parameter optimisation, especially for more complex model specifications. Genetic programming (GP) offers more potential for the creation of models, and it is to this subject that attention is now turned. A brief description of the techniques involved is presented, followed by an example application, in an attempt to build new spatial interaction models. Finally, some thoughts on the future of such techniques are provided.

9.2 Implementing genetic programming techniques

9.2.1 The basic concepts of evolutionary algorithms

Genetic algorithms are an adaptive search technique developed initially by Holland (1975). Despite being a relatively new field of artificial intelligence research, they are already well established and have been successfully used in a wide range of disciplines, perhaps most notably as an optimization technique (Bethke, 1981; Fogel, 1994; Schwefel, 1995). This is due partly to their assumption-free properties, whereby the only information required about the problem used in the algorithm involves an appropriate performance measure. This makes them both powerful and robust.

Genetic algorithms use strings of bits to represent solutions for a problem, as an analogy to chromosomes which occur in nature, upon which the processes of evolution operate. After initially generating a random set of such strings, these are decoded into an appropriate set of values for use in the evaluation function and their performance determined. Individuals are then selected on the basis of this performance measure. Good solutions are evolved to create offspring for the next generation, through the simulation of genetic operators. The next generation is evaluated and the process continues until some termination criteria are met.

Genetic programming is a relatively new hierarchical extension of genetic algorithms, pioneered by Koza (1992, 1994). It utilises the same principles of evolution (natural selection, survival of the fittest, etc.) but exhibits one important difference (especially in the context of spatial interaction model breeding), namely, the representation used consists of symbolic mathematical expressions. Whilst these are evolved in a similar manner to the bit strings which represent the population of a genetic algorithm, GP yields the advantage in a model breeding context of eliminating the need for an arbitrary transformation between a binary bit string and a model, which allows a greater degree of flexibility since the representation is no longer rigid. Like the genetic algorithm, however, the method is extremely robust, with only fatal errors halting a run, whilst minor programming errors can exist undetected as the method is error tolerant and will do its best to cope (Openshaw and Turton, 1994). In consequence, there is a need to rigorously check the algorithm throughout its development.

Among the numerous problem domains in which genetic programming possesses potential (Koza, 1992), empirical discovery and forecasting is the relevant theme for the development of spatial interaction model breeding software. This involves

> '. . . finding a model that relates a given finite sampling of values of the independent variables and the associated (often noisy) values of the dependent variables for some observed system in the real world.'
>
> (Koza, 1992, p. 12)

Once such a model is successfully identified, it can then be used in a predictive capacity, an activity that is especially useful in the context of spatial interaction modelling. The task is to identify a computer program which utilises a series of independent variables (such as origin and destination characteristics, distances, population data, etc.) to produce a mathematical description of the dependent variable (interaction volume). In this case, the programs which are evolved are not programs in the true sense (lines of code), but symbolic mathematical expressions.

9.2.1.1 Representation

Fixed length binary strings are normally used to represent problem solutions in genetic algorithms, but these are limited in potential for many applications, severely constraining the search process. These shortcomings have been further highlighted in early attempts at program induction using genetic algorithms (for example Cramer, 1985). Numerous attempts have been made to overcome such limitations, as outlined by Koza (1992, pp. 64–68), which include many domain-specific representation schemes as well as higher level representations (Antonisse, 1991; Antonisse and Keller, 1987) and hierarchical structures (Wilson, 1987). Goldberg *et al.* (1989) implemented what they termed a 'messy' genetic algorithm, which manipulates variable length strings. Diplock (1996) implemented a variety of genetic algorithm based model builders, but the binary representation yielded insufficient power for the task.

Koza (1992) utilises a LISP representation for his genetic programming algorithm, which utilises symbolic expressions (or S-expressions) to represent equations in a prefix notation. These expressions are represented as parse trees, which consist of node linking operators (such as basic arithmetic operators) and the nodes themselves (variables or constants). An example is given below; an infix (normal) expression (equation (9.1)) can be represented as a prefix expression, as illustrated by equation (9.2), whereby the function is specified first, followed by the required arguments.

$$(X - Y)$$ (9.1)

$$(- X\ Y)$$ (9.2)

In this case a subtraction calculation is being performed. This is termed a function list. Multiple lists can be used to represent full expressions being processed in a recursive fashion, starting from the left. Given the infix expression of equation (9.3)

$$((X - Y) + Z)$$ (9.3)

this becomes the prefix expression

$$(+ (- X\ Y)\ Z)$$ (9.4)

This identifies an addition operator as the 'root' operation, with the two arguments being $(- X\ Y)$ and Z, so the former is evaluated and the result used as the first argument of the addition operator, which has Z added. These structures can also be represented as 'trees' (with equation (9.4) illustrated in Figure 9.1), which is helpful for visualising the structures being manipulated and the operators being applied. Koza (1992) chooses LISP for his genetic programming algorithm since it uses such structures although, surprisingly, virtually any programming language (such as FORTRAN) is capable of manipulating them.

When constructing models it is also necessary to incorporate constants into the specifications, which then have to be optimised for the data to which the model is being applied (constants being the main mechanism for calibrating a model for a specific dataset). A strategy for this involves simply increasing the terminal set specified to introduce a special terminal called the 'ephemeral random constant' (Koza, 1992). During the initialisation of a population, if a terminal is identified as an ephemeral random constant, a random value within a suitable range is generated and inserted into the tree at the appropriate point. A different value is generated for each constant identified in a model structure, with the genetic processes operating to identify 'good' constant values as well as good variable and function combinations. New parameter values are generated by manipulating these ephemeral constants.

9.2.1.2 Composition of individuals

The individuals within a genetic programming framework consist of a set of functions (arithmetical, mathematical, logical, etc.) and a set of terminals

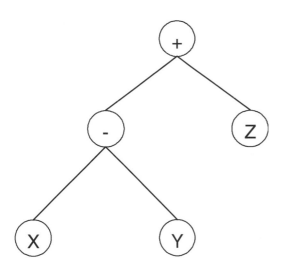

Figure 9.1 Example LISP tree structure.

(variables, constants, etc.). The set of possible structures which can be represented is the set of all possible combinations of these (with varying length and complexity). However, an important difference between the evolved structures in genetic programming and those associated with a conventional genetic algorithm is that the former are hierarchical, whereas the latter are not (Koza, 1992).

The composition of the function and terminal sets is obviously application-specific, but Koza (1992) identifies two general requirements that should be satisfied irrespective of the specific problem domain. These are the closure and sufficiency of the sets. Closure requires that

> '...each of the functions in the function set be able to accept, as its arguments, any value and data type that may possibly be returned by any function in the function set and any value and data type that may possibly be assumed by any terminal in the terminal set.'
>
> (Koza, 1992, p. 81)

Examples include dealing with occurrences which break mathematical rules (such as divide-by-zero, logarithms of negative numbers, etc.), which can be relatively easily dealt with by protecting the function. A related issue is that of exceptions and errors that may occur through the evolution of these structures; those giving rise to mathematical errors need to be discouraged from mating so their genetic material becomes extinct within the population. Diplock (1996) discusses this issue in greater depth. Sufficiency requires that

'. . . the set of terminals and the set of functions can be capable of expressing a solution to the problem.'

(Koza, 1992, p. 86)

although this is a problem which is common to most methods of quantitative analysis in a wide variety of disciplines.

9.2.1.3 Initialising a population

In a conventional genetic algorithm, individuals can easily be randomly initialised. For symbolic expressions, the process is more complex, since functions and their arguments need to be identified correctly, as does the size of an individual.

A function is chosen at random. For each argument of this function, another function or a terminal is selected. If further functions and terminals are selected, the process continues recursively until a completely closed tree structure is obtained (with each branch ending in a terminal). Two methods can be used to determine the length of the individuals generated (Koza, 1992, p. 92). The 'full' method creates branches with trees extending to the maximum specified depth (so initially, they are all the same size), whereas the 'grow' method initialises structures of various lengths, up to the maximum, since there is no explicit finishing point. Given the relatively smaller number of possible structures (especially with small function and terminal sets), the initial population is usually checked for duplicates, since this reduces diversity; if one is found, an alternative individual is generated to replace one of the duplicates. Such a strategy is not normally required in a genetic algorithm, since the number of possible individuals is much larger, although it is sometimes used (see Davis, 1991). The seeding of the population with sub-structures or whole expressions is another strategy for initialisation, although caution must be taken, since a few 'good' individuals inserted into an otherwise random population will soon dominate. Koza (1992) recommends that if such a strategy is to be implemented, the population needs to be completely seeded with individuals possessing a similar level of performance, to avoid this problem.

9.2.1.4 Fitness and selection

The evaluation of individuals usually requires some form of comparison between observed data and the results yielded by that particular solution, or a test to see if a solution meets certain criteria or constraints. In an optimisation problem, for example, a sum-of-squares function could be used, utilising a transformation function so that the higher the value, the 'fitter' the individual, as this makes the genetic algorithm implementation more straightforward (Davis, 1991). The error values need to be converted into a fitness value, to determine the selection pressure for the generation whilst also preventing 'super-fit' individuals dominating the population (Beasley *et al.*, 1993; Oie *et al.*, 1991).

The selection pressure within an algorithm is a critical parameter (Hancock, 1994) since it drives the algorithm towards a solution. Too much and the search will terminate prematurely; too little and progress will be slower than potentially possible. The selection procedure differentiates between good and bad solutions, and in a simple genetic algorithm (Holland, 1975), strings are selected proportional to their fitness relative to the whole population. One good way of doing this is by utilising an intermediate gene pool which consists of chromosome copies based on their fitness. A chromosome of average fitness has one copy of itself automatically placed into the gene pool. Chromosomes with fitness levels less than average have no copies made automatically, whilst above average individuals have the number of chromosomes corresponding to the truncated integer part of their fitness copied into the gene pool. Therefore, a chromosome with a fitness of 3.4 has three chromosomes placed in the gene pool, a value of 1.9 means one copy, etc. These integer values are then subtracted to leave all chromosomes with a fitness of between zero and one (3.4 becomes 0.4, 1.9 becomes 0.9, etc.). These are then ranked and the best chromosomes used to fill the gene pool so that its size is equal to that of the real population.

This method creates a problem of scaling (Hancock, 1994) with a tendency for premature convergence to occur, if relatively good solutions exist in the initial population. As a consequence, various modifications have been developed to this basic technique. Windowing involves subtracting the worst fitness value found over a set of previous generations. Sigma scaling utilises standard deviations to control the fitness variance and avoid wide ranges (between the best and worst value) affecting the selection pressure adversely. Linear scaling adjusts fitnesses so that the best individual gets a set number of offspring and others are scaled to produce the required number. Ranking methods apply a fitness value according to a predefined sequence, such as an exponential function. More details can be found in Beasley *et al.* (1993), Blickle and Thiele (1995) and Hancock (1994), all of which encourage experimentation in the identification of the best procedure, a principle which applies to most aspects of a genetic algorithm specification. Tournament selection is an alternative method to that outlined previously for obtaining an intermediate gene pool. A series of individuals are chosen randomly from the population and the best of these is placed in the intermediate population and the process repeated until the required number of chromosomes have been selected.

It should be noted that the two procedures described here are most suited to a generation-based genetic algorithm, which proceeds in steps after a population of a given size has been evaluated. An alternative strategy is to only replace a few strings at a time, after an initial population has been evaluated. Such a method is termed a steady-state genetic algorithm. A single population is maintained as before, but only a few chromosomes are evolved at each time step and inserted into the population, and another few strings evolved. Such a scheme was first used by Whitley (1989, 1993), although there is some debate on which is the best strategy to use (Hancock, 1994), and this is determined by the application and the hardware on which it is being implemented. A further possibility is to utilise a steady-state genetic algorithm without duplication, which checks that offspring do not already exist within the population, and if they do, new offspring are created instead (Davis, 1991; Syswerda, 1989). If in doubt, experiment!

9.2.1.5 Reproduction

Once some form of selection procedure has been completed, simulated genetic operators are applied to the survivors in an attempt to evolve the existing population into a set of better performing individuals, since the 'selected' individuals should represent a collection of good genetic material. The two most common operators used are called crossover and mutation. In a simple genetic programming scheme, the application of operators is usually probabilistic. A 'well performing' string is selected, and a probability test applied.

The principle behind crossover is to select subsections of parents and combine them to create offspring, or simply copy a string unaltered into the next generation. If a crossover operation is to be applied, two parents are selected, proportional to fitness. A crossover point is then chosen (independently for each parent), which is a subtree of the parents, consisting of the entire structure 'below' the point identified for the crossover (which can be any size up to the maximum). The process is best illustrated using an example. Suppose equations (9.5) and (9.6) have been selected as parents and are to be evolved into two offspring (represented as infix equations for the sake of simplicity).

$$((A + B) + C) \tag{9.5}$$

$$((X - Y) - Z) \tag{9.6}$$

The associated parse trees are given in Figure 9.2, which also indicates (i) the extent of the structures to be swapped (the dotted line), which are at the same location in this example, and (ii) the expression fragments.

The left-hand fragment is inserted into the right-hand tree in place of fragment two and vice versa, which yields the two trees shown in Figure 9.3, which represent equations (9.7) and (9.8) below.

$$((X - Y) + C) \tag{9.7}$$

$$((A + B) - Z) \tag{9.8}$$

These are similar structures, but with some swapped elements.

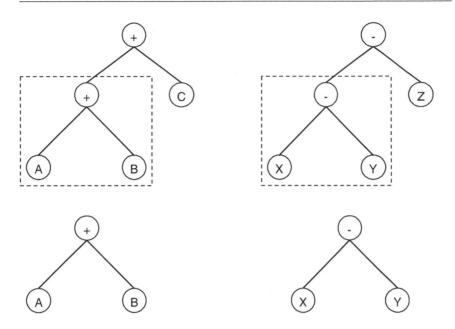

Figure 9.2 Parent trees, before crossover, with crossover fragments; similar structure, same crossover points.

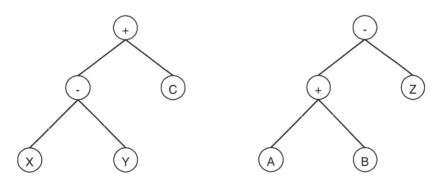

Figure 9.3 Offspring trees, after crossover.

However, if the crossover points were located in different subsections of the parents, the equations yielded are quite different (which also usually occurs if two parents of different size and shape are selected). Using equations (9.5) and (9.6) as parents, if the crossover points were identified in different locations, as shown in Figure 9.4, the fragments are swapped to yield the two markedly different trees shown in Figure 9.5, which decode to the following equations.

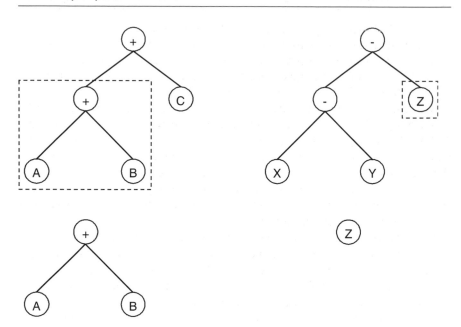

Figure 9.4 Parent trees, before crossover, with crossover fragments; similar structure, different crossover points.

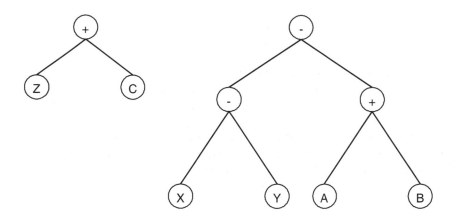

Figure 9.5 Alternative offspring trees, after crossover.

$$(Z + C) \tag{9.9}$$

$$((X - Y) - (A + B)) \tag{9.10}$$

It is in this process that the real power of genetic programming lies, since not only are valid models also produced because of the choice of crossover locations, but dynamic (rather than fixed-length) structures are manipulated; similarly structured parents produced two very different offspring, which would be problematic to replicate within a binary representation. It is also advisable to restrict the maximum tree length, or exceptionally large and computationally intensive strings can be produced.

Secondary operators are also available to a genetic programming implementation, although these are less effective and not often utilised (Koza, 1992). Mutation generally plays a secondary but necessary role in a genetic algorithm, but is not necessary in a genetic programming algorithm since the chromosomes are not fixed, there are fewer functions and terminals than bits in a chromosome, and rarely do functions disappear completely from a run. If it is used, a point is chosen, and a randomly created subtree of a maximum specified length inserted.

Permutation is analogous to the inversion operator of a genetic algorithm, but instead of simply reversing the selected sub-section, any possible combination of the elements can be chosen. However, like inversion, its role is unclear, and its full use is yet to be identified. Editing is utilised to eliminate duplicate or redundant features from a string (for example, (* 2 3) is replaced by (6)). It is time consuming given its recursive nature and can also result in the loss of genetic information and variety between individuals.

Termination criteria often consist of a set of generations or the attainment of a predetermined level of fitness (as with a genetic algorithm) and the results obtained utilise similar techniques (store the best value and report the best result at the end). The processes described here can now be combined with the basic algorithm described earlier to yield a simple genetic programming algorithm.

Crossover provides the main search mechanism for the algorithm, and usually has a high probability of occurring (typically in the region of 0.7 or more), whilst mutation maintains diversity but has a significantly lower probability of happening (typically in the region of 0.01). The former is more important in the early stages of an algorithm execution, when looking for an area of the search space which contains the global optimum, whilst the latter is more important towards the final stages of the algorithm, providing a mechanism for fine tuning the results.

9.2.1.6 Termination versus convergence

Although genetic programming is being increasingly used, there has been little research concerned with convergence properties (Rudolph, 1994). An evolutionary algorithm needs to be told when to stop. An obvious choice is to terminate when the function is optimal, but this may never be achieved in many problems, or may not be known beforehand. A satisfactory level of performance can be specified, but this restricts the algorithm and requires the specification of a termination level. Alternatively, the run is determined by the size of the population and the number of generations (which can be altered through a process of experimentation); the algorithm runs to completion. The algorithm can also be halted when no improvement has been made for a specified number of evaluations or when genetic diversity is low.

Another consideration is that of premature convergence, whereby a partial or complete 'super-fit' solution propagates through a population since its fitness level is significantly higher than the other chromosomes. If this is not the global optimum then the algorithm loses its search power and with mutation operating, the algorithm is reduced to a slow random search (Whitley, 1993, discusses this concerning genetic algorithms). This highlights the importance of a good selection method within a genetic algorithm. Fitness scaling is a method used to prevent premature convergence, by smoothing fitness values across the population and reducing the risk of one individual becoming dominant. This is mostly due to the random initialisation of strings; as the algorithm progresses the difference between the best and the worst solutions decreases, but at the start of a run, a good individual could rapidly dominate the population, hence the need for careful fitness scaling (see also Beasley *et al.*, 1993; Grefenstette, 1986). Given enough care and experimentation, the risk of premature convergence can usually be reduced.

9.3 Parallel genetic programming

Undoubtedly, one of the most impressive technological advances of recent decades concerns high performance computer hardware, and the appropriate software developed for it. The real cost of computer hardware is rapidly decreasing, whilst storage capabilities and processing speeds of these machines are increasing at a very rapid rate. The development of parallel computer architectures stands out as a major step in the advancement of computer technology. This has led to radical new techniques in the implementation of computer-based research and genetic programming is an ideal application.

The principal difficulty is the need to convert code developed on a conventional serial machine for parallel hardware. Fortunately, there is now a very good set of portable software that greatly eases this task (see Chapter 3 by Ian Turton). The best way to program a parallel supercomputer is via message passing and there is now a *de facto* world standard called the message passing interface (MPI).

The fundamental mechanism behind MPI is that each processor possesses a local memory to which no other processor has direct access, and hence there is no globally addressable memory. The only way of communicating between processors is to explicitly send and receive messages. Programs are generally written as normal, but a communicator is used to connect processors, one of which is designated as a 'master' process, the others being termed 'slaves', and regions of code are designated as being executable by either the master or a slave. The message passing parallelisation strategy allows the dynamic allocation of work at run-time, but also gives the programmer control over the distribution strategy of the work, as it can be explicitly defined, rather than being implemented by the compiler.

The major benefit of a message passing strategy in the context of genetic programming is that load-balancing is easier to achieve. A single processor is designated as the 'master', and the rest of the processors are designated as 'workers' or 'slaves'. The master initialises the population, allocates the work and controls the evolutionary stages of the evolutionary algorithm whilst the slaves are used to evaluate the population. Initially, each slave is sent a member of the population to evaluate. When this population member has been evaluated, the slave sends the performance back to the master, and that slave is sent another population member. This process is repeated until a population of a predetermined size has been evaluated.

When the master receives a performance value, it compares this with the current population. If the performance value is worse than the current worst value, it is left to extinction and a new population member created from the existing population. If the performance value is better than that of members of the current population it is inserted and the worst member left to extinction, and a new member is then created from this modified population. This newly created population member is then sent to the idle slave. A byproduct of this method is that there is also an implicit element of parsimony in the algorithm mechanism, whereby the more complex an equation the longer it takes to evaluate and the less likely it is to be selected for reproduction. This form of evolutionary algorithm is termed asynchronous, since the population is fully dynamic, being constantly updated rather than progressing in generation steps, and therefore can also be considered a lot more representative of the mechanisms existing in nature. It is this method which is utilised for the model breeding exercises in this chapter.

9.4 Breeding new spatial interaction models

Computational model building based on data provides a new way of model formulation and derivation which can potentially yield radically different models from those derived using more traditional methods. Diplock (1996) and Diplock and Openshaw (1996) demonstrate that genetic algorithms can be used to calibrate spatial interaction models, but the real challenge is to utilise the technology in an innovative way, and to demonstrate the real potential of genetic programming in a model breeding capacity. Openshaw (1988) implemented a genetic algorithm based automated modelling system (AMS) and Diplock (1996) used genetic algorithms for a series of 'model breeding machines', both of which were experimental.

This section outlines a series of experiments concerned with developing a more flexible and thus potentially more powerful automated spatial interaction model building tool. The algorithm to be implemented is discussed, and a series of model breeding experiments are then reported. Openshaw and Turton (1994) and Turton *et al.* (1996, 1997) present other genetic programming strategies.

The algorithm to be utilised is the steady-state version which replaces population members as better ones are identified, rather than progressing through distinct generations. The following describes the representation, selection and evolutionary mechanisms of the model breeding genetic programming algorithm. The parallelisation issues are discussed fully in Diplock (1996).

9.4.1 Model breeding genetic programming algorithm

9.4.1.1 Representation

LISP S-expressions are used to encode the model structures and provide the genetic material for the evolutionary stages of the algorithm. The data have been vectorised (the arrays have been unrolled) to ensure maximum computation efficiency.

9.4.1.2 Initialisation, evaluation and selection

The initial population is selected through the methods outlined previously to ensure a diverse initial population with respect to both composition and size of individuals (Diplock, 1996). The initial population is also checked for duplicates before the strings are evaluated, and if any are detected, further random equations are created until each individual in the initial population is unique.

The selection procedure utilised in the model breeder is based on average population fitness, supplemented by a linear scaling transformation function. The method outlined here is described by Goldberg (1989). It has been chosen because it is suited to the steady-state algorithm which is to be used, and it avoids domination of the population by 'super-fit' individuals whilst encouraging competition between chromosomes yielding a similar level of performance in the latter stages of execution (Goldberg, 1989, p. 79).

The error function is again calculated using a trip-pair sum-of-squares function, as given in equation (9.11), adjusted to handle the data vectors.

$$f_p = \sum_{k=1}^{m \times n} (S_k - T_k)^2 \tag{9.11}$$

where f_p represents the error of an individual, S_k and T_k are the vectors of observed and predicted trips respectively with m denoting the number of origins and n the number of destinations. This error is used as the raw fitness value (F_p), which is then transformed using the function

$$E_p = (a \times F_p) - b \tag{9.12}$$

where E_p is the transformed fitness of an individual, with a and b calculated for each selection event as follows:

$$a = \frac{0.2 \times F_{avg}}{F_{best} - F_{avg}} \tag{9.13}$$

$$b = \frac{F_{avg} \times [F_{best} - (1.2 \times F_{avg})]}{F_{best} - F_{avg}} \tag{9.14}$$

F_{avg} is the average raw fitness value of the current population and F_{best} the best individual raw fitness of the current population. This fitness derivation is described in more detail by Goldberg (1989). Once these values have been adjusted, cumulative fitness values for each member of the population are calculated and a selection procedure applied.

9.4.1.3 Evolution

The evolutionary mechanism of crossover is that outlined by Koza (1992), and described previously. No other evolutionary operators are to be implemented as Koza (and others) argue that these do not yield significant extra potential since crossover is the dominant force. A stack of unevaluated individuals is maintained, with an individual being passed out to a processor as and when one becomes idle. If the stack becomes empty new individuals are created. Two parents are selected. If the crossover probability test is passed (in this case the probability of a crossover event is 0.9), a random point on each parent is identified. From this point in the string, a backwards sweep is performed, considering each character in turn, until an opening bracket is identified. This sweep is then reversed until the matching bracket is found, which yields the fragment to be crossed. An example is given in Figure 9.6. The process is necessary to ensure valid models are always produced by a crossover event.

9.4.2 Initial experiments

Two datasets were used for initial experimentation purposes: Durham journey-to-work data and Seattle car sales data:

a. The Durham data set is a 73×73 matrix for journey-to-work flows.
b. The Seattle data set represents car sales in the city for a single Japanese car manufacturer. It includes the most popular market segments (e.g. Fiesta/Corsa, Escort, Mondeo/Laguna, Scorpio).

S-expression and initial point, marked with ‖

$$\left(* \;\; (- \;\; (X) \;\; (Y)) \;\; (+ \;\; \| \;\; (Z) \;\; (/ \;\; (Y) \;\; (Z))))\right)$$

Opening bracket identified, marked with [

$$\left(* \;\; (- \;\; (X) \;\; (Y)) \;\; [\;\; (+ \;\; (Z) \;\; (/ \;\; (Y) \;\; (Z))))\right)$$

Closing bracket identified, marked with]

$$\left(* \;\; (- \;\; (X) \;\; (Y)) \;\; [\;\; (+ \;\; (Z) \;\; (/ \;\; (Y) \;\; (Z))) \;\;] \;\;)\right)$$

Crossover fragment

$$\left(+ \;\; (Z) \;\; (/ \;\; (Y) \;\; (Z)))\right)$$

Figure 9.6 Identification of a crossover fragment.

9.4.2.1 Model pieces

A simple set of terminals and functions were selected, as listed in Tables 9.1 and 9.2 respectively. The identifier (ID) in Table 9.1 is the notation used for reporting the results, to avoid the use of subscripts. Variables V1, V2 and V5 are self-explanatory. Variables V3 and V4 involve the inversion of origin and destination characteristics, so that O_j represents all trips that terminate in origin j and D_i represents all trips that start from destination i. The intervening opportunities term, V6, is expressed as the percentage of intervening destinations between each trip-pair and V7, the competing destinations term, is defined as the sum of competing destination attractiveness divided by their distance from alternative destinations (Fotheringham, 1983). Finally, the intra-zonal trip flag, V8, identifies when a trip starts and finishes in the same zone. The 'Symbol' column in Table 9.2 gives the operator notation. Most of the operators are self-explanatory, being standard arithmetic, trigonometric or mathematical functions. The two exceptions are the 'less-than' and 'greater-than' operators. These compare the arguments they receive and set the result accordingly (a value of 2.0 for true and 1.0 for false); if 1.0 and 0.0 were used, this could lead to division-by-zero in model

Table 9.1 Initial variables included in GP runs

ID	Notation	Description
V1	O_i	Origin total
V2	D_j	Destination total
V3	O_j	Inversed destination total
V4	D_i	Inversed origin total
V5	C_{ij}	Distance measure
V6	X_{ij}	Intervening opportunity
V7	Z_{ij}	Competing destinations
V8	I_{ij}	Intra-zonal trip flag

Table 9.2 Initial operators included in GP runs

Symbol	Argument	Description	Form
+	2	Addition	$R = X + Y$
−	2	Subtraction	$R = X - Y$
*	2	Multiplication	$R = X \times Y$
/	2	Division	$R = X/Y$
^	2	Power	$R = X^Y$
abs	1	Absolute	$R = |X|$
sqrt	1	Square root	$R = \sqrt{X}$
log	1	Natural logarithm	$R = \log(X)$
exp	1	Exponential	$R = e^X$
<	2	Less than	$if\ (X < Y, R = 2, R = 1)$
>	2	Greater than	$if\ (X > Y, R = 2, R = 1)$
sin	1	Sine	$R = \sin(X)$
cos	1	Cosine	$R = \cos(X)$

Notes
R = result of operation; X = first argument; Y = second argument (if applicable).

specifications or large chunks of redundant model pieces if multiplied by the intra-zonal variable when set at zero.

This set of operators and variables has been chosen since it represents a good starting point for experimenting with various model breeding schemes. The set represents not only the components of numerous traditional models, but also some new components which might be applicable. In general it is only through investigation and experimentation that a good set of variables and operators may be identified. Finally, all of the results described in this chapter concern origin-constrained models.

9.4.2.2 Preliminary investigations

Initial experimentation was performed using the Durham journey-to-work dataset. A variety of population and generation sizes were utilised, ranging from a few hundred to several thousand, although the effects of variable population and generation sizes are not investigated here. In the course of some early runs, the 'redundancy' of the absolute function was discovered. Whilst not relatively computationally expensive, its inclusion reduced the overall efficiency of the algorithm, since positive parameters were being included as the absolute value of a negative number, or the absolute value of a variable in which all values were already greater than zero. The absolute operator was therefore removed from all further model breeding exercises.

Initially, the GP model breeder was run with the complete set of operators and variables, but without any other mechanisms to guide the search. The best two equations were:

$$T = \left[V2 \left(12.0 V1 \; V2 \; V6^{1.82} V2^{1.82} + 3.34 + 2.0 V1 \right. \right.$$

$$\left. \left. - V1 \; V5^{-1.82} + V1 \; V4^{1.82} + \frac{V6 \; V5^{-1.82}}{V2} \right) \right]^{-V5}$$

$$\times \left[\left(V2 + V6 + 3.34 + \frac{V6 \; V5^{-1.82}}{V2} \right) \right]^{-V5} \times V2 \log(1.67 + V6^{1.67})$$

$$- 3.0 V6^{1.82}) \times [V6^{1.67} - 2.0 V6 + V1 \log(V4^{1.82} \; V5^{-1.39})$$

$$+ V1 \; V5^{-1.82} + V2] \times \left(V2 + \frac{V2}{2.0 V2^C} \right) \tag{9.15}$$

and

$$T_{ij} = V2 \log(V1) \frac{V6^{-1.7} + V6^{V2 V6 \log(V1)}}{V2}$$

$$+ V1 \log(V1^2 \; V6^{-1.2}) \sqrt{V2} + V2 \; V8 + V2 \; V6^{\log(V1)}$$

$$\times [\log(V2) + V1 \; V2 \; V6 \times (\log(V2) \; V2^{1.5} \; V6^{-1.7}(V6 + V8))^{-1.4}]$$

$$+ \left[\frac{\frac{\sqrt{V2 \; V6}}{V5^{1.31}} + \frac{V2 \; V6 + V8}{V6^{3.2}} + \frac{\log(V1) \; V2^2 + V8}{V6}}{V2 \log(V2^2 \; \log(V2)) \times (V6 + V8)} \right]$$

$$\times \frac{\sqrt{V2} + V8}{V5^{1.17}} \right] \tag{9.16}$$

The error function values for these two equations were 136.6 and 156.0 respectively. The former relates to probability based data whilst the latter utilised volume interaction data. The performance of these compared very favourably with that of conventional models, given that the sum of squares error was almost half the value yielded by conventional models for the probability-based model, and almost two thirds for the volume-based model (a conventional model yields a sum-of-squares error of approximately 260.0). However, the model structures are incredibly complex. The models obtained by the GP algorithm have been simplified using Maple V (Release 3), which is a mathematical package that can simplify the equations using a series of rules which allows terms to be removed or combined accordingly. This identifies repetitive or redundant terms, combines constants, etc. The problem is that these models are tending towards the maximum size allowed by the program, and despite the significant performance improvement of a factor of two, it is desirable to also produce a more interpretable model.

9.4.2.3 Breeding simpler models

To address the problem of model complexity, a mechanism was introduced that is designed to incorporate a notion of parsimony and generate less complex model structures. This is achieved by analysing the model structure after a fitness value has been assigned; each variable or arithmetic operator 'scores' one point, whilst mathematical functions and parameters 'score' two points, the summation of which yields the equation's complexity score. If the total for a model exceeds the specified threshold complexity score, then the fitness value is increased. A threshold of 20 was set as it was felt that this permitted adequate components without restricting model development. For example, a conventional model (as given in equation (9.17)) would score as follows: the three variables (O_i, D_j and C_{ij}) score a point each, as do the three linkage multiplication operators, the exponential two points and the β parameter a further two points, giving a total of 10 points (which is well inside the threshold value). The error increase is simply the score minus the threshold (which is added to the error value).

$$T_{ij} = O_i \times D_j \times \exp(-\beta \times C_{ij}) \tag{9.17}$$

The resulting best two equations for each of the data-types are given in equations (9.18) (probability-based) and (9.19) (volume-based), with the corresponding error values 180.2 and 205.0 respectively.

$$T_{ij} = V1 \left[\frac{V2}{-2.97\exp\left(V8 + \dfrac{-31.40 + V5^{-0.008}}{V4} - 4.63V6\right)(-0.78 - \sqrt{V6})} \right.$$

$$\left. \times \frac{1}{\exp\left(-9.41V5 + \dfrac{V8 + 20.23}{V2}\right) + V8} + V8 \right] \tag{9.18}$$

$$T_{ij} = \frac{V2^{2.0} \exp(-0.05V5)}{V6^{1.2}} \tag{9.19}$$

Whilst the error functions seem less impressive compared with the results of the first experiment, the modelling equations are significantly less complex. More importantly, however, the two error functions incorporate the parsimony measures for the model structures. Whilst this does not affect the volume-based model as this is a sufficiently small structure, given the size of the probability-based model the 'real' error function value (not including the parsimonious measure), is 126.2, which is the best model so far.

9.4.2.4 Seeding the starting population

The final stage involved two further modifications to the GP algorithm. First, a duplicate check was introduced, as it had become apparent that bred models might duplicate those already in the population; if this is found to be the case, then another equation is bred, so as to improve the efficiency of the evolutionary search. Second, provision was made to seed the initial population with structures from existing equations to see if this could help improve the search. The seeds were conventional model structures incorporating elements of the function and terminal set (from basic models to intervening opportunity and competing destination versions). The models are given in equations (9.20), (9.21), (9.22) and (9.23); the first two utilised probability-based data, whilst the last two utilised volume-based data.

$$T_{ij} = V2 \exp(-16.7V5 + 11.2 \exp(V2) - 27.8 \ V4 + V8) \tag{9.20}$$

$$T_{ij} = \exp\left(\frac{2V5 - 35.0}{2V4^{V2} - 35.0}\right) \times \exp(V2 \ (V5 - 13.8) + V8) \tag{9.21}$$

$$T_{ij} = \frac{(V4^{50})^{-1.9V5} \times (V8 + 0.7)}{V2} \tag{9.22}$$

$$T_{ij} = \frac{V2 \ V6^{-0.6} \ \exp(-0.08V5 + V8 - V4^{\exp(-1.65)})}{V2 + (V8 + 2.4)^{1.1}} \tag{9.23}$$

The error function values for these models were 150.2, 151.4, 190.4 and 178.6 respectively. However, as in the previous experiments, equation (9.23) was penalised in the GP because of its size, and the true error value was 143.6.

The performance of the models is still better than conventional specifications, but demonstrates no significant improvement over the last stage of experimentation. However, the influence of the initial population seeding can be seen, in that the models take on a more recognisable form, or at least possess certain components that can be associated with conventional model specifications. The models are relatively simple, and possess structures similar to the seeding equations used (multiplicative-based modelling equations, for example).

This series of experiments was repeated for the Seattle car sales data. The initial results for the basic model breeder implementation are given in equations (9.24) (probability-based) and (9.25) (volume-based), with the error function values being 6.3 and 7.5 respectively. In contrast to the Durham models, these equations are extremely complex, and their overall level of performance is disappointing in comparison to the conventional model benchmarks.

$$T_{ij} = V1^{V5} \ V2 \ \exp\left[\left(\frac{V6 \ V5^{0.2938} \ \exp(V7)}{V1 \ V2^2}\right)\right.$$

$$+ \left(V1 + \frac{V7^{1.81} \ \exp(V8) \ \log(V5)}{\exp(V2 \ V5^{0.99} \ V7^{0.29})}\right)\right] \times \exp\left[(V7^{1.81} \ \log(V5) - V1)\right.$$

$$\times \frac{V1 \ V2}{V5}\left(V7 + \frac{V6 \ V7}{\exp(V8) \ V5}\right)^{0.99}{}^{V7^{1.81}\log(V5)}\right] - \left(\frac{V6}{V2}\right)^{-0.21V5} \times \frac{V1^2}{V5}$$

$$\times \left[\frac{V5^{\log(V5)} \ V7^{1.99}}{V2 \ \exp(V7)} + \frac{V5^{0.09} \ (V2 \ \exp(V8))^{0.09}}{\exp(V5^{0.29} \ V6) \ \exp(-2.04V5)}\right]^{V7 \ \log(V5)} \tag{9.24}$$

$$T_{ij} = \left[V2 + \left(\frac{V6 \ V8 \ \log(V6)}{V2 \ V5^{0.73}}\right) \times \left(V2 + \frac{2V2 \ V6 + V8 - 1.87}{V2 + V8 \ (V2 + V5)}\right)\right]$$

$$+ \left[V2 + \frac{3V2 \ V8}{V6 \ V5^{0.84} - 1.87}\right] \times \left[\frac{V5^{-0.73}}{V2} + \frac{2V8 + V2 + \log(V6)}{V5}\right]$$

$$+ \left[\frac{V2 \; V6 \; (V7 + V8)}{V2 \; V6 + 3V2 \; V8 + \log(V6) - 3.74} \right]^{-0.32}$$

$$+ \left[2\frac{V8 + \log(V6)}{V5^{0.73}} \right] \tag{9.25}$$

With the parsimony check incorporated into the GP algorithm, much simpler equations were generated, but again the level of performance improvement was disappointing. Equation (9.26) represents the probability-based version (with an error of 7.1) whilst equation (9.27) represents the volume-based version (with an error of 7.5).

$$T_{ij} = V2 \exp(V6 - V2) \exp\left(\frac{-12.7}{V5} \right) \tag{9.26}$$

$$T_{ij} = \frac{V5^{-0.2V5}}{\sqrt{V7 + V2}} \tag{9.27}$$

Finally, the initial population was seeded with the same series of conventional model components, to see if this improved the performance of the bred models. The results are given in equations (9.28), (9.29), (9.30) and (9.31) (the first two probability-based and the last two volume-based). The error function values for these equations were 6.9, 7.0, 7.0 and 6.9 respectively.

$$T_{ij} = V2 \exp\left(V6 + 2.7V2 - 0.54 + \frac{14.2V5}{V7^{0.06}} \right) \tag{9.28}$$

$$T_{ij} = V2 \exp[\exp(-3.55V2) + 1.1V6] \exp(-13.5V5) \tag{9.29}$$

$$T_{ij} = \exp(V6) \exp\left(\frac{-0.1}{V5} \right)(V2 + 46.9) \tag{9.30}$$

$$T_{ij} = \left[V2 \left(\frac{-0.1V5}{V4^{0.03}} \right) + V6 + 0.9 \right]^{0.91} \tag{9.31}$$

A significant improvement has been obtained for one of the datasets, but a less marked improvement for the other. Therefore, these results have to be examined more closely, to attempt to explain this, to enable further development of the model breeding software. In summary, for the Durham data the best result for a basic model breeding algorithm was 136.6, and 126.2 for the parsimony version, compared to the conventional model benchmark of 260.0.

Table 9.3 Summary: best-performing model errors

Data	Model	Train	Unseen
Durham	Conventional	260.78	7.67
	Evolutionary	126.23	8.15
Seattle	Conventional	7.67	260.78
	Evolutionary	6.36	268.14

Notes
Train = data used for model breeding; Unseen = data not used for model breeding.

For the Seattle data, the best results were 6.4 and 7.0 respectively, compared with the benchmark of 7.7.

It would be interesting to determine why this method is performing more successfully for one dataset than the other. Diplock (1996) identified that the two datasets possess different characteristics which are significantly influencing the success of the model breeding exercises. The Durham data appears more suited to spatial interaction modelling, driven more by distance factors; the Seattle data may not necessarily need spatial interaction-based model components. Perhaps no significant improvement has been seen because the choice of the functions and terminals is inappropriate, as they are based around spatial interaction principles. The nature of car sales data may be more suited to data on dealership availability, price, dealer characteristics and a host of other potentially useful predictor variables.

9.4.3 Cross-validation

It has been demonstrated that models can be bred for a specific dataset that can outperform a conventional model specification, but a major issue for such bred models is their generality. Do the models perform well when applied to data that have not been used to breed them? This issue is investigated using the models bred from both the Durham journey-to-work data and the Seattle car sales data. The models are applied to the 'other' dataset which corresponds to the data type used for the model breeding exercise (either volume or probability based) and the parameters optimised accordingly.

The results of this cross-validation exercise are summarised in Table 9.3. For the training data, models which were approximately twice as good as conventional specifications (with respect to their error function values) were discovered for the Durham data, but only small improvements could be made over conventional model specifications for the Seattle data; the ratios of the best bred to conventional error function values are 2.07 and 1.21 respectively. In the case of the unseen data, the best bred models were yielding slightly larger sum-of-squares error values than their conventional counterparts. What is important is the fact that in the case of the Durham data, the best bred model was also the best performing model when applied to the Seattle data. In the case of the Seattle based models, the best performing model on the unseen data was the third best performing model on the original data.

One reason for these findings could be the nature of the data being used here. The Seattle car sales data presents a potentially more difficult problem concerning model breeding, since the 35 car dealers in the set are located in only a few census tracts, which means that large trips tend to dominate the data and can obscure the mechanisms of the spatial system. This is further emphasised by the nature of the datasets; the Durham dataset is a square 73×73 matrix which emphasises intrazonal trips explicitly, whereas the Seattle dataset is an 86×35 where the destinations demonstrate no explicit relationship with the origins.

The nature of the two datasets is also markedly different, one representing journey-to-work flows, and the other car sales (of many different types) which raises questions as to the compatibility of models bred on the datasets; given these differences can it be expected that models can be applied to unseen data successfully, given that the original dataset was the driving force behind the construction of the model structure. These results are discussed more fully in Diplock (1996). Nevertheless, the genetic models would appear to be surprisingly robust, although clearly far more research is needed to investigate this aspect.

9.4.4 Issues for developing GP-based model breeders

The experiments outlined previously have demonstrated one of the fundamental issues concerned with breeding models inductively; namely the question of data utilisation. There are two broad strategies that can be adopted when attempting to develop automated model building strategies. The first was the utilisation of a single dataset to develop a model specifically for that data, which can then be applied to another dataset and recalibrated accordingly. The other concerns the utilisation of multiple datasets for breeding models; a model is applied to several datasets and each data-specific error function value combined to yield the overall performance measure. This presents the potential for developing more generally applicable spatial interaction models. It is not possible to use multiple datasets because: (1) the Cray T3D used here is not fast enough – runs of six hours would increase by a factor of k, where k is the number of datasets used, and (2) the share of the Cray T3D time used for this project exceeded all other social science uses of the machine in 1995–6 and used up the entire 5% share officially allocated for social science usage.

The choice of strategy ultimately reflects upon the purpose of the model breeding exercise; utilising a single dataset yields a potentially good level of performance for that dataset, but the model might not be as successful when applied to other data. On the other hand, whilst multiple datasets yield models which can perform adequately over a series of datasets, there is no guarantee that this trend will extend to unseen data, and there is also the problem that for any given dataset, the model will perform less well than a model bred using solely that dataset. There is also the question of how many datasets to use.

It can be argued that the former approach is more suited for an applied perspective, where individual performance is the top priority, given that model predictions could provide the basis for decision making or policy tasks, whereas the latter is more suited to a theoretical perspective, whereby a general understanding of the spatial systems is more of a priority than the predictive capability of the models. To a certain extent, this could be achieved by breeding numerous individual data-based models and then considering the structures that appear in the models. The utilisation of multiple datasets creates issues concerning the measure of performance when combining the data, how best to parallelise the task and the number and types of dataset to use (similar characteristics or a variety of spatial interaction types), and is beyond the scope of this chapter.

The compatibility of models might also be improved if spatial interaction datasets with similar characteristics are considered. This concerns not only the type of interaction (migration, retail, etc.) but also the characteristics of the data (square or rectangular matrix, intra-zonal trips included or excluded, etc.) and this is another issue that needs to be investigated.

A further problem concerns the parsimony mechanism outlined previously. The identification of complex model structures works well enough, but the problem concerns the variable error function values between datasets; if an increase of 5.0 is specified, then this would affect a model based on the Durham data (with a best error value of approximately 126.0–127.0) but would severely impact on the Seattle data (the best error being approximately 6.3–6.4). This effect is clearly seen when Durham-based models have been reported as the best for a run, even though they have had their error values increased, but this is not the case for the Seattle-based models.

To overcome this, the actual error increase is determined by taking the difference between the score and the threshold, and converting this into a percentage of the model's error function value, as shown in the equation below.

$$f_{new} = f_p + \left[\frac{f_p}{200.0} \times (N_{pen} + N_{thr}) \right] \qquad \text{if } N_{pen} > N_{thr} \qquad (9.32)$$

where f_{new} is the adjusted error value, f_p the original error value, N_{pen} the 'complexity score' of the equation and N_{thr} the threshold complexity score value. The value 200.0 was used simply to reduce the magnitude of the penalty percentages. Thus a complexity score of 30 yields a fitness increase of 5% (the difference between the score and the threshold being 10). If the complexity score is lower than the threshold, then the error value remains unaltered. This parsimony mechanism will be used in all subsequent model breeding runs.

9.4.5 Further experimentation

Given the issues discussed in the previous section, a final experiment concerns the investigation of model breeding for single datasets of similar characteristics, to investigate not only the cross-data compatibility of such models but also the theoretical potential possessed by the model breeding methods used in this investigation. Three datasets are to be used in these experiments, each consisting of a square matrix of trips comprising: (1) a different journey-to-work dataset this time for England and Wales (1991 data of 55 counties) and two national migration datasets from (2) Japan (1985 data of 46 prefectures) and (3) the United States (1970–75 data of 58 standard metropolitan statistical areas). Each of these has had the intra-zonal trip diagonal values set to zero, so that these do not dominate the model breeding process. Bred models are first compared with conventional model benchmarks. They are then applied to the other two datasets to see if their cross-data performance is better than the previous results and, finally, the equations are considered to see if there is any theoretical potential for the utilisation of automated modelling procedures.

A similar set of functions and terminals is to be used as outlined in Tables 9.1 and 9.2, with the exclusion of the intra-zonal trip flag ($V8$), since there are no intra-zonals in the datasets, and the following functions (sin, cos, <, > and abs), which have not appeared in any of the 'good' model specifications produced so far. The conventional model benchmarks for these datasets are given in Table 9.4.

Table 9.4 Conventional model benchmarks

Data	Error
England & Wales	107 527.43
Japan	2 300 424.78
United States	4 304.69

The experiments were again divided into two stages. The first utilised a basic model breeding GP algorithm, whilst the second incorporated the parsimony measure outlined in the previous section. Both are discussed in turn. Cross-validation was also undertaken but omitted here for the sake of brevity; for more information, see Diplock (1996).

For the England and Wales journey-to-work data, the best model is given in equation (9.33). This yielded an error value of 88 911.73, which is an improvement over the benchmark value of 107 527.43, but not as significant as might be hoped for. Using the Japanese migration data, the best bred model yielded an error function value of 8886.03, with equation (9.34).

$$T_{ij} = V7^{-0.5} \times$$

$$\left[V3 + \cfrac{V2 \sqrt{V7}}{V6\left(V2\, V5^{-0.92} + \left(\frac{V5}{-5.08} + 1.0 \right)^{V2} \right)} + 0.55 \right]^{-1.2751} \tag{9.33}$$

$$T_{ij} = \left[V3 + \cfrac{V4}{V3 + \sqrt{V3} + 0.18} \right] \times V5^{\sqrt{\frac{V2^2\, V4}{V4\, V5^{-0.12}}} \times V7^{-0.21}}$$

$$\times \left[V3^{-1.0} \times \left(-1.05 V3 + \sqrt{\frac{V7}{V4}} + V2\, \frac{V5^{-0.62}}{V6^{-0.75}} \right) \right] \tag{9.34}$$

This represents a greater improvement, with the benchmark error value being 2 300 424.78. Finally, for the United States migration data, the best error function value obtained through model breeding was 923.91, with the model being given by

$$T_{ij} = V2 \times \left[V5^{-2.24} + \frac{V4}{8.53} + \left(V5^{-1.92} + V6^{-2.63} - \frac{V2}{V5^{-2.73} + 0.47} \right) \right.$$

$$+ V5^{-0.69} - V7 - \frac{V1}{0.03} \times \left(\frac{V5^{-0.30} + V3 + \log(V4) + V4}{V7} \right.$$

$$\left. \left. - \frac{V2}{V4} + V5^{-3.47} \right) \right] \tag{9.35}$$

This also represents a significant improvement over the benchmark model, which yields an error value of 4 304.69.

Whilst these models have been produced without any form of parsimony measure being incorporated into the error function value, it is interesting to note that the models are relatively less complex than the basic model specifications obtained in the previous set of experiments, which is perhaps a result of the removal of the intra-zonal interactions, which could make the model breeding process more difficult. The specifications are still more complex than is perhaps useful, so as a final experiment, the GP model breeder was run again, but this time with the parsimony measure incorporated into the model evaluation.

For the England and Wales data, the best model is given in equation (9.36) with an error value of 101 701.25.

$$T_{ij} = V2^{-1.0} \times \left[\frac{V5^{-0.75} - V5}{V2\ V3} - \log(V6) \right] \qquad (9.36)$$

The Japanese data yielded a model with an error function value of 960 418.93, with the specification being

$$T_{ij} = \frac{(V4\ V5^{1.04} + V2) \times \left(\dfrac{V2}{V7} + 0.03 \right)^{1.30}}{V4\ V5^{1.04}\ V7} \qquad (9.37)$$

Finally, the United States data saw the discovery of the model given in equation (9.38), with an error value of 908.95.

$$T_{ij} = V5 \times \left[V4\ \{\exp(V5^{-0.51}) - \exp(V5^{-0.56})\} \right.$$

$$\left. + \left(\frac{V2}{V5^{V6+2.96} + V7} + V4 - V6 \right)^{-1} \right] \qquad (9.38)$$

However, these performance measures do not yield the 'true' result for the last two equations, due to the effect of the parsimony measure. The equation (9.36) is sufficiently small not to have been punished, but improves on the benchmark model only slightly and is worse than the previous model. But equation (9.37) actually yields an error function value of 937 057.06 which represents a significant improvement over the benchmark model performance, but not the previous model, and the third a value of 830.08 which is an improvement over both the benchmark and previous models. After the findings of the first set of experiments, it is perhaps not surprising to discover that the models obtained are relatively simple in their specification, but in the case of the England and Wales and Japanese data, this simplicity has been obtained at the expense of model performance.

9.5 The way forward

The genetic programming algorithms identified models that performed significantly better than the conventional model benchmarks, and sometimes the improvements were in the order of several times. Such improvements were achieved for all except two of the datasets examined; the Durham journey-to-work data yielded a smaller improvement, and the Seattle car sales data saw little improvement, although for this dataset, this could be partly explained by the characteristics of the data, with a small error function value

Table 9.5 Example processing times (hours) for a GP run on a single Sun UltraSparc system

Batch	PEs	Total	Time
I hour	128	128	21.3
I hour	256	256	42.7
I hour	512	512	85.3
6 hour	128	768	128.0
6 hour	256	1536	256.0
6 hour	512	3072	512.0

Notes
PEs = number of processing elements; Total = total execution time, in hours, including all processors; Time = equivalent execution time, in hours, for a single processor run.

and an apparent dominating importance of destination characteristics over distance.

9.5.1 The potential of automated model breeding

Although it is desirable to be able to run the algorithms on a highly parallel supercomputer, it is not absolutely essential. For example, a processor in a Sun UltraSparc workstation is about six times better than the DEC Alpha chips in the T3D, due primarily to the limited cache on the Alpha chip (Openshaw and Schmidt, 1996). Table 9.5, given this information, demonstrates the execution times for typical GP runs over a series of different system configurations. The batch time is the 'real' time limit for the job; after this all of the processors are shut down, whereas the total time represents the 'full' processing time (the number of processors multiplied by the batch time). It can be seen that, for the smaller scale jobs (one hour batch runs), it is still feasible to run a GP algorithm for a similar size problem as would require 128 or even 256 processors on the T3D, as this would only take approximately one and two days respectively. The 512 processor job is longer (around four days), but is still feasible. However, the six-hour batch jobs are more problematic, taking 256 or 512 hours to complete.

Further, given the fact that the algorithm is written using MPI and has a fairly coarsely grained degree of parallelisation, it could easily be ported onto a multiple workstation farm, which would operate in a similar manner to the multiple processors of the T3D. This would reduce the amount of time taken for each run, allowing either more runs to be performed or larger individual runs to be implemented. If a network of ten workstations were available, then a significant amount of time is saved, and even large scale runs become feasible, as illustrated in Table 9.6.

Table 9.6 Example processing times (hours) for a GP run on multiple processor systems

Batch	1 PE	10 PEs	20 PEs
1hr 128PE	21.3	2.13	1.07
1hr 256PE	42.7	4.27	2.14
1hr 512PE	85.3	8.53	4.27
6hr 128PE	128.0	12.80	6.40
6hr 256PE	256.0	25.60	12.80
6hr 512PE	512.0	51.20	25.60

For example, all of the one hour jobs take a few hours, and even the larger jobs become more realistically usable; an equivalent 256 processor six hour batch job takes only a day, whilst a 512 processor version still only takes two days. Similarly, if 20 workstations were available, then these times are further reduced, again illustrated in Table 9.6. It must be acknowledged that these are only approximate figures, since the algorithm is not fully parallel, and whilst similar speedups might be possible, they are theoretical maximums.

9.5.2 A wider perspective

If the hardware requirements for a GP based approach to model building is not a serious problem, then the major issue becomes a question of whether building such models is a worthwhile task. Again, this must be viewed as an application-specific issue, with two broad perspectives, namely theoretical and applied.

In the case of applied research (encompassing work involving both spatial analysis and GIS application), the main potential for such an approach would lie in applications which could potentially save money, such as the development of models for transportation planning. If the model breeder can identify a better model than exists currently for predicting traffic flows through transportation systems then this could potentially save a considerable amount of money. Similarly, in retail applications, bred models which predict the flows of consumers to retail outlets of whatever type more accurately than conventional models could provide more powerful and successful decision making tools for companies to utilise.

From a theoretical perspective, the potential offered by the model breeding approach is less clear, since the benefits are at present virtually impossible to quantify. From a research perspective the attraction is more obvious. The use of automated model breeding techniques is potentially useful for a variety of spatial analysis tasks, ranging from developing a better theoretical understanding of a spatial system for a particular application to attempts to gain a more general understanding of complex spatial systems.

In its current form this model breeding software would require a set of model breeding runs for each dataset, and the results could be brought together and any similarities in the yielded models may suggest areas for further study. For example, in some cases, such as the Durham data, it might be more worthwhile to simply breed the distance decay function, whereas in others, such as the Seattle data, consideration of the destination attractiveness may be more beneficial, with the other pieces of a conventional model kept constant. This is an issue for further investigation.

This experimentation may take two broad forms. A series of datasets for a single spatial interaction type (such as migration, journey-to-work, food retailing, etc.) could be chosen and a series of models bred for each, and the results considered, or a mixture of datasets could be utilized, in an attempt to obtain a more general understanding of spatial interaction. Of course, the method used is determined by the precise nature of the research, but either could potentially yield useful results for spatial analysis specifically, or geographical modelling more generally, given a research subject that can be quantified adequately to permit mathematical description through models.

9.5.3 Problems of automated model building

Throughout the course of this research, numerous issues have been highlighted concerning the implementation of an automated model building approach for the development of spatial interaction models. Some of these have been solved, but others are more problematic. They are presented briefly below, followed by a discussion of possibilities for developing the model breeding algorithm and further uses for it.

There is still a need for the user to provide the model breeder with adequate descriptive and model components to represent the system, which could prove to be difficult. The importance of specifying an adequate set of functions and terminals was illustrated in the first set of GP model breeding experiments. Given a set of potentially applicable functions and terminals, these worked well for one dataset, but not the other, where it became apparent that a different set of variables and perhaps even functions was required if the data were to be modelled successfully.

The initial problem of obtaining complex model specifications was overcome by the incorporation of a parsimony mechanism, although sometimes at the expense of model performance. Parsimony was also implicit in the parallelisation strategy of the algorithm, whereby more complex model specifications took longer to evaluate and therefore were given less opportunity to reproduce. However, much more research is needed here.

Concerning the problems relevant to traditional modelling, the issue of floating point exceptions was dealt with satisfactorily, resulting in a strategy which ensured that exceptions did not interfere with the model breeding process other than in the forced extinction of any model specification which gave rise to them. The optimisation of model parameters was also accommodated through the use of a hybrid optimisation technique which utilises the strengths of both conventional and evolutionary strategies.

9.5.4 The development of model breeding machines

If the utilisation of automated model breeding techniques is to be further developed, there are several other issues that need to be addressed. The specification of the sets of functions and terminals need to be considered closely, not just concerning the choice of functions and variables, but also perhaps the task of developing a series of application-specific functions which may be better suited to the model building task being undertaken. This is a potentially hard problem, with no clear guidance as to how one might initially proceed, except for that provided from existing models and a limited theory of spatial interaction processes.

An obvious new approach is the breeding of models on multiple datasets. This ties in with the task discussed earlier, but instead of using individual runs and pooling the results, the model breeder could be provided with all of the datasets and try to breed a model that best describes them. Whilst the models will probably perform less well on an individual dataset than a model bred using just that data, it may yield models with more satisfactory performance over unseen data, given that it should not have been moulded to a single dataset, as long as one of those used does not dominate the model breeding process (through the standardisation of error values).

Ultimately, the relevant approach is determined by the aims and objectives of the application for which the automated model breeding methodology is being utilised. Thus research has demonstrated the potential of such a technique in the context of spatial interaction modelling from both a theoretical and an applied perspective, as well as the potential of evolutionary algorithms in general. There is much to gain from the adoption of new and innovative approaches to geographical research, and little to lose. The approach is not intended as a replacement for more conventional model building techniques, but rather as a complement, to assist model development in areas where conventional methods have proved inadequate, or simply to approach the problem from a different perspective. If the results of this research are an indication, geographical modelling will surely lose out if such approaches to spatial analysis and modelling (and perhaps geographical modelling more generally) are ignored, especially since they offer such an exciting opportunity for the development of new spatial analysis tools and methods.

Bibliography

The following contains references from this investigation and also several texts for further information. Numerous evolutionary computing based bibliographies exist.

Antonisse, H. J. (1991) 'A grammar-based genetic algorithm', in Rawlins, G. (ed.) *Foundations of genetic algorithms*, California: Morgan Kaufmann.

Antonisse, H. J. and Keller, K. S. (1987) 'Genetic operators for high-level knowledge representations', in Grefenstette, J. J. (ed.) *Proceedings of the Second International Conference on Genetic Algorithms*, New Jersey: Lawrence Erlbaum Associates.

Beasley, D., Bull, D. R. and Martin, R. R. (1993) 'An overview of genetic algorithms. Part 1; fundamentals', *University Computing*, 15, pp. 58–69.

Bethke, A. D. (1981) *Genetic Algorithms as Function Optimisers*, Technical Report, Logic of Computers Group, Univerity of Michigan.

Blickle, T. and Thiele, L. (1995) *A Comparison of Selection Schemes Used in Genetic Algorithms*, Zurich: TIK report, Swiss Federal Institute of Technology.

Chabris, C. F. (1987) *A Primer of Artificial Intelligence*, London: Kogan Page.

Cramer, M. L. (1985) 'A representation for the adaptive generation of simple sequential programs', in Grefenstette, J. J. (ed.) *Proceedings of the first International Conference on genetic algorithms*, New Jersey: Lawrence Erlbaum Associates.

Davis, L. (1991) (ed.) *Handbook of Genetic Algorithms*, New York: Van Nostrand Reinhold.

Diplock, G. J. (1996) 'The application of evolutionary techniques to spatial interaction modelling', Unpublished PhD thesis, School of Geography, University of Leeds, Leeds.

Diplock, G. and Openshaw, S. (1996) 'Using simple genetic algorithms to calibrate spatial interaction models', *Geographical Analysis*, 28, pp. 262–279.

Fogel, D. B. (1994) 'An introduction to simulated evolutionary optimisation', *IEEE Transactions on Neural Networks*, 1, pp. 3–14.

Fotheringham, A. S. (1983) 'A new set of spatial interaction models; the theory of competing destinations', *Environment & Planning A*, 15, pp. 15–36.

Goldberg, D. E. (1989) *Genetic Algorithms in Search, Optimisation and Machine Learning*, Reading, MA: Addison Wesley.

Goldberg, D. E., Korb, B. and Deb, K. (1989) 'Messy genetic algorithms: motivation analysis and first results', *Complex Systems*, 3, pp. 493–530.

Grefenstette, J. J. (1986) 'Optimisation of control parameters for genetic algorithms', *IEEE Transactions on Systems, Man and Cybernetics*, 16, pp. 122–128.

Hancock, P. J. (1994) 'An empirical evaluation of selection methods in evolutionary algorithms', in Fogarty, T. C. (ed.) *Evolutionary Computing; Lecture Notes in Computer Science; Vol. 865*, Berlin: Springer Verlag.

Holland, J. H. (1975) *Adaptation in Natural and Artificial Systems*, Cambridge, MA: MIT Press.

Kinnear, K. (1994) (ed.) *Advances in genetic programming*, Cambridge, MA: MIT Press.

Koza, J. R. (1992) *Genetic Programming: On the Programming of Computers by Means of Natural Selection*, Cambridge, MA: MIT Press.

Koza, J. R. (1994) *Genetic Programming II: Automatic Discovery of Re-usable Programs*, Cambridge, MA: MIT Press.

Oie, C. K., Goldberg, D. E. and Chang, S-J. (1991) *Tournament Selection, Niching and the Preservation of Diversity*, ILLiGAL Report no. 91011, Illinois: Illinois Genetic Algorithm Laboratory.

Openshaw, S. (1988) 'Building an automated modelling system to explore a universe of spatial interaction models', *Geographical Analysis*, 20, pp. 31–46.

252   Gary Diplock

Openshaw, S. and Schmidt, J. (1996) 'A social science benchmark (SSB/1) for serial, vector and parallel supercomputers', *School of Geography Working Paper*, Leeds: University of Leeds.

Openshaw, S. and Turton, I. (1994) 'Building new spatial interaction models using genetic programming', *Proceedings of the AISB workshop on Evolutionary Computing*, Leeds.

Rudolph, G. (1994) 'Convergence analysis of canonical genetic algorithms', *IEEE Transactions on Neural Networks*, 5, pp. 96–101.

Schwefel, H. P. (1995) *Evolution and Optimisation Seeking*, New York: Wiley.

Syswerda, G. (1989) 'Uniform crossover in genetic algorithms', in Schaffer, J. D. (ed.), *Proceedings of the Third International Conference on Genetic Algorithms*, California: Morgan Kaufmann.

Syswerda, G. (1991) 'Schedule optimisation using genetic algorithms', in Davis, L. (ed.) *Handbook of Genetic Algorithms*, New York: Van Nostrand Reinhold.

Turton, I., Openshaw, S. and Diplock, G. (1996) 'Some geographical applications of genetic programming on the Cray T3D supercomputer', in Jesshope, C. R. and Shafarenko, A. V. (eds) *UK Parallel '96: Proceedings of the British Computer Society Parallel Processing Specialist Group Annual Conference*, Berlin: Springer Verlag, pp. 135–150.

Turton, I., Openshaw, S. and Diplock, G. (1997) 'A genetic programming approach to building new spatial models relevant to GIS', in Keme, Z. (ed.), *Innovations in GIS 4; Selected Papers from the Fourth National Conference on GIS research UK*, London: Taylor & Francis, pp. 89–102.

Whitley, D. (1988) 'GENITOR: a different genetic algorithm', *Proceedings of the Rocky Mountain Conference on Artificial Intelligence*, Denver.

Whitley, D. (1989) 'The GENITOR algorithm and selection pressure', in Schaffer, J. D. (ed.) *Proceedings of the Third International Conference on Genetic Algorithms*, California: Morgan Kaufmann.

Whitley, D. (1993) *A Genetic Algorithm Tutorial*, Technical report CS-93-103, Department of Computer Science, Colorado State University.

Wilson, A. G. (1971) 'A family of spatial interaction models and associated developments', *Environment and Planning A*, 3, pp. 1–32.

Wilson, S. W. (1987) 'Hierarchical credit allocation in a classifier system', in Van Steenwyk, E. (ed.), *Proceedings of the Tenth International Joint Conference on Artificial Intelligence*, California: Morgan Kaufmann.

Chapter 10

Visualization as a tool for GeoComputation

Mark Gahegan

10.1 Introduction

Visualization is an emerging science that draws from a rich and diverse research literature including computer graphics, computational geometry and theories of human perception from cognitive science and psychology. From a GeoComputation (GC) perspective, visualization is an advanced form of display technology offering the potential for improving the efficiency and effectiveness of data communication between the machine and the user. From the perspective of human–computer interaction, visualization can be viewed as one member of a larger family of methods, where interaction is restricted to the visual domain only and specifically by forming images, scenes or virtual realities to graphically portray data. In sighted humans, vision is the dominant sense, making it the strongest candidate on which to base interaction with the computer (McCormick *et al.*, 1987).

Visualization is used in a wide variety of different ways, ranging from the generation of photo-realistic images to the interactive study of highly abstract data-spaces. For many tasks in scene understanding or vision (such as perception of structure, segmentation and generalization), the human visual system can outperform even the most sophisticated of computer vision systems. Recognition of this fact has led to a shift in emphasis away from the automation of such tasks and towards a collaborative mode of interaction with the user. Consequently, there is a trend within the computer vision and artificial intelligence community to now regard the human as an essential component within many systems, and interest in developing collaborative approaches that can utilize the strengths of both computers and people working together[1]. Visualization can provide an approach to data analysis that is based around such collaboration. It is also usually typified by being highly interactive, allowing the user to move around the scene to explore different perspectives on the underlying data.

10.1.1 Definitions

Visualization is a means of interaction between computers and people based around a graphical environment. It involves the production of graphical representations of data, often termed as scenes. A scene may attempt to depict (render) data as it might appear to a human observer (e.g. photo-realistic), or alternatively, transform data values which do not have a true visual appearance (such as depth, soil pH and rainfall) into a pictorial form where they can be readily perceived. This transformation is known as visual encoding.

A scene is composed of a number of primitive graphical objects which are synthesized together, rather like the base types in a conventional programming language. Example objects might be a point, a line, an arrow or a mesh (lattice); these are sometimes termed marks, with a collection of graphical objects being known as a mark set. The visualization environment allows a user to assign data to marks in such a way as to determine appearance. A simple example, common in cartography, involves using the size of a population to control the area of a circle used to represent a city. More complex examples are given later.

10.1.2 Emergence of visualization

Buttenfield and Mackaness (1991) describe the emergence of visualization in terms of several needs (here paraphrased): to search through huge volumes of data for relevant information, to communicate complex patterns or structures, to provide a formal framework for data presentation and for exploratory analysis of data. In addition, the recent popularity of visualization is also due in part to technological developments. Like so many other computer-based technologies, it can be considered in terms of hardware, software and data, with recent progress in these areas contributing to the current interest.

Visualization has, since its inception, required sophisticated (and expensive) hardware to provide accelerated graphical performance (rendering speed). For handling geographic datasets, large quantities of random access memory (RAM) are also usually required. However, the ever-improving cost/performance ratio of computer hardware has ensured that suitable graphics capabilities are available now for many platforms, including personal computers (PCs), at a much lower cost and indeed are considered standard equipment on many entry-level workstations (such as the Silicon Graphics O_2). On the software side, the emergence over the last few years of commercial visualization systems and libraries, such as ENVI/IDL, NAG-Explorer[2], IBM's Visualization Data Explorer, ERDAS Imagine and AVS[3] has provided comprehensive development environments that can greatly ease the task of engineering a visualization, often by the use of visual programming techniques based on the flow of data. Slocum (1994) provides a useful overview of a variety of these environments and ranks their ease of use and utility for geographic visualization. Finally, strong motivation is provided for the use of visualization by the increasing amounts of data becoming available for many types of application in the geosciences. For example from whole new families of geo-physical and thematic imaging tools, with large numbers of data channels, and often providing high spectral, spatial and temporal resolution. Such datasets give rise to new challenges relating to data presentation, some of which can be addressed by using the enhanced functionality and finer control that visualization environments offer. Care is needed to deal with the ensuing information overload (O'Brien *et al.*, 1995).

10.1.3 Nature of geographic data

Geography is an holistic discipline, so the gathering and analysing of additional (relevant) or more detailed datasets may increase the understanding of some geographic process or system. Thus, the datasets used in many geographic and geo-scientific studies are often large and heterogeneous, in terms of areal extent, number of distinct layers (or channels) of data, and the geometry and content of each layer (e.g. images, point samples, defined objects, contours and so forth)[4]. For example, a demographic study using census and medical data might require 20 or more variables including point incidence data, administrative boundaries and demographic surfaces whereas a geological mapping exercise might use multi-channel, geo-physical imagery to supplement drill-hole samples, elevation surfaces and aerial photographs. Many traditional forms of analysis can become prohibitively complex or unreliable when dealing with such large and diverse multi-variate databases. The same can be said of artificial intelligence techniques (though perhaps to a lesser extent) which can become difficult to configure and slow to converge as the complexity of the analysis task increases. Visualization and virtual reality offer some hope of providing an environment where many data layers can be viewed and understood concurrently without recourse to statistical summarization or data reduction – both of which may mask or even remove the very trends in the data that might be of greatest interest.

10.1.4 Visual attributes

A visualization environment usually provides a high degree of control over the visual attributes that together define the appearance of the graphical marks. For example, the following visual attributes are often separately configurable for each mark:

*   position, usually in three dimensions;
*   colour, often as three separate variables representing red, green and blue or hue, saturation and intensity;

- transparency, as a single value, referred to as alpha (α);
- material properties, to describe the light reflectance and absorption characteristics of the objects in a scene.

Table 10.1 Some example descriptions of graphical marks in terms of the visual attributes over which the user has control (from NAG Explorer)

Geometric mark	Visual attributes
3D mesh	x, y, z, red, green, blue, alpha
2D sphere	x, y, z, size, colour
3D arrows	x, y, z, x-offset, y-offset, z-offset, colour

In addition, further control is usually available over the entire scene, including the position from which the scene is observed (viewpoint) and the location and characteristics of light sources to choose how the scene is illuminated[5].

Depending on the system and the type of graphical marks, the properties available may differ. As an example, Table 10.1 shows the visual attributes of some example marks from the NAG-Explorer visualization environment in terms of the visual attributes over which the user has control. The appropriateness of a particular mark is determined by various statistical properties of the data that are to be visually encoded and this is discussed in section 10.2.1.2. In many visualization environments, it is possible to extend the visual library by adding in new mark descriptions.

By contrast, the degree of control over visual appearance available in many commercial GIS is rather limited, often being restricted to a small number of predefined spatial primitives, with few visual attributes available to the user, and able to show only a limited number of concurrent displays or layers. Whilst many GIS are capable of producing high-quality output in a traditional cartographic sense, they are simply not up to the task of visualization[6]. Such shortcomings are being addressed in one of two ways; either the visual capabilities of the packages are increased or links to visualization software are incorporated (Hartmann, 1992; Tang, 1992).

10.1.5 Visualization effectiveness: the study of perception and psychophysics

To understand why visualization can be effective, and to improve this effectiveness, it is necessary to understand something of the process of visual perception in humans. Although it is difficult to be definitive about perception, much research and experimentation has led to the establishment of some useful models of perception. For example, it is known that perception of certain visual properties occurs earlier in the task of overall cognition than others. The early stage of visual processing is often termed 'pre-attentive' or 'early' vision and can be considered as a low-level cognitive process in humans. Experiments with the separability of a visual stimulus (Treisman, 1986; Treisman and Gormican, 1988) show that shape and colour and also light-ness and size are to a large extent separable, pre-attentive, and occur in parallel, hence they are good candidates for using together to visually en-code a multivariate dataset. Additional research has identified major neural pathways through which visual information is received, including colour, form (or structure) and movement (Livingstone and Hubel, 1988). To a large extent, perception of these three different types of information appears to be performed orthogonally, thus allowing different types of information to be separately encoded within the overall visual stimulus. The hypothesis is that, on perception, an observer may be able to maintain this separation, thus perceiving many different data values simultaneously. By exploiting these pathways, and also some heuristics that determine how accurately and at what rate human subjects can assimilate these various types of visual attributes, it becomes possible to design visualization strategies that are effective, making good use of both the available hardware and the cognitive abilities of the user.

The fact that some tasks in vision require more effort and take more time has led to a good deal of research into ranking perceptual tasks, according to their difficulty. Early work in this field that has had an influence on researchers in both cartography and perception was conducted by Bertin (1981), who categorized visual attributes (referred to as retinal variables) according to whether they could be observed separately when combined into a visual stimulus. Working with a similar (but not identical) set of visual attributes, Mackinlay (1986) synthesized a rank ordering of the diffi-culty of perception. In the absence of a definitive framework, these guide-lines can be adopted as a starting point from which to construct a visualization, as discussed below. As Table 10.2 shows, the orderings actu-ally differ according to the statistical scale of the data under consideration (nominal, ordinal or quantitative). It is interesting to note that position is rated as the most important visual attribute for all types of data. Section 10.2.1.2 introduces some other metadata that can also be useful in the visualization design process.

10.1.6 Designing a visualization

In order to ensure that a visualization achieves its goal, the data should be encoded to the available visual attributes bearing in mind the perceptibility of these attributes. For example, humans are poor performers at quantify-ing differences in colour hue (and many are also red-green colour-blind), so using colour to highlight differences between two similar datasets might be less effective than using position or movement (animation). The total band-width available across all visual attributes effectively sets an upper bound

Table 10.2 Ranking of elementary perceptual tasks according to statistical scale after Mackinlay (1986) (bracketed items not applicable).

Quantitative	Ordinal	Nominal
Position	Position	Position
Length	Grey saturation	Colour hue
Angle	Colour saturation	Texture
Slope	Colour hue	Connection
Area	Texture	Containment
Volume	Connection	Grey saturation
Grey saturation	Containment	Colour saturation
Colour saturation	Length	Shape
Colour hue	Angle	Length
(Texture)	Slope	Angle
(Connection)	Area	Slope
(Containment)	Volume	Area
(Shape)	(Shape)	Volume

on the volume of information that can be conveyed concurrently. However, even within this limit, the use of a greater number of visual attributes does not necessarily increase the effectiveness of a scene, since many combinations of visual attributes are known to interfere with each other (for example, intensity and transparency). The use of such knowledge in the construction of a scene is described later.

Methodologies for the design of visualization have been proposed by a number of researchers including Senay and Ignatius (1991), Beshers and Feiner (1993) and Duclos and Grave (1993). Turk (1994) provides a useful taxonomy of approaches, reviewing a number of different ways by which the assignment of data to visual attributes can be achieved.

10.1.7 Useful metaphors for visualization

The human visual system appears to be specifically 'tuned' for various perception tasks, making it more adept at assimilating some types of visual information than others. To exploit this bias, Chernoff (1973, 1978) proposed the use of human faces to encode data. By assigning the shape of the mouth to one variable, the size of the nose to another, and so on, it is possible to convey many data values simultaneously in a way that humans can easily interpret, because we are skilled at interpreting facial expression. Dorling (1994) shows some excellent examples of Chernoff faces used to visualize trends in political and social conditions in the form of a cartogram. Plate 10.1 shows a visualization of health-related variables imposed on cartograms for districts in England and Wales. Each image encodes a single health variable via the use of colour (source: Dr Daniel Dorling, Department

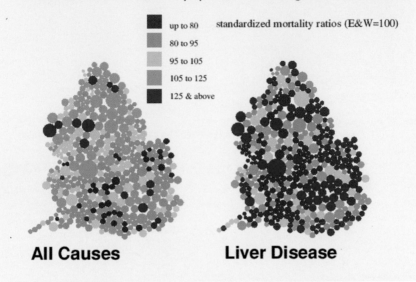

Plate 10.1 Two cartograms showing a comparison of two different health related variables for population districts in England and Wales. Causes of death before age 65, 1981–1989.

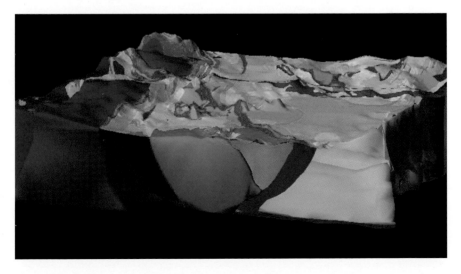

Plate 10.2 A 3D solid geological model, interpreted from a 1:100 000 mapsheet and rendered to support visual interaction with the user.

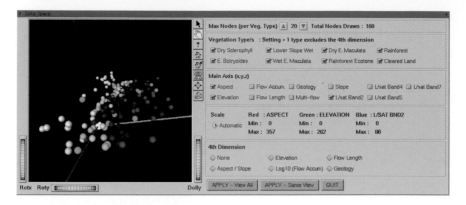

Plate 10.3 A visualization tool for viewing dataspaces as scatterplots. The user can compose one or many graphs to explore relationships between datasets.

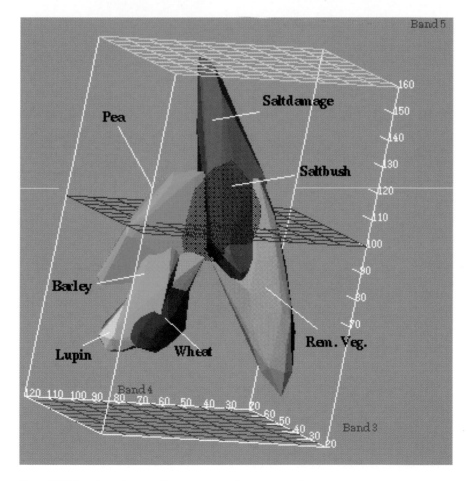

Plate 10.4 Display of spectral training data from a classification exercise using remotely-sensed imagery.

Plate 10.5 An example of mark and axis composition for visual exploration of a complex dataset. Refer to the text for an explanation.

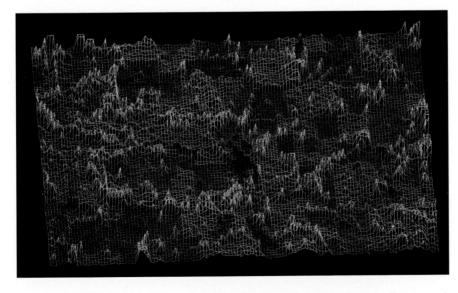

Plate 10.6 Visualization used to convey classification uncertainty; vertical offset is used to represent the probability of a pixel being assigned a particular class.

Plate 10.7 Visualization used for the simulation of alternate realities with the emphasis on visual realism.

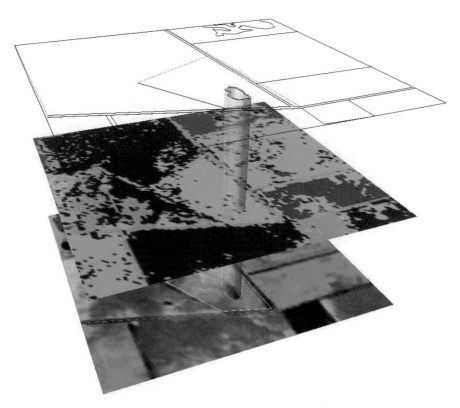

Plate 10.9 The use of visual interactors to study positional relationships between layers of data. A full description is given in the text.

of Geography, University of Bristol, England). Landscapes are another example of a visual metaphor which seems to be effective for humans to work
with (Rheingans and Landreth, 1995), as shown by Robertson (1990), who
used various properties of a natural vista to visually encode a dataset.

10.2 Uses of visualization within the geosciences

One obvious use of visualization is to enable three-dimensional (3D) models
to be constructed and viewed. Plate 10.2 shows a 3D rendering of a solid
geological model, interpreted from a 1:100 000 mapsheet. The user can view
and slice through the model from any angle (source: Fractal Graphics
Geoscientific Consultancy, Nedlands, Perth, Australia). In this section, some
further visualization techniques and their uses within the geosciences are
described under three main headings: exploratory analysis, visual realism
and animation.

10.2.1 Exploratory visual analysis

Data can tell many stories; a comprehensive geographic dataset contains a
huge wealth of different relationships, both within and between the different
layers or themes. When faced with all this diversity and variance the first
question posed might be 'What should I ask?'

It is the goal of exploratory visual analysis (EVA) to present data to a
user in such a way that relationships and structure contained therein are
made apparent (Cleveland and McGill, 1988; Tufte, 1990; Monmonier,
1990; MacDougall, 1992). The recognition of some underlying pattern or
structure is known colloquially as visual pop-out. For pop-out to occur,
the target must be identifiable via the combined effect of its visual properties, which implies that the stimulus is not shared with any distracters,
i.e. irrelevant patterns or structures that produce a similar response in the
observer.

Exploratory visual analysis is one tool in a larger arsenal for performing
data mining or knowledge discovery that could be grouped under the heading of exploratory data analysis (EDA). Other tools attempt to recognize
or discover knowledge by computing the strength of relationships within
and between layers of data (for example, using correlation or co-variance),
possibly distilling out some sort of model as a result. Such tools are usually
computer intensive. Haslett et al. (1991) describe interactive tools for identifying statistical anomalies in data.

Where there is no associated task or goal to a rendering, then the task of
EVA is largely unstructured. Where a specific task or goal is identified, EVA
should encourage perception of this. EVA is an example of a collaborative
mode of interaction in which the visualization environment produces a stimulus which is then interpreted by the user. It is important to note that the

visualization software does not 'find' or evaluate any relationships in the data, but instead aids the user in identifying such items for themselves. It is therefore the goal of EVA to help increase understanding of a dataset to the point where the user might feel confident to formulate an hypothesis.

10.2.1.1 Scatterplots and dataspaces

Perhaps the most well-known exploratory visualization technique is the scatterplot. With 2D systems or paper output, scatterplots are restricted to showing relationships between a small number of variables (usually two). With the use of interactive visualization, a third axis can also be included, so three variables may be graphed together. By employing some further cartographic devices, yet more channels of data can be added. Where a large number of independent variables are to be studied, it may first be necessary to employ some form of dimensionality reduction (e.g. Noll, 1967) or rotation (Cook *et al.*, 1995).

Plate 10.3 shows an application developed to display a number of concurrent data spaces as scatterplots[7]. The dataset used here is the Kioloa NASA Pathfinder data of a coastal region in New South Wales, Australia[8]. In the plate, known values for certain vegetation species are plotted on the three axes and the colour of each icon represents the vegetation type. For each additional scatterplot, a similar number of data values can be added. A further example, shown in Plate 10.4, depicts a spectral dataspace in which a convex hull has been formed around sets of training data for an agricultural land-use classification. The overlap of the various land-use classes indicates a measure of their inseparability.

The construction of scatterplots involves the following steps:

1. Scaling of data. This is required to ensure that axes are visually comparable. In some cases, it may be best to use non-linear scaling (such as logarithmic or exponential) depending on the distribution of values in the data.
2. Assignment of data channels to a specific scatterplot property (*x*-axis, *y*-axis, *z*-axis, colour, size, etc.). This might be carried out by a user or an expert system. The goal of assignment must first be identified. One common goal is to look for combinations of channels that give the best separation of the data within the graph, for example to explore the possibilities of classifying the data. There is no reason why the same data channel should not appear in many plots.
3. Selection of which scatterplot(s) to visualize. The size of the viewing area and the performance of the hardware often dictate an upper limit on the number of graphs that can be viewed concurrently.
4. Rendering of the scatterplot(s).

The scatterplot is often used to graph variables directly against each other, effectively removing locational information. Hence, the data is transformed from a spatial frame of reference to one that is based on some arbitrary dataspace, devised by the user. These artificial dataspaces can be difficult to comprehend. The scatterplot functions well when the number of variables to be considered can be contained in a single graph. In this case, attention can exclusively focus on a single dataspace, which significantly reduces the cognitive difficulty in studying the data. If the data extends over two or more plots, the user's focus of attention must then continually shift to a different dataspace, within which the axes and spatial relationships contained are not the same (Aspinall and Lees, 1994). This makes the study of relationships between graphs difficult, and has led to the development of interaction techniques to re-establish relationships fractured by the division of the data amongst the graphs. For example, a brush can be applied in one plot to highlight data points in another. The embedding of such interactors within complex scenes may provide some useful possibilities for studying cause and effect, as shown later in section 10.2.3.

10.2.1.2 Compositional approaches to visual exploration

Rather than try to study individual data values and their patterns we may instead construct a visual stimulus that combines several different visual attributes, where each attribute encodes a different data value. Ideally, such a combined stimulus would be observable only when the underlying data layers contain regions or patterns of interest. Such an approach is in keeping with integrated theories of perception, as described by Treisman (1986). Compositional approaches to exploratory analysis provide an environment in which the user can search for various different types of (visual) relationship between the datasets, such as correlation, by combining layers of data within the same scene.

If we assume that certain processes might be better understood by exploring a larger number of data layers concurrently then matters concerning the composition of data layers to create a single scene must be addressed. To produce effective visualizations using composition requires the use of several different types of knowledge and meta-data including: (1) relevant psychophysical principles; (2) preferences of the observer; (3) the task to be undertaken; (4) the data in question, its statistical scale, its distribution, localization and dynamic range; (5) facilities provided by the visualization environment (available graphical marks and the degree of control provided); and (6) limitations imposed by the hardware, such as rendering speed, spatial and colour resolution, or available RAM. As an example, Table 10.3 describes some of the meta-data required in order to select appropriate graphical marks according to certain properties of the data.

Table 10.3 The definition of some example marks in terms of meta-data requirements for the visual assignment process.

3D mesh	Bound variables: *x*, *y*, *z* Free variables: red, green, blue, alpha Dimensionality: 3D Coverage: local, global
2D sphere	Bound variables: *x*, *y* Free variables: *z*, size, colour Dimensionality: 2D Coverage: point
3D arrows	Bound variables: *x*, *y*, *z* Free variables: *x*-offset, *y*-offset, *z*-offset, colour Dimensionality: 3D Coverage: point, local

Senay and Ignatius (1991, 1994) propose different mechanisms by which data may be combined within a scene whilst maintaining overall effectiveness of the complete visualization. Two commonly used mechanisms are axis composition and mark composition.

Axis composition combines datasets as a series of layers (a stacking paradigm) providing a fixed spatial framework within which positional relationships are preserved. When using axis composition, the locational properties of the marks must be bound to geographic position. This is basically a process of dataset registration. By enforcing such a constraint, we disallow other forms of assignment (such as might be achieved in a scatterplot) because the meaning of the axes has been dictated. This constraint need only be used when position is of prime importance, as is often the case in geographical analysis.

Mark composition recognizes that some of the symbols and surfaces used in visualization are capable of encoding many variables simultaneously (like an iconograph). Spare capacity within a mark set is used up by encoding further data to the unassigned visual attributes. Obviously, the data to be added must conform to the same location as the original data and be capable of representation using the same geometry. Plate 10.5 shows an example of mark composition using a set of arrows, whose properties are shown above in Table 10.3. The scene depicts various properties of certain ground truth samples: colour indicates dominant vegetation type and position gives the ground location. Orientation of the arrows is determined by the spectral response in three channels of Landsat imagery, each defining the offset of the arrow on three different directions, *x*, *y*, and *z*. So the arrow encodes three variables for orientation, two for position of the origin and one for its colour. The resulting scene can appear rather confusing at first and requires some time for the observer to become oriented to the visual encodings used. Once this is achieved some interesting insight into the spectral data is revealed, notably: (1) sample points from the same class are not always characterized in a similar way (arrows of the same colour have differing orientation); and (2) sample points from different classes may be characterized in a similar way (arrows of differing colour with similar orientation). We may safely conclude that in this case, regions with the same landcover class are not always distinguishable using spectral data alone. Plate 10.5 is also an example of axis composition, the lower layer represents a Landsat false colour image draped over an elevation model of the same region. A detailed discussion of iconographic visual displays is given by Pickett *et al.* (1995).

By themselves, these paradigms have some cognitive limitations, caused by the necessity to separate data into different layers, to avoid over-cluttering in any one layer. When additional layers of data are required then the user's focus of attention must 'shift' between layers in order to assess their inter-relationships (to 'see' pattern or structure). As with the scatterplots previously described, this attention shifting is undesirable as it leads to a weakening of the overall stimulus at any given point in space, since it is now divided amongst n layers.

10.2.1.3 Visualizing data validity and uncertainty

One use of visualization that seems to have captured the interest of the GIS community is in the presentation of uncertainty information. The extra visual attributes that a visualization environment provides can be used to add a further dimension to a map, in order to communicate measures such as uncertainty, error, validity or probability. For example, the reliability of a classification process (labelling uncertainty) can be used to change the transparency of a region or to form a surface where the vertical offset denotes some measure of confidence. Animation can also be used by dynamically changing colour mappings or moving boundaries in accordance with uncertainty values. Hunter and Goodchild (1993) and Wood (1994) describe the use of visualization for depicting interpolation accuracy and uncertainty in elevation data, and in the same vein, Fisher (1994) applies visualization to convey the uncertainty of a viewshed, and also to soils classification (Fisher, 1993). An ongoing research issue is to determine the most effective ways to communicate this type of additional data without detracting from the visual impact of the remainder of the map's content.

As an example of labelling uncertainty, Plate 10.6 depicts a false colour satellite image fragment of an agricultural area where vertical offset is used to represent the probability (as determined by a classifier) of a pixel being classified as 'wheat'. (Cropped areas appear as red and remnant vegetation and eroded areas as blue.)

10.2.2 Visual realism

Visualization can be used to create photo-realistic scenes, in order to plan or evaluate a certain course of action. In the geographic realm, photo-realism has been used for many diverse applications, including the evaluation of forestry planting and harvesting strategies (Thorn *et al.*, 1997), planning for developments such as road construction, and for visualizing streetscapes or buildings (Bishop, 1994). Obviously, there is a major advantage in being able to provide a realistic view of some proposed development ahead of time (an alternative reality), in order for a better evaluation of impact to be made.

The key to success here is in the authenticity that can be achieved, since the observer must be convinced that they are seeing a realistic scene in order for the simulation to work. This may involve a blending of observed data with a virtual model that exists only in the computer[9]. Visual realism in geographic visualization is discussed by Bishop (1994). Emphasis must usually be placed on lighting and surface absorption and reflection properties, using ray-tracing techniques to ensure that the modelled objects have a realistic appearance (Whitted, 1980). Such sophisticated lighting models are also useful for blending together existing photographs and imagery with proposed models, since the models can be 'lit' in a way that copies the lighting in the imagery. For realism in the natural world, it is possible to include texture mapping from satellite, photographic or video imagery.

The production of visually realistic scenes can be extremely demanding in terms of computing resources. Depending on the complexity of the scene, it may not be possible to achieve rendering at a speed that permits user inter-action in real time. In such circumstances, fly-overs and walk-throughs have to be computed off-line and then assembled into video sequences. Plate 10.7 depicts one frame from an animated walk-through of suburban Melbourne, Australia. The geometry of the simulated environment contains approximately 100 000 polygons and was derived primarily from aerial photographs. Trees and other surfaces are defined by 72 separate texture maps captured on-site. Modelling, animation and rendering all used the Alias/Wavefront software. (Source: Dr Ian Bishop, Centre for GIS and Modelling, University of Melbourne, Australia).

10.2.3 Animation and the use of interactors

Animation techniques provide a powerful and visually effective means of studying the relationships between higher objects and their defining data (Keller and Keller, 1993). Movement has been shown to have a high visual impact, and its detection in humans uses significantly different neural pathways to the perception of 'retinal' variables (see Section 10.1.5). Animation is therefore highly complementary to techniques based around shape, colour and position. Rex and Risch (1996) describe a query language for the animation of geographic data.

The link between animated cartography and visualization is described by Dorling (1992) and a comprehensive account of the different ways that temporal variation can encode geographic data is provided by MacEachren (1994). Another possibility, geared towards visual data exploration, is to use animation to establish connections or investigate relationships. Examples include the movement of surfaces, one through another, and the projection of interactors between surfaces. A number of different geographical interactors can be used to communicate a variety of different types of relationship between data layers. Types of interaction include the projection of objects, pixels, lines and points between data layers to describe processes such as interpolation, object extraction, classification, edge detection, and so forth. To be useful for exploratory analysis these techniques must facilitate perception of the structural and positional relationships between specific regions in the data. The example screen shot in Figure 10.8 shows the specification of an object (polygon) interactor describing a geological region. This particular interactor is implemented via a software module, effectively adding to the geometric base objects within the NAG Explorer environment and is animated to 'move' vertically through layers or surfaces to establish a visual connection, rather like brushing in a scatterplot. Some examples, in the form of movie clips, are available on the accompanying website.

Animated interactors have proven useful for studying cause and effect relationships between different data layers to address questions such as: 'What evidence is there to support a particular (hypothesized) structure in the data?' or, conversely, 'How does a particular (known) structure appear in the data?' Both of these questions refer to the direction of the interaction, being either from or to some primary data source. As a practical example, Plate 10.9 shows an interactor describing the extent of a salt scald (outbreak of surface salination) based on three data layers. The lowest layer shows the source data from Landsat TM as a false colour image fragment, the middle layer is a landcover theme produced from the Landsat image by classification, and the top layer contains some geographic objects known from ground truth. If the salt scald is known then the interactor shows how the scald is manifested in the thematic and image domain. If it is hypothesized, then the interactor allows the user to visually assess the appropriateness or plausibility of the salt scald as a defined landcover object. Some further examples are given by Gahegan (1998).

10.3 Major research issues in visualization

Visualization, being a relatively new discipline, has a number of outstanding issues that have as yet been only partially addressed. Discussion here is restricted to issues that relate to the use of visualization as it applies to

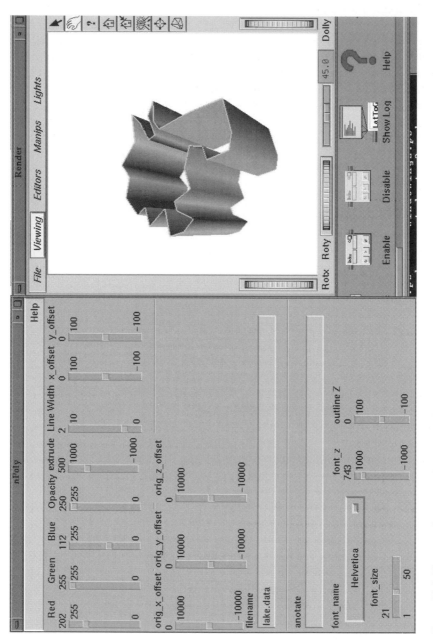

Figure 10.8 The specification and control of a polygon interactor within the NAG Explorer environment.

geography. The development of related hardware and software to support visualization activities is not covered since it is being independently pursued by different research communities.

Many research questions relate to the use of visualization for visual exploration and analysis and, specifically, how to make the best use of the available functionality to communicate a better understanding. This is a familiar problem to cartographers, but the large increase in visualization possibilities raises some difficult issues. It should be recognized that the development of a visualization (specifically the building of a scene) is often a highly under-constrained problem. This means that there are a vast number of assignment possibilities, with too few rules or constraints to narrow down the set of 'good' solutions to one or a few scenes. To illustrate this point, some of the major decisions to be taken when constructing a scene are as follows:

- Which of the layers or channels of data should be included, and which should be left out?
- For each included layer, which visual attribute(s)[10] should be assigned to it for visual encoding?
- For each assignment, how should the data be mapped to the visual attribute?

Each of these questions hides deeper research issues (Senay and Ignatius, 1994). For example, when deciding on which data to include or exclude, we must consider limitations on how much information can be assimilated by the user and rendered by the system. In choosing the visual encoding, we must also bear in mind that some visual attributes are more effective than others and some interfere with each other. Finally, in selecting a mapping function we must understand that human perception of visual attributes is generally non-linear, but may follow a specific activation function (e.g. Robertson and O'Callaghan, 1988). Similarly, the mappings used by particular display and printing devices are also not linear, so may need to be accounted for.

Highly relevant to the successful adoption of visualization tools for geographic data are the following three problems:

1. It is difficult to define the psychophysical principles that a 'good' visualization should follow. In very simple scenarios, existing perceptual and cartographic knowledge can be applied with relative ease. However, the richness of the geographic domain (including the spatial and temporal dimensions) and the large number of variables that might be required concurrently point to the need for this existing knowledge to be expanded.
2. There is the engineering problem of using existing perceptual and cartographic knowledge to good effect, i.e. a system to which this knowledge

can be applied effectively in the building of a scene. Apart from requiring that suitable visualization techniques are available and that control over visual attributes is possible with the desired flexibility, the visualization environment also requires a degree of interaction with a geographic information system or data store. At the very least, it is necessary to provide data import facilities for the visualization environment and some meta-data descriptions.

3. The task of evaluating the effectiveness of a scene is problematic. Ideally, some quantification of the utility of a visualization should be available, as feedback to the first and second problems above. Effectiveness evaluation is complicated because the results are ultimately evaluated by humans, whose judgements are subjective, and which will of course vary between individuals. However, this problem should not be overlooked since it is necessary to provide an objective means of discerning the 'good' from the 'bad'[11]. At a higher level, evaluation is also necessary to ensure progress in research. Without evaluation the 'scientific loop' of implementation, testing and the analysis of results is never closed (Leedy, 1993), which can lead to directionless or bad science.

Bertin (1981, 1985), Mackinlay (1986) and Rheingans and Landreth (1995) address the first of these problems, providing useful guidelines from the science of visual perception. Freidell et al. (1992) and Duclos and Grave (1993) show how such an approach may be automated using rules and grammars. Gahegan and O'Brien (1997) describe such a rule base, designed around the needs of geographic datasets. The problem of evaluation can be tackled by ranking each scene according to how assignments are made, using some of the many types of guidelines available (such as those shown in Table 10.3). This approach is taken by Jung (1996). The effect of combining many different perceptual devices in one scene is often not clear. The use of guidelines often assumes that each assignment can be evaluated in isolation, which may conflict with the ethos of data combination.

10.3.1 Knowledge-based visualization

Within the cartographic discipline, a good deal of effort has been expended towards automated mapping, including the capturing of map making expertise in rule-based or knowledge-based systems. This is a rich source of experience from which visualization scientists can draw. However, the target platform for output has shifted from the paper map to an interactive environment, capable of dynamic behaviour and with a correspondingly richer set of primitives for use in data encoding. There are now some examples of visualization production systems in the literature (Senay and Ignatius, 1991; Beshers and Feiner, 1993; Gahegan and O'Brien, 1997). These are aimed at automating the generation, or collaborating with the user, to produce a scene.

Such systems could potentially be of great benefit within the geosciences since they can encapsulate expertise in cartography, visualization and psychophysics that the user would otherwise be expected to supply. A comprehensive list of rules for visualization production that could be automated is given by Senay and Ignatius (1998).

10.4 Future directions

As the hardware required for visualization becomes more powerful, easier to use and cheaper to buy then the more sophisticated virtual reality tools, such as virtual workbenches and immersive environments, will become affordable options. However, research has yet to quantify the increase in effectiveness that such devices can offer. Certainly, commercial visualization systems will continue to develop, and the ease of inter-operation with existing GIS and remote sensing packages is set to improve with the continued development of better exchange and inter-operation standards.

Perhaps the most significant developments currently relate to the emergence of the Virtual Reality Modelling Language, VRML (ISO, 1997) and JAVA 3D. These offer a new generation of web-oriented visualization environments, with a rich object-based structure and good facilities for customization. The significance is that a common web browser (such as Netscape) can contain (via a plug-in) all that is required to render and interact with a scene, thus bringing visualization capabilities to almost any desktop. VRML, now in its second major revision, is based on the longstanding OpenInventor graphics definition. Whilst there are some quite severe performance limitations at present (current browsers work from an interpreted form of data description rather than a more efficient compiled form) these will surely be overcome. VRML may well become a *de facto* standard for scene graphs, since it has been designed to run across different platforms and is already supported (albeit slightly differently) by a variety of developers.

Further reading

The book *Visualization in Geographical Information Systems* (Hearnshaw and Unwin, 1994) contains a useful collection of position papers that introduce many of the important aspects of visualization from the perspective of the geosciences. Likewise the volume on *Visualization in Modern Cartography* (MacEachren and Taylor, 1994). For those interested in the perceptual aspects of visualization, the work of Mackinlay or Treisman is a good place to start. An excellent primer is the book *Perceptual Issues in Visualization* (Grinstein and Levkowitz, 1995). A more computational perspective can be gleaned from *Scientific Visualization: Advanced Software Techniques* (Palamidese, 1993). Two journal special editions also include a good account of very recent work in this field: *Computers & Geosciences Special Issue on Exploratory Cartographic Visualization* (MacEachren and Kraak, 1997) and the *International Journal of Geographic Information Science* special issue of papers from the 1998 meeting of the International Cartographic Association working group on Carto-graphic Visualization (MacEachren and Kraak, 1999).

   The images used in this chapter and some further examples of scenes, animated sequences, and VRML applications to download are available from an accompany-ing web site: *http://www.geovista.psu.edu/members/mark*. The author would like to acknowledge the following research students who contributed some of the scenes that are used in this paper: Julien Flack, Vilya Harvey, David O'Brien, Scott Pleiter and Nathan Shepherd.

Notes

1. This is not to say that computers should not be used for complex data analysis. On the contrary, there are many tasks within GC that can be better performed by an algorithm than by humans. For example, classification and the search for minima and maxima in high-dimensionality spaces (that form the basis of many of the artificial intelligence techniques discussed in this volume, including neural networks, decision trees and genetic algorithms) are usually best performed using a computational approach.
2. NAG Explorer is a visualization environment developed originally by Silicon Graphics and now enhanced and marketed by the Numerical Algorithms Group, Oxford, England.
3. AVS is a visualization environment developed and marketed by Advanced Visual Systems Inc. Waltham, MA, USA.
4. Whilst this leads to fascinating datasets, it also makes the performance of visu-alization systems on even highly specialized hardware quite frustrating at times.
5. These scene properties can be used as a basis for interaction. For example, a fly-through or walk-through can be simulated by incrementally changing the viewpoint.
6. It seems rather ironic that, as users of GIS, we expend much effort to free our data from the constraints of a 2D map in order to force it back into this form for all types of display and output.
7. The application was developed using the IRIS Inventor graphics library as part of the geographic visualization project at Curtin University. It can be made available to other researchers (along with some suitable data) through the author. The code was developed to run on Silicon Graphics hardware and requires the OpenInventor graphics libraries.
8. The Kioloa dataset is used by kind permission of Dr Brian Lees, Department of Geography, Australian National University.
9. Such techniques are used extensively in the film industry to enable dinosaurs, aliens and volcanoes to be seen on the streets of contemporary America.
10. It is possible, and sometimes desirable, to doubly encode data to two visual attributes for added emphasis. A common example is the use of both z-offset (height) and colour, so called redundant assignment.
11. For example, how do we establish that one paradigm works better than another for a given task? Ideally, we would measure effectiveness as experienced by human operators, requiring that we perform experiments, use control groups, questionnaires and so forth, in the tradition of psychological research. Whilst it

then might be simple to establish if a visualization is 'good' or 'bad', such terms are hard to quantify with respect to the visual attributes used, the data assigned to them and other qualities of the scene.

References

Aspinall, R. J. and Lees, B. G. (1994) 'Sampling and analysis of spatial environmental data', in Waugh, T. C. and Healey, R. G. (eds) *Proceedings, 6th International Symposium on Spatial Data Handling*, Edinburgh, Scotland. University of Edinburgh, Vol. 2, pp. 1086–1098.

Bertin, J. (1981) *Graphics and Graphic Information Processing*. Berlin: Walter de Gruyter.

Bertin, J. (1985) *Graphical Semiology*, Madison, Wisconsin: University of Wisconsin Press.

Beshers, C. and Feiner, S. (1993) AutoVisual: Rule-based design of interactive multivariate visualizations, *IEEE Computer Graphics and Applications*, 13, 4, pp. 41–49.

Bishop, I. (1994) 'Visual realism in communicating spatial change', in Hearnshaw, H. M. and Unwin, D. J. (eds) *Visualization in Geographical Information Systems*, Chichester: Wiley, pp. 60–64.

Buttenfield, B. P. and Mackaness, W. A. (1991) 'Visualization', in Maguire, D. J., Goodchild, M. F. and Rhind, D. W. (eds) *Geographical Information Systems*, Harlow: Longman, pp. 427–443.

Chernoff, H. (1973) 'The use of faces to represent points in k-dimensional space graphically', *Journal of the American Statistical Association*, 68, pp. 361–368.

Chernoff, H. (1978) 'Graphical representations as a discipline', in Wang, P. C. C. (ed.) *Graphical Representations of Multivariate Data*, New York: Academic Press.

Cleveland, W. S. and McGill, M. E. (1988) *Dynamic Graphics for Statistics*, Belmont, CA: Wadsworth & Brookes/Cole.

Cook, D., Buja, A., Cabrera, J. and Hurley, C. (1995) 'Grand tour and projection pursuit', *Computational and Graphical Statistics*, 4, 3, pp. 155–172.

Dorling, D. (1992) 'Stretching space and splicing time: from cartographic animation to interactive visualization', *Cartography and Geographic Information Systems*, 19, 4, pp. 215–227.

Dorling, D. (1994) 'Cartograms for human geography', in Hearnshaw, H. M. and Unwin, D. J. (eds) *Visualization in Geographical Information Systems*, Chichester: Wiley, pp. 85–102.

Duclos, A. M. and Grave, M. (1993) 'Reference models and formal specification for scientific visualization', in Palamidese, P. (ed.) *Scientific Visualization: Advanced Software Techniques*, pp. 3–14.

Fisher, P. F. (1993) 'Visualizing uncertainty in soil maps by animation', *Cartographica*, 30, 2, pp. 20–27.

Fisher, P. F. (1994) 'Probable and fuzzy models of the viewshed operation', in Worboys, M. (ed.) *Innovations in GIS*, London: Taylor & Francis, pp. 161–175.

Freidell, M., Kochhar, S., LaPolla, M. V. and Marks, J. (1992) 'Integrated software, process, algorithm and application visualization', *Journal of Visualization and Computer Animation*, 3, pp. 210–218.

Gahegan, M. N. (1998) 'Scatterplots and scenes: visualization techniques for exploratory spatial analysis', *Computers, Environment and Urban Systems*, 21, 1, pp. 43–56.

Gahegan, M. N. and O'Brien, D. L. (1997) 'A strategy and architecture for the visualization of complex geographical datasets', *International Journal of Pattern Recognition and Artificial Intelligence*, 11, 2, pp. 239–261.

Grinstein, G. and Levkowitz, H. (eds) (1995) *Perceptual Issues in Visualization*, Berlin: Springer Verlag.

Hartmann, J. L. (1992) *Visualization Techniques in GIS. Proc. EGIS '92*, Vol. 1, Utrecht, Netherlands: EGIS Foundation, pp. 406–412.

Haslett, J., Bradley, R., Craig, P., Unwin, A. and Wills, G. (1991) 'Dynamic graphics for exploring spatial data with application to locating global and local anomalies', *The American Statistician*, 45, 3, pp. 234–242.

Hearnshaw, H. M. and Unwin, D. J. (eds) (1994) *Visualization in Geographical Information Systems*. Chichester: Wiley.

Hunter, G. J. and Goodchild, M. F. (1993) 'Mapping uncertainty in spatial databases: putting theory into practice', *Journal of Urban and Regional Information Systems Association*, 5, 2, pp. 55–62.

ISO (1997), ISO/IEC DIS 14772, The Virtual Reality Modeling Language: (VRML97), URL: http://vag.vrml.org/VRML97/DIS/index.html.

Jung, V. (1996) 'A system for guiding and training users in the visualization of geographic data', in *Proceedings of the 1st International Conference on Geo-Computation*, University of Leeds, UK, 2, pp. 470–482.

Keller, P. R. and Keller, M. M. (1993) *Visual Cues: Practical Data Visualization*, Los Alamitos, CA: IEEE Press.

Leedy, P. (1993) *Practical Research Planning and Design*, 5th edn, New York: MacMillan.

Livingstone, M. and Hubel, D. (1988) 'Segregation of form, colour, movement and depth: anatomy, physiology and perception', *Science*, 240, pp. 740–749.

MacDougall, E. B. (1992) 'Exploratory analysis, dynamic statistical visualization and geographic information systems', *Cartography and Geographical Information Systems*, 19, 4, pp. 237–246.

MacEachren, A. M. (1994) 'Time as a cartographic variable', in Hearnshaw, H. M. and Unwin, D. J. (eds) *Visualization in Geographical Information Systems*, Chichester: Wiley, pp. 115–130.

MacEachren, A. M. and Kraak, M.-J. (eds) (1997) *Exploratory Cartographic Visualization*, Special issue of *Computers & Geosciences*, 23, 4.

MacEachren, A. M. and Kraak, M.-J. (eds) (1999) *Exploratory Cartographic Visualization*, Special issue of *International Journal of Geographic Information Science*, 13, 4.

MacEachren, A. M. and Taylor, D. R. F. (eds) (1994) *Visualization in Modern Cartography*, Oxford: Pergamon, pp. 91–122.

Mackinlay, J. D. (1986) 'Automating the Design of Graphical Presentations of Relational Information', *ACM Trans. Graphics*, 5, 2, pp. 110–141.

McCormick, B. H., Defanti, T. A. and Brown, M. D. (1987) 'Visualization in scientific computing', *SIGGRAPH Computer Graphics Newsletter*, 21, 6.

Monmonier, M. S. (1990) 'Strategies for the interactive exploration of geographic correlation', in Brassel, K. and Kishimoto, H. (eds) *Proceedings of the 4th International Symposium on Spatial Data Handling*, Dept. of Geography, University of Zurich, 1, pp. 381–389.

Noll, A. M. (1967) 'A computer technique for displaying *n*-dimensional hyperobjects', *Comm. ACM*, 10, 8, pp. 469–473.

O'Brien, D., Gahegan, M. N. and West, G. A. W. (1995) 'Information overload – the visualization of multiple spatial datasets for mineral exploration', in *Proc. GISRUK '95*, Department of Geography, University of Newcastle.

Palamidese, P. (ed.) (1993) *Scientific Visualization: Advanced Software Techniques*, Chichester: Ellis Horwood.

Pickett, R. M., Grinstein, G., Levkowitz, H. and Smith, S. (1995) 'Harnessing pre-attentive perceptual processes in visualization', in Grinstein, G. and Levkowitz, H. (eds) *Perceptual Issues in Visualization*, Berlin: Springer Verlag, pp. 59–69.

Rex, B. and Risch, J. (1996) 'Animation query language for the visualization of temporal data', in *Proceedings of the Third International Conference Workshop on Integrating GIS and Environmental Modelling*, Santa Fe, NM, January 21–26, 1996. Santa Barbara, CA: National Center for Geographic Information and Analysis. *http://www.ncgia.ucsb.edu/conf/SANTA_FE_CD_ROM/main.html*

Rheingans, P. and Landreth, C. (1995) 'Perceptual principles for effective visualiza-tions', in Grinstein, G. and Levkowitz, H. (eds) *Perceptual Issues in Visualization*, Berlin: Springer Verlag, pp. 59–69.

Robertson, P. K. (1990) 'A methodology for scientific data visualization: choosing representations based on a natural scene paradigm', in *Proc. IEEE. Visualization '90*, New York: ACM Press, pp. 114–123.

Robertson, P. K. and O'Callaghan, J. F. (1988) 'The application of perceptual colour spaces to the display of remotely sensed imagery', *IEEE Trans. Geoscience and Remote Sensing*, 26, 1, pp. 49–59.

Senay, H. and Ignatius, E. (1991) 'Compositional analysis and synthesis of scientific data visualization techniques', in Patrikalakis (ed.) *Scientific Visualization of Phys-ical Phenomena*, Hong Kong: Springer Verlag, pp. 269–281.

Senay, H. and Ignatius, E. (1994) A knowledge-Based System for Visualization Design, *IEEE Computer Graphics and Applications*, pp. 36–47.

Senay, H. and Ignatius, E. (1998) Rules and principles of scientific data visualization. URL: http://homer.cs.gsu.edu/classes/percept/visrules.htm.

Slocum, T. (1994) 'Visualization software tools', in MacEachren, A. M. and Taylor, D. R. F. (eds) *Visualization in Modern Cartography*, Oxford: Pergamon, pp. 91–122.

Tang, Q. (1992) 'A personal visualization system for visual analysis of area-based spatial data', Proc. *GIS/LIS' 92*, Vol. 2, Bethesda, MD: American Society for Photogrammetry and Remote Sensing, pp. 767–776.

Thorn, A. J., Daniel, T. C. and Orland, B. (1997) 'Data visualization for New Zea-land forestry', in Pascoe, R. (ed.) *Proc. Goecomputation'97*, Dunedin, New Zealand: University of Otago, pp. 227–239.

Treisman, A. (1986) 'Features and objects in early vision', *Scientific American*, Nov. 1986, 255, 5, pp. 114B–125.

Treisman, A. and Gormican, S. (1988) 'Feature analysis in early vision: evidence from search asymmetries', *Psychological Review*, 95, pp. 15–48.

Tufte, E. R. (1990) *Envisioning Information*, Cheshire, Conn.: Graphics Press.

Turk, A. (1994) 'Cogent GIS visualizations', in Hearnshaw, H. M. and Unwin, D. J. (eds) *Visualization in Geographical Information Systems*, Chichester: Wiley, pp. 26–33.

Whitted, T. (1980) 'An improved illumination model for shaded display', *Communications of ACM*, 23, 6, pp. 343–349.

Wood, J. (1994) 'Visualizing contour interpolation accuracy in digital elevation models', in Hearnshaw, H. M. and Unwin, D. J. (eds) *Visualization in Geographical Information Systems*, Chichester: Wiley, pp. 168–180.

Chapter 11

Spatial multimedia

Antonio Camara, Henk J. Scholten and
J. Miguel Remidio

11.1 Introduction

Consider the following quotation from a Dutch daily newspaper

'Using modern information technology for design and simulation of projected – public or private – plans (e.g., the positioning of a new landing stage) in combination with technology of multimedia and virtual reality will enhance the quality of decision making, because one works from the very beginning towards support and acceptance of the plans'.
(Didde, Volkskrant, 7 June 1997)

Placed under the headline of 'computer-democracy', this quotation points to the direction of the latest development in information technology. An extra dimension can now be added to a conventional system using modern multimedia and virtual reality technologies. As a result multimedia has become a 'buzzword' of the 1990s. Laurini and Thompson (1992) defined multimedia as 'a variety of analogue and digital forms of data that come together via common channels of communication', while Furht (1994) describes multimedia in the bigger picture as a 'merging of three industries: computing, communication and broadcasting'.

Multimedia software applications are computer-based tools concerned with the simultaneous display and processing of several different types of multimedia data, e.g. electronic games, hypermedia browsers, interactive videodiscs, multimedia authoring tools, and desktop conferencing systems. These applications can be standalone or distributed over a network. With the popularity of distributed systems, networks must also match the communication characteristics of multimedia distributed applications. Multimedia networks are specifically designed for digital multimedia traffic, especially audio and video. Distributed multimedia systems have many advantages and can enable 'telepresence' through the use of live cameras and real-time information updates. In this chapter we will examine in detail the use of multimedia in spatial information systems (SIS).

The five common data types which it is now possible to manipulate and display using multimedia tools are according to (Kemp, 1995):

1. text of infinitely variable size, font and structure;
2. still images, such as bitmaps and rasters, either generated, captured, or digitized;
3. still and animated computer-generated graphics;
4. audio, whether synthesized or captured, and replayed sound;
5. video or moving frames.

Text is traditionally the most widely used type of media and text-based information will always be integrated into multimedia applications. Text is the least space-intensive data type in terms of data storage. Most current network protocols (e.g. SMTP, NNTP and HTTP) are also text based (ASCII) and use different binary to ASCII encoding methods for data transfer.

Still images are the second most widely used data type in multimedia applications. Their storage requirements depend on image size, resolution and colour depth. Given that images are more storage-space-intensive than text, image compression techniques play an important role in multimedia developments. Audio is an increasingly popular data type that is being integrated into mainstream multimedia applications. There are a number of different audio file formats all of which are quite storage-space intensive. One second of digitized sound can consume several tens of kilobytes of storage. Compression reduces these storage requirements but they are still large. Video is the most storage-space intensive multimedia data type. The video objects are stored as sequences of frames and, depending on resolution and size, a single frame can consume 1 megabyte or more of storage-space. In order to have realistic video playback, a transfer rate of about 30 frames/second is required. Interleaved structures that incorporate timed sequencing of audio and video playback are now popular. Quicktime from Apple and AVI formats from Microsoft are examples of such structures. However the continued rapid increase in network bandwidth (e.g. a gigabit ethernet standard has been defined), the rapidly falling cost of disc storage, and their rapidly increasing capacity suggest that the hardware obstacles to spatial multimedia developments will be short lived.

Nevertheless, until recently the representation of spatial information using computer tools was based on the use of conventional maps. Although being very efficient tools for communicating certain types of spatial information, traditional maps have a number of limitations in their portrayal of spatial information, regardless of their storage media (i.e. paper or digital via a computer screen). Parsons (1992) points out that the communication of spatial information within SIS has used traditional cartographic techniques which have therefore limited the type of spatial information it is possible to represent. The same author mentions as limitations of traditional maps the two-dimensional representation of a three-dimensional reality, the difficulty in representing dynamic features (such as tidal limits), the use of fixed-scale representation and generalization, and the difficulty in representing temporal information. But the information which these traditional maps convey is nevertheless characterized as quantitative, in that the information expresses the relationships between spatial objects, in absolute, and often numerical, terms. Traditional maps are also characterized by high levels of abstraction resulting in the extensive use of symbols, which although often well designed, can be confusing to inexperienced map users. Maps can also be easily manipulated to tell whatever story the map maker wishes to communicate.

Spatial data are increasingly becoming available from sources such as remote sensing, airborne videography, video, aerial photographs, and statistical archives (Raper and McCarthy, 1994). Most of this data appears to be qualitative and in order to convert it into traditional maps, a certain degree of abstraction is required. In the process of abstraction, some of the information will be lost or can be biased. In order to realize the full value of this information, users must therefore be provided with the tools that allow them to consult it, analyse it and create customized reports to support their everyday activities. The use of multimedia data types therefore appears to offer an ideal solution to the various problems of representing qualitative spatial data. However for the maximum level of understanding or cognition to be achieved, the data must be structured in an efficient and appropriate way, which allows for a high degree of user interaction.

11.2 Historical perspective

The development of multimedia follows closely the evolution of computing. Back in 1945 the Director of the Office of Scientific Research and Development of the US Government foresaw a device 'in which one stores all his books, records, and communications, and which is mechanized so that it can be consulted with exceeding speed and flexibility'. These devices would also need to be associative so that related items could be easily located. More than 20 years later the concept and first applications of hypertext appeared, linking associated data for easy access. A hypertext editing system on an IBM 360 was developed at Brown University in the late 1960s by Ted Nelson. At the same time, in 1967, Nicholas Negroponte formed the Architecture Machine Group in the Architecture Department at MIT. The aim of this group was to make it easier for people to use computers. Sensorama, a device created by Morton Heilig in 1960, is also a classic reference on multimedia and is considered the first cyberspace machine. Similar to an arcade machine, it was outfitted with handlebars, a binocular-like viewing device, a vibrating seat, small vents which could blow air when commanded, stereophonic speakers and a device for generating odours specific to the events viewed stereoscopically on film (Pimentel and Teixeira, 1993).

During the 1970s much research was done and significant technological developments were achieved using computers. At the end of that decade, the Aspen Project took place at MIT, in which film shots taken from a moving vehicle travelling through the town of Aspen were stored on videodiscs, and then accessed interactively to simulate driving through the town. The MIT Media Lab, the Olivetti Research Group and the Apple Computer Multimedia Lab also carried out a number of innovative multimedia projects during the 1980s and contributed enormously to the development of this field. However, until a few years ago, the computer storage space and processing speed needed to manage and manipulate large image files was only available on expensive specialized systems in the best-funded departments.

The integration of multiple data types and SIS have as a classic pioneer example, the BBC Doomsday Project (Openshaw et al., 1986; Rhind et al., 1988). Using a videodisc one could consult a map of Great Britain and open windows containing video clips, aerial photographs, ground images and natural sounds from certain localities. More recent developments in the field of multimedia and SIS include the implementation of applications to urban planning, where sound and video manipulation are explored to improve the analysis of traffic problems, and for noise assessment. In these instances, speech recognition and voice annotation are used to facilitate public participation during the planning process (Shiffer, 1993). Moreover, making qualitative data available to the general public in a form that is easy to understand can significantly improve public participation and provide some level of transparency to the task in hand.

11.3 Current status

The exploration of multimedia capabilities within SIS involves two main topics: the data sources used and the integration of, and access to, different data via a common interface. Video and sound can be used within SIS in different contexts (Fonseca et al., 1993): for illustration purposes; as a source of information; to enrich the visualization of some SIS operations; to support the definition of the criteria to be used in spatial analysis; to support the implementation of models within SIS; and/or to help in the evaluation of some SIS results. Some of the explored capabilities related to the addition of video and sound to SIS include (Cassetari and Parsons, 1993; Fonseca et al., 1993; Shiffer, 1993):

1. the use of video to show backgrounds, point scenes, or transitions;
2. the superimposition of synthetic video on video images of natural sights;
3. the use of navigation images allowing the user to fly over the studied area;
4. the use of stereo sound to provide the notion of space;
5. the use of sound icons or music to create movement or illustrate point scenes;

6. the use of digital video as a source of information e.g. for use in simulation modelling.

One other important characteristic of a multimedia SIS concerns the possibilities that exist for creating multiple representations of the same phenomena. This multiple representation of a problem enables the user to view information in several different contexts, thus providing numerous potential opportunities for the generation of alternative solutions to a problem, which can be a very important item in the planning process (Shiffer, 1993).

Current spatial multimedia applications have been developed for use on two different types of platform: stand-alone computers, such as the PC or Mac (e.g. Fonseca *et al.*, 1995; Blat *et al.*, 1995, Raper and Livingstone, 1995); and computer networks, such as local area networks (LAN) or the World Wide Web (WWW) (e.g. Shiffer, 1995). This last medium offers advantages for projects that involve the participation of large audiences from different geographical locations. Numerous products make use of multimedia technology or maps and in most cases such items are distributed on CD-ROM. Common examples include digital atlases (e.g. *Encarta 96 World Atlas* which is produced by Microsoft Corporation, Redmond, WA), tools that provide a spatial dimension to tourist, historical and environmental information (e.g. *ExplOregon* which is produced by the University of Oregon, Eugene, Oregon, USA; or *Explorer French Camping Guide* which was produced by a European consortium for the ANWB, The Netherlands), and several other products that support education in a diverse range of subjects (e.g. *Interactive Geography*, which is produced by Pierian Spring Software, Portland, OR).

The extensive use of aerial photographs in SIS projects was until recent times quite rare. But recent advances in dedicated compression software and hardware, image tiling and multiple pixel size techniques and low-cost storage media such as CD-ROM with fast access disc drives have now made possible the efficient management of the massive amounts of data involved in digital imagery based SIS projects. For example, one such application is described by Romão *et al.* (1999), who developed a multimedia system for coastal management in the Netherlands based on aerial photograph mosaics.

To date SIS developers have been slow to take advantage of new multimedia technologies and appear confused as to whether to extend existing SIS or to add spatial functionality to multimedia. A very good review of this subject can be found in Raper (1997). Additionally, many GIS builders have been slow to appreciate the opportunities for even simple map animation capabilities. Quantitative geographers too have been slow to realize the potential offered by spatial multimedia for showing scale effects or for animating spatial analysis and modelling algorithms. A visual spatial and multimedia approach to geo-analysis is something now seriously worth considering.

Many of the historic hardware and software barriers to these developments have now evaporated.

11.3.1 Key hardware developments for spatial multimedia

The hardware developments reviewed here are divided into 'multimedia facilitating' and 'hardware–software integration'.

11.3.1.1 Multimedia facilitating developments

The uptake of multimedia technology has been facilitated above all else by a huge rise in the performance of low cost processors such as Intel's Pentium (and more recently the MMX) or Motorola/IBM's PowerPC, and the availability of cheap computer memory, both of which have been incorporated into microcomputers, workstations, games machines and (now) television receivers. Dedicated processors have also been developed which are capable of real-time data compression, decompression, sampling, filtering and buffering, and the availability of cheap fast memory has facilitated high speed caching. These hardware technologies have made it possible to develop robust digital audio and video handling systems which are error free and which can be incorporated in microcomputers as add-on boards. Recent experiments with digital TV signals and cameras preface a move towards fully integrated digital systems that extend from data capture through recording and broadcast to editing and compilation.

From an applications perspective, there are four main multimedia platforms that are used in multimedia SIS: the multimedia PC as defined by a manufacturers group led by Microsoft known as the Multimedia PC marketing council; the multimedia Apple Macintosh systems; the Intergraph systems; and the Silicon Graphics Indy Unix workstation. Spatial datasets are stored on large, fast, hard discs or compact discs (CD). Most current spatial data products on CD are based on the CD-ROM Yellow Book format (ISO 9660). Many users still have the original CD drives with a data transfer rate of 150 kbps (1×) and a relatively slow seek time compared to that of a modern hard disc. As yet few if any GIS datasets have been written in the White Book Video CD format allowing full motion video to be played back from MPEG encoded video on an accelerated CD drive.

There has been limited progress towards developing a library of spatial multimedia data based on information gathered using digital audio and video capture techniques. However, experiments have been conducted with the capture of digital video from aerial videography (Raper and McCarthy, 1994; Green and Morton, 1994), and on the use of touch screens or pen-based command systems to update maps stored in portable computers such as the MGIS system (Dixon *et al.*, 1993). What is still rare will soon be commonplace.

11.3.1.2 'Hardware–software integration' developments

There are also a variety of new developments which integrate dedicated hardware and software. The main systems with spatial applications will be described briefly below:

1. Compact Disc-Interactive (CD-I) from Phillips (launched in 1991) uses the Green/White Book formats to allow the use of a CD as a source of alphanumeric data, images, animation, high-quality audio and video. Authoring is carried out for example using MediaMogul running under OS9. Modern CD-I players are designed to plug into a television or VCR and be used with a remote controller. Typical spatial applications include museum guides and interactive golf!
2. Data DiscMan from Sony is an extension of the audio DiscMan portable player. The Data DiscMan uses 8 cm discs to display graphics on an LCD screen and to play sound, for reference work, and to provide business information such as city maps.
3. Kodak PhotoCD is a system for scanning colour prints and storing the images on a CD. To view images in the ImagePac format requires PhotoCD hardware attached to a television and uses a further extension to the CD-ROM standard known as CD-ROM XA (Extended Architecture). Although this system is intended to appeal to anyone regularly developing colour print film, there are already spatial applications in the archiving of earth surface images and maps.
4. Personal Digital Assistant (PDA) is a generic term for a hand-held computer with a small LCD screen and pen-based input using pointing or handwriting. A number of such PDAs, such as Apple's Newton, have now appeared and have been used for field data entry and the update of stored maps.
5. Known as 'active badges', ultra-small computers which can be worn by a user, are being developed to track the location of individuals within buildings or sites and to route phone calls to the nearest phone. This work has involved the development of new forms of spatial database. Real-time spatial tracking, analysis, and prediction is now a distinct possibility.
6. Games machines, such as Sega Megadrive, Nintendo Gameboy and 3DO, have developed simulated spatial scenarios which are based on dedicated hardware (sometimes incorporating a CD drive) with 'plug-in' game software cartridges. The same tools have obvious geographical applications.

11.3.2 Key software architectures for spatial multimedia

Although multimedia developments have progressed through a series of self-reinforcing cycles, comprising hardware and software innovation in which developments in one field fuel the other, the software developments appear to be pre-eminent in the multimedia GIS field at the moment. This arises from the need to serve a market that is only equipped with standard hardware systems. Development of software outside the dedicated hardware environments reviewed above has therefore been based on extended operating systems offering support for the 'multimedia' data types, specifically video and sound.

All the main operating systems for personal computers have been extended to offer support for specialized multimedia-supporting hardware such as video capture and for software-only handling of video and sound. Microsoft Windows has offered multimedia extensions since version 3.1, IBM's OS/2 first offered the Multimedia Presentation Manager/2 extensions with version 2.1, Apple Macintosh System 7 has supported the Quicktime extensions since 1991, and the Silicon Graphics Irix operating system has offered support for video since the release of version 4.0. In addition to multimedia support at the operating system level, the software aspects of data storage have also seen major developments, largely driven by 'information superhighway' requirements which called for high-speed multimedia servers capable of handling 'media objects' as well as the traditional database data types. Further developments of this kind can be expected as the draft version of the relational database query language SQL3 is finalized. This will permit relational databases to store and query 'abstract' data types especially multimedia GIS data (Raper and Bundock, 1993). Built on top of the extended operating systems, many new tools have been developed for visualizing multimedia spatial data (e.g. for use in Exploratory (Spatial) Data Analysis), for the animation of spatial processes, and for map-to-video 'hot-links' (which can be developed using tools such as Visual Basic).

Java adds a future critical dimension here as a universal software environment that may, if it survives and develops its full potential, be the software environment within which spatial multimedia comes of age. The hardware and OS free nature of Java-based multimedia has much to commend it and it could well provide the ideal mechanism for delivering spatial multimedia over the internet in the early decades of the new century. The re-usable and building block features have much to commend them and could well dramatically speed-up many future software developments in this area.

Multimedia authoring in the field of spatial applications has become dominated by proprietary personal computer based tools. These systems can be classified by the informational structures and metaphors they use for the development of their materials. The simplest informational structures are based on page-turning electronic books, scenario-building based on user interaction, or movie making. The sophistication of some of the latter systems has been extended using scripting languages. More complex informational structures allow for linked multimedia data types grouped as abstractions within semantic nets. Such options are provided with systems that use the hypertext model, which have proved popular in the development of spatial applications. Hypertext or hypermedia systems have also been used to develop hypermaps which are 'browsable hypermedia databases integrated with co-ordinate-based spatial referencing' (Raper and Livingstone, 1995). The hypermedia model is particularly useful for heterogeneous collections of data where many abstractions have unique properties and cannot be typed or named in advance, or where arbitrary aggregations are meaningful, such as in spatial applications.

Open hypermedia systems have begun to emerge in order to reduce the dependence on package-specific file structures and to provide operating system level support for the associative linking of multimedia resources. Microcosm is an example of one such system that has been developed for Windows 3.1, Macintosh and Unix. Microcosm was experimentally coupled to the SPANSMAP GIS (Simmons *et al.*, 1992) in order to link spatial and non-spatial data, of various different forms, with reference to a development site. Work has also begun to design intelligent agents capable of locating and retrieving spatial information, generating spatial model templates, monitoring task execution in a GIS, and managing collaboration between GIS and other systems (Rodrigues *et al.*, 1995).

11.3.3 Compression

Digital data compression has also been of crucial importance to the effective storage and high-speed delivery of multimedia data types. Compression methods can be divided into lossless methods (images compressed are recovered exactly) and lossy methods (images compressed are recovered approximately). Lossless methods are appropriate for the compression of data which must be preserved exactly, such as earth surface images and maps, and can generally achieve on average a 3:1 reduction in data volume for single frames. However, lossy techniques can be used to compress images-to-be-viewed (such as photographs), since human perception of the high-frequency variations in an image is poor. Most systems now use the Joint Photographic Experts Group (JPEG) standard for still image compression, which in lossy mode can achieve on average a 20:1 compression, whist still managing to retain excellent reconstruction characteristics. However, the JPEG format introduces noise in areas of high contrast (which is a common feature of maps), and does not allow point and click interaction or image mapping. Many World Wide Web (WWW) sites compress maps and images using the CompuServe Graphic Interchange Format (GIF) which is a compressed image format for 8-bit graphics.

For video, a further reduction can be obtained over and above that of the individual image frames compressed with JPEG using the Motion Picture Experts Group (MPEG) standard. The MPEG standard designates some frames of the video as reference frames and other frames are then encoded in terms of their differences from the nearest reference frames. Due to the limited implementation of MPEG in software and hardware at the time of writing most work carried out with spatially-referenced video has used proprietary video compression such as Apple's Quicktime format. However, this is likely to change very rapidly.

11.3.4 Networked multimedia

Internet delivery of digital imagery is becoming very common. Currently, raster formats are the most common way of publishing geospatial information across the internet. CompuServe's Graphic Interchange Format (GIF) and the Joint Photographic Experts Group Image Format (JFIF, or JPEG) are the two standard image formats that have web browser support. These formats are great for displaying raster data, such as digital orthophotos. However, they are not adequate for storing vector files, such as contour lines or ones created by sketching operations, and our ability to display vector maps and drawings at different display sizes without losing resolution is thus severely limited. Postscript (PS) and Data Exchange Format (DXF) formats are widely used standards for vector graphics but most web browsers do not support them due to their complexity. However, although web browsers still do not have direct support for any standard vector formats, viewing and manipulating vector spatial data on the internet is progressively becoming possible through the development of plug-in components that extend web browser capabilities. Softsource recently released their first plug-in for Netscape Navigator, called Vdraft internet CAD Tools which allows users to embed 2D and 3D CAD drawings as inline drawings in HTML documents. Corel Corp. and Numera Software Inc. are developing a Visual CADD plug-in which provides drawing information retrieval and layer management through the internet, for some common drawing formats, such as DXF and DWG.

Although multimedia systems were initially developed on standalone platforms, the development of the high-bandwidth global internet into a single heterogeneous wide area network (WAN) has focused much development on the delivery of multimedia data over large distances. Throughout much of the developed world, governments, universities and corporations are developing digital networks while telecommunication utilities and cable television companies are developing access to homes and small businesses. This is being achieved either by offering digital services based on integrated services digital network (ISDN) connections transmitted over copper wires, or by replacing analogue services over copper wires with optical fibre networks, capable of handling digital data. The huge increase in the transmission of multimedia data over the internet has generated an urgent need to update existing infrastructures and research is underway to revise the open systems interconnect (OSI) protocols and to develop multimedia conferencing using the International Telecommunication Union T.120 standards. Telecommunication utilities and cable television providers are also experimenting with 'video-on-demand' services transmitted over existing analogue connections such as home shopping, videoconferencing, interactive television and long distance collaborative game playing. High-bandwidth networks also facilitate computer-supported cooperative work (CSCW) which permits physically separated colleagues to work together exchanging data and speaking to each other, as and when necessary, perhaps using videoconferencing systems, such as CU-SeeMe.

At the time of writing probably the most important implemented multimedia network 'service' is the World Wide Web (WWW). Many spatial data resources have been 'published' on the WWW using JPEG/MPEG compression of images, graphics, video and sound. Shiffer (1995) reviews some of the resources available and shows how this kind of spatially referenced data can be linked to the collaborative planning system developed for Washington's National Capital Planning Commission enabling spatially-oriented browsing over the internet. Raper and Livingstone (1995) describe a spatial data explorer called SMPViewer designed to facilitate extraction of spatial information from geo-referenced image and map data resources obtained from the internet, for example using the Xerox PARC Map Viewer application. Important related technological developments are the virtual reality (VR) plug-ins for web browsers that have now started coming onto the market. Virtual reality solutions, which will not require graphics acceleration to work fast, will facilitate much better fly-over capabilities. Moreover, the recent VRML 2.0 specification has several new features that already make possible the incorporation of the missing geographic meaning in VRML files, specifically through the use of its interaction, scripting and prototyping capabilities in association with Java.

To manage these huge and distributed resources of data, many organizations are now starting to create and maintain 'digital libraries' which are designed to give network access to these collections of data. The first major attempt at establishing a digital library for spatial data is the Alexandria Project (Smith and Frew, 1995). The Alexandria Project is a consortium of researchers, developers, and educators, spanning the academic, public, and private sectors, who are exploring a variety of problems related to a distributed digital library for geographically-referenced information. Distributed means the library's components may be spread across the internet, as well as coexisting on a single desktop. Geographically referenced means that all the objects in the library will be associated with one or more regions ('footprints') on the surface of the earth. The key aim of the Alexandria Project is 'to provide geographically dispersed users with access to a geographically dispersed set of library collections. Users will be able to access, browse and retrieve specific items from the data collections of the library by means of user-friendly interfaces that integrate visually based and text-based query languages.' The Alexandria Project will initially include access to maps, orthophotos, AVHRR, SPOT and LANDSAT images as well as geo-demographic data. Other networked spatial multimedia services include the US Geological Survey (USGS), NASA's EOSDIS (a program that will allow users to graphically query global climate change using the WWW), and the UC Berkeley Digital Library Project.

With the proliferation of software tools in the multimedia field, the need has been recognized for data exchange standards capable of translating one proprietary system into another. There are several standards at an early stage of development. HyTime is a Draft International Standard (DIS) 10744 for the representation of information structures for multimedia resources through the specification of document type definitions and is based upon the Standardized Generalized Mark-up Language (SGML) (ISO 8879). HyTime uses the concepts of 'frontmatter', 'infoblurb' and 'pool' as funda-mental units of multimedia data representation, and these units, and the links between the units, are grouped into 'techdoc' abstractions. The Multi-media and Hypermedia information encoding Experts Group (MHEG) is a draft standard for representing hypermedia in a system independent form. It is optimized for run time use where applications are distributed across dif-ferent physical devices or on networks. The PResentation Environment for Multimedia Objects (PREMO) is a draft computer graphics standard which will define how different applications present and exchange multimedia data. It will define interfaces between applications so that different applications can each simultaneously input and output graphical data. But each of these standards is developing a different concept of space and time. The PREMO and MHEG working groups are discussing the use of a four-dimensional space-time which provides a comprehensive frame of reference for all events in a multimedia presentation. However, HyTime separates space and time permitting the definition of multiple time reference systems, for example permitting both 'video frame rate time' (30 frames per second in NTSC / 25 fps in PAL and SECAM) and 'user playback control time' describing how frames are actually played-back by a user who is responsible for pauses and fast-forward events. The resolution of these debates and the concepts of space and time which are implemented have important implications for the spatio-temporal analysis of geo-referenced multimedia data.

11.4 Illustrative examples

The main applications of spatial multimedia include education, professional and tourism products. In this section we will present some illustrative ex-amples of spatial multimedia applications to show what can be achieved.

11.4.1 Interactive Portugal –
http://helios.cnig.pt/~pi/pub/cg97/java/sketch/help.html

In the summer of 1995, Portugal was photographed using colour infrared film at a scale of 1:40 000 (approximately). The corresponding 4500 aerial photographs are now being scanned using a high-accuracy scanner (Zeiss/Intergraph PhotoScan PS1), rectified for tilt and topographic displacement, and organized into 1:25 000 scale regional mosaics by the Geographic Institute of the Portuguese Army (IGeoE). These digital orthophotos are available for public use through the National System for Geographic Information (SNIG, Sistema Nacional de Informação Geográfica) network (http://snig.cnig.pt). The SNIG network was launched in 1995, and represents the first national geographic information infrastructure in Europe that connects all georeferenced product information via the internet. The central node and coordinating body of the SNIG network is the National Center for Geographic Information (CNIG). This organization was also given the task of developing WWW applications for the management and exploration of geo-referenced information.

The main objective of Interactive Portugal is the development of a distributed digital spatial library based on mosaics of digital orthophotos, that will enable users both to retrieve geospatial information from the SNIG WWW server, and to develop local databases that can be connected to the main system. Interactive Portugal will therefore benefit from these two developments: the existence of an important and updated base imagery covering all of Portugal; and the opportunity to disseminate products that connect those images with existing distributed spatial databases over the SNIG network. Relevant information to be linked to the digital orthophotos, that can already be retrieved from the SNIG WWW server, includes administrative boundaries at a local level, toponomy placement, and elevation data. Being photographically based, Interactive Portugal will contain more interpretable and visible detail than any of the vector databases currently available in Portugal. Compared to traditional vector maps, instead of conveying details by lines and symbols, previously interpreted by someone during the mapping process, an orthophoto depicts a photographic image of real-world objects as they were at the time of the photography. Compared to aerial photographs, a digital orthophoto shows the exact location of objects, since distortions due to tilt and relief have been compensated for, thus allowing the precise measurements of areas, distances and bearings (Paine, 1981).

Interactive Portugal proposes the use of sketching and dynamic sketching in the exploration of digital orthophotos (Figure 11.1). Three different levels of interactive experiences are possible with these tools: exploratory, simulation and non-immersive virtual reality. Exploratory functions in Interactive Portugal are mostly associated with the point and click feature found in common browsing operations. In addition, the user can perform sketching

Figure 11.1 Interactive Portugal main screens.

operations to determine spatial measurements. A higher level of interactivity, typical of simulation, is brought to the exploration of digital orthophotos using the dynamic sketching tool. In the context of environmental applications this tool enables the creation and comparison of different scenarios and the implementation of 'what–if' analysis of spatial phenomena such as floods, forest fires or in air and water pollution accidents. Finally, the fly-over metaphor chosen for the navigation process aims at providing the highest level of interactivity, normally associated with non-immersive VR applications. In fact, users will be able to freely explore information connected to the digital orthophoto database in a non-structured fashion. Here, we intend to extend the sketching tool to enable the user to fly in a preferred direction. Thus, sketching and dynamic sketching are key tools that facilitate the use of Interactive Portugal both for exploration and analysis.

Several contributions can be expected from this proposed distributed system. Not only does it bring to the general public a valuable source of geospatial data that has never been so easily accessible, but these digital images and associated vector layers can also be used for various professional and educational activities. Particularly in the educational context, the

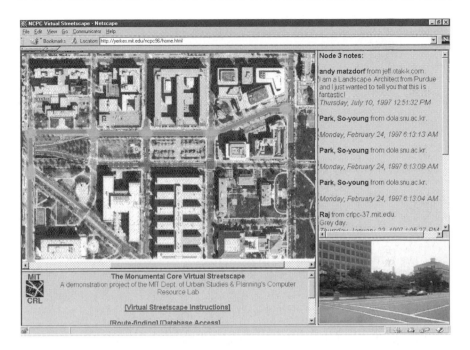

Figure 11.2 Monumental Core Virtual Streetscape.

proposed system can be the basis for earth sciences projects in geology, geography or the environment. Another important aspect of the project is the expected collaborative work performed by system users in the development of information layers. From an educational standpoint, it will enable the creation of a spatial multimedia database covering the whole country, made by high school students in a collaborative fashion. This feature is only possible using an internet-based system, with sketching tools such as those proposed in this project.

11.4.2 The monumental core virtual streetscape – http://yerkes.mit.edu/ncpc96/home.html

This is another very interesting generically applicable example of spatial multimedia. The main web page of the Monumental Core Virtual Streetscape presents a hypermap of downtown Washington with active nodes and paths (Figure 11.2). Clicking on a hypermap node shows a Quicktime VR of that node in the bottom right window frame and people's comments on that node in the top right window frame. A Quicktime VR presentation is a 360 degree, panoramic view of a place. One may look left, right, up, or down, by holding the mouse button down and dragging in the direction one wants to look. One may also move forwards by pressing the option key and backwards by pressing the control key. When a node is viewed, a corresponding list of notes appears in the window frame above the node. At the bottom of this list is a form for adding a new note to the list. Anyone who accesses the node will see all the notes that have been entered, so this is a place for people to record their observations and insights on the node and pose questions and topics for debate. Clicking on a hypermap path plays a video in the bottom right window frame which takes you on a drive along the indicated route.

11.5 Conclusions

Spatial multimedia is a rapidly developing and expanding topic of consider- able practical significance. It provides a way of seeing and sensing both real and artificial worlds. It can be used to look at visible spatial entities or to look at invisible ones that have never been seen before. The geo-cyberspace that the GIS revolution is creating for us contains many thousands of potential spatial multimedia applications. The nice feature is that most of them are as yet undefined. The worst aspect is that the copyright aspects are as yet still largely undefined. From a litigational perspective, spatial multimedia based on multiple data sources creates the prospect of multiple simultaneous copyright violation unless the thorny issues of the ownership and licensing of derived products can be resolved.

References

Blat, J., Delgado, A., Ruiz, M. and Segui, J. M. (1995) 'Designing multimedia GIS for territorial planning: the ParcBIT case', *Environment and Planning B*, 22, pp. 665–678.

Cassetari, S. and Parsons, E. (1993) 'Sound as a data type in a spatial information system', in Harp, J., Ottens, H. and Scholten, H. (eds) *Proceedings of the European GIS Conference*, Genoa, Italy, pp. 194–202.

Dixon, P., Smallwood, J. and Dixon, M. (1993) 'Development of a mobile GIS: field capture using a pen-based notepad computer system', in *Proceeding of the AGI 95 Conference*, Birmingham: AGI, 16, pp. 1–6.

Fonseca, A., Gouveia, C., Raper, J. F., Ferreira, F. and Câmara, A. (1993) 'Adding video and sound to GIS'. *Proceedings of the European GIS Conference*, Genoa, Italy, pp. 187–193.

Fonseca, A., Gouveia, C., Câmara, A. S. and Silva, J. P. (1995) 'Environmental impact assessment using multimedia spatial information systems', *Environment and Planning B*, 22, pp. 637–648.

Furht, B. (1994) 'Multimedia systems: an overview', *IEEE Multimedia*, 1, 1, pp. 12–24.

Green, D. and Morton, D. (1994) 'Acquiring environmentally remotely sensed data from model aircraft for input to GIS', *Proceedings of the AGI 94 Conference*, Birmingham: AGI, 15, 3, pp. 1–27.

Kemp, Z. (1995) 'Multimedia and spatial information systems', *IEEE Multimedia*, 2, 1, pp. 68–76.

Laurini, R. and Thompson, D. (1992) *Fundamentals of Spatial Information Systems.* London: Academic Press.

Openshaw, S., Wymer, C. and Charlton, M. (1986) 'A geographical information and mapping system for the BBC Domesday optical disks', *Transactions of the Institute of British Geographers*, 11, pp. 296–304.

Paine, D. P. (1981) *Aerial Photography and Image Interpretation for Resource Management*, New York: Wiley.

Parsons, E. (1992). 'The development of a multimedia hypermap', *Proc. of AGI 92*, 2, 24, pp. 1–3.

Pimentel, K. and Teixeira, K. (1993) *Virtual Reality*, New York: McGraw Hill.

Raper, J. (1997) 'Progress in spatial multimedia', in Craglia, M. and Couclelis, H. (eds) *Geographic Information Research*, London: Taylor and Francis, pp. 525–543.

Raper, J. and Bundock, M. (1993) 'Development of a generic spatial language interface for GIS', in Mather, P. M. (ed.) *Geographical Information Handling*, Chichester: Wiley, pp. 113–143.

Raper, J. and Livingstone, D. (1995) 'The development of a spatial data explorer for an environmental hyperdocument', *Environment and Planning B*, 22, pp. 679–687.

Raper, J. and McCarthy, T. (1994) 'Virtually GIS: the new media arrive', in *Proceedings of the AGI'94 Conference*, Birmingham: AGI, pp. 18.1.1–18.1.6.

Rhind, D. W., Armstrong, P. and Openshaw, S. (1988) 'The Doomsday machine: a nationwide GIS', *Geographical Journal*, 154, pp. 56–68.

Rodrigues, M. A., Raper, J. and Capitao, M. (1995) 'Implementing intelligent agents for spatial information', in *Proceedings of the Joint European Conference on GIS*, Vol. 1, pp. 169–174.

Romão, T., Câmara, A. S, Molendijk, M. and Scholten, H. (1999) 'CoastMAP: aerial photograph based mosaics in coastal zone management', in Camara, A. S. and Raper, J. (eds) *Spatial Multimedia and Virtual Reality*, London: Taylor & Francis, pp. 59–70.

Shiffer, M. (1993) 'Implementing multimedia collaborative planning techniques', *Proceedings of the Urban and Regional Information Systems Association Conference*, pp. 86–97.

Shiffer, M. (1995) 'Interactive multimedia planning support: moving from stand-alone systems to World Wide Web', *Environment and Planning B*, 22, pp. 649–664.

Simmons, D., Hall, W. and Clark, M (1992) 'Integrating GIS with multimedia for site management', *Proc. Mapping Awareness 92*, pp. 303–318.

Smith, T. and Frew, J. (1995) 'Alexandria digital library', *Communications of ACM*, 38, 4, pp. 61–62.

Chapter 12

Fractal analysis of digital spatial data

Paul Longley

12.1 Introduction

The purpose of this chapter is to describe and reflect upon the emergence of fractal measurement and simulation of geographical structures. There is now a very wide range of basic introductions to fractal geometry, and it is not the intention here to revisit in detail the basic concepts of scale dependency and self-similarity and their application to real world objects: for a good introduction to this field see, for example, Peitgen *et al.* (1992). Rather, in the spirit of a research-oriented text-book, we will focus upon the application of fractal concepts to the modelling and simulation of social systems, with specific reference to the achievements and limitations of a ten-year-long project in which the author has been centrally involved. This is the 'fractal cities' project (Batty and Longley, 1994) which has counterparts in the urban modelling and dynamics literatures of a number of countries (e.g. Frankhauser, 1993; Pumain *et al.*, 1989). More generally, however, we will reflect upon some of the ways in which digital geographical data may be assembled and used in quantitative urban analysis. As such, this chapter is about GeoComputation (GC) in that it consolidates a range of experiences in the development of digital statistical, analogue and computational models of urban systems. However, it differs in approach from some of the other contributions to this volume in that it is less about the 'power' of computing in the sense of intensive numerical processing, and more about the use of emergent digital technologies and digital datasets to depict and analyse the shape and form of urban settlements. There is also a specific substantive emphasis upon the form and functioning of urban systems. Here we will emphasize the development of data-rich representations of real-world systems using measures that have been developed within conventional urban geography, or are analogous to them. This work should also be seen as arguing for a return in urban geography to the measurement of physical and socio-economic distributions within cities, as a prerequisite to improved theorising about the form and functioning of urban systems.

12.2 Background

In broad terms the emergence of fractal models of urban structure may be viewed as an evolving response to the computer graphics and digital data handling revolutions, their impacts upon the ways in which we build and interact with urban models, and the ways in which we seek to confirm our conjectures about what we believe is going on in the real world. More specifically, the origins of the 'fractal cities' project may be traced to the innovation of large-scale graphical forecasts, and their use to complement inductive statistical models. Related work has used fractal geometry, through the medium of computer graphics, to reinvigorate statistical and analogue urban model-building in geography; here, in particular, we will review the use of analogue models to simulate the processes of urban growth and the evolution of urban morphologies. Thirdly, and most recently, such approaches are now being enriched by GIS-based data models in empirical analysis, and as such are leading to advances in the empirical modelling of real world urban systems.

Until the end of the 1980s, urban models were overwhelmingly theory rich, but information poor. These are characteristics that have prevailed elsewhere in quantitative analysis, perhaps most notably in econometrics.

A theme of this chapter is to illustrate how changes in the supply and processing of digital data are at last making this a dated perspective. The development of new, powerful and disaggregate datasets is making possible data-rich depictions of the morphologies of urban settlements and the human functions that such entities perform. Such datasets may now be customized to particular applications through a 'horses for courses' approach to data modelling, which is important since the way in which urban phenomena are measured quite tightly prescribes the likely outcome of analysis. However, 'data rich' inputs should be complemented by more than 'data led' thinking and, conscious of concerns that the latter are generating within geography (e.g. Johnston, 1999), it is necessary also to mesh data-rich morphological models with prevailing spatial and locational theory. We will describe how fractal measurement, simulation and modelling has been used to provide a complementary framework for the analysis of information-rich data models. Together, this approach is creating new prospects for quantitative comparison of real world urban morphologies across space and time. In sum, the data processing revolution has profound implications for the analysis of urban structure, and fractal geometry should be seen as providing one path for analysis in the quest to create data-rich models of the form and functioning of geographical reality.

This optimistic view stands in contrast to those of some detractors of GIS (e.g. Curry, 1995), in that GC is viewed as clearly pushing forward geographical analysis and understanding in a theoretically coherent way. As such, geographical model-building is seen as responding to technological changes, as well as changes in the way that statics and dynamics are conceptualized, in order to create avowedly subjective, yet detailed and plausible portraits of reality. In sum, harnessed to developments in data handling and computation, fractal geometry is enabling us to see a very great deal of geographical reality differently.

12.3 Statistical and analogue urban modelling using fractal geometry

Many of the current concerns with GC can be traced to long-established uses of quantitative methods in geography and, within this (and notwithstanding continuing work in the spatial statistical field, e.g. Getis and Ord, 1996), the shift away from statistical methods towards computational and numerical techniques. They are also closely associated with the development of GIS, a field which has, until quite recently, been viewed as being technology led (Martin, 1996) and of only limited importance for the development of spatial analytic methods (Openshaw, 1991). Also, GIS itself has developed in an inter-disciplinary context, drawing in particular upon developments in computer-assisted cartography and remote sensing which themselves now provide increasingly important sources of information for urban analysis (e.g. Barnsley and Donnay, 1999). In this respect, the development of GC too may be viewed as data led, in that interest has developed out of the quest to structure and understand the large, detailed and multiattribute datasets that are now becoming available.

It is salutary to note that the urban modelling tradition in geography has to date taken rather few of these developments on board, beyond incorporation of aspatial graphs and charts to improve the specification, estimation and testing of models (e.g. Wrigley *et al.*, 1988; Longley and Dunn, 1988). Whilst improved algorithms, in harness with the precipitous fall in the cost of computer processing power, have undoubtedly played a marginal role in maintaining interest in analytical urban models amongst the *cognoscenti*, such approaches nevertheless now account for a much reduced share of intellectual activity in geography compared to, say, 20 years ago. The shift towards exploratory data analysis has also done something to arrest waning interest in urban models, but an important structural weakness has been that a very great deal of quantitative urban analysis has remained resolutely aspatial in nature. For example, housing market modelling (often based upon hedonic approaches) has developed using imprecise specifications of physical and social spaces, coarse and often inappropriate areal geometries, and crude distance metrics. Moreover, city structures have been resolutely viewed as monocentric, consumption patterns and lifestyles taken as spatially uniform and spatial externality effects ignored (but see Orford, 1999; Can, 1992). These kinds of manifold and evident shortcomings have likely contributed not a little towards disillusionment with approaches based upon analytical model-building in geography (as described in Bertuglia *et al.*, 1994). In short, urban modelling during the 1980s and early 1990s has been characterized by a lingering asymmetry, in that inductive generalization of parameter estimates and suchlike has not been complemented by large-scale deductive spatial forecasting.

There is a paradox in this, in that interest in urban problems, such as the economic efficiency of cities and the sustainability of their urban forms, remains a vigorous, exciting research area and indeed is one which is undergoing a renaissance of interest. Many of the central themes of the rash of current 'cities' initiatives by research councils (in the UK, at least) can be directly addressed only through generalization across space and time, and the urban modelling tradition would at first sight appear well-placed to develop enhanced, data-rich models and forecasts of urban form and functioning. Over a longer time period, it is to these goals that the origins of the 'fractal cities' project can be traced. In what the authors (at least!) consider to be an important paper written ten years ago (Batty and Longley, 1986), the results of a conventional discrete choice model were used for the first time to inform a large-scale spatial fractal simulation of urban structure. Figure 12.1 illustrates how inductive generalization about the structure and form of the city might be complemented by deductive spatial partitioning of urban space using appropriate fractal generating rules.

The detailed motivation for this was the recognition that competing variable specifications in discrete choice models (Wrigley, 1985; Wrigley and Longley, 1984) were frequently found to result in near-identical goodness-of-fit statistics, and that forecasting of what such competing specifications mean 'on the ground' provided a more important means of discriminating between them than rudimentary mapping of residuals and potential leverage values alone. More generally, the paper noted the dearth of urban modelling applications that used inductive models to fuel deductive spatial generalizations, and the simulation of the occurrence of different physical housing types across Greater London was the first time (to our knowledge) that this 'inductive-deductive loop' had been joined. The spirit of this work chimed with the views of the arch-popularist of fractals, Benoit Mandelbrot's (1982) that '. . . the basic proof of a stochastic model of nature is in the seeing: numerical comparisons must come second', and aired such a view in the geo-graphical sciences. The outcome for the work was an illustration of what is now developing into the 'virtual city' representations of urban structures that were purportedly implicit in many urban models, but which had never been articulated through any explicit land use geometry.

That said, the graphics used by Batty and Longley (1986) to articulate the simulations were crude (reflecting PC graphics technology of the time) and the assumptions invoked heroic; in particular, there was no justification for our use of fractal geometry beyond allusions to the self-similar, hierarchical nature of central place theory. Even if physical land use patterns could

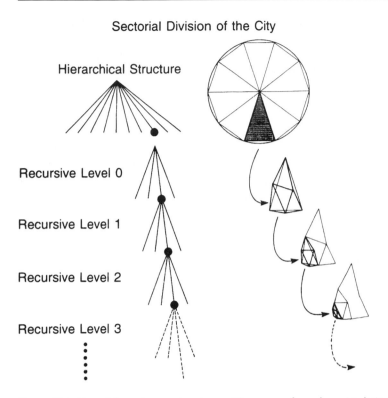

Figure 12.1 Use of fractal geometry to partition space for urban simulation (Batty and Longley, 1986).

plausibly be assumed to be fractal, there was no justification for the degree of fractal recursion used, simply because no relevant graphical statistics of the morphology of urban land-use existed, beyond the crude shape measurements carried out in the 1960s (see Haggett *et al.*, 1977, pp. 309–312). The land use geometry was also independent of the rest of the modelling process. As such the simulations were an adjunct to the modelling process rather than an integral part of them.

It is a truism that measurement tightly prescribes the course of empirical analysis, and it was to such measurement that the 'fractal cities' project next turned. A short series of papers (Batty and Longley, 1987a, b, 1988) developed and applied a range of perimeter-scale and area-perimeter relations, in order to demonstrate whether fractal measurement indeed provides a robust and defensible means of quantifying the irregularities of land use. The rudiments of a fractal measurement are illustrated in Figure 12.2 using the so-called 'structured walk' method of deriving scaled measures of length. This method mimics the classic coastline conundrum developed by Mandelbrot (1967), and has been set out by Batty and Longley (1994) amongst others.

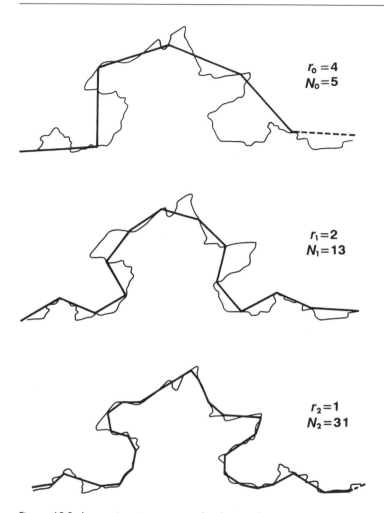

Figure 12.2 Approximating an irregular line and measuring perimeter length at three adjacent scales (Batty and Longley, 1994). See text for explanation of terms.

There is a range of measures of the ways in which fractal phenomena fill space (Batty and Longley, 1994), but the following measure sets out to do so by relating the number of parts into which a line can be divided, and its length, to some measure of its scale.

Scaling relations may be derived with respect to an irregular line of unspecified length R between two fixed points, such as that illustrated in Figure 12.2. We begin by defining a scale of resolution r_0, such that when this line is approximated by a sequence of contiguous segments or chords each of length r_0, this yields N_0 such chords. Next we determine a new scale of resolution r_1 which is one-half r_0, i.e. $r_1 = r_0/2$. Applying this scale r_1 to the line yields N_1 chords. If the line is fractal, then it is clear that 'halving the interval always gives more than twice the number of steps, since more and more of the self-similar detail is picked up' (Mark, 1984). Formally this means that

$$\frac{N_1}{N_0} > 2, \quad \text{and} \quad \frac{r_0}{r_1} = 2 \tag{12.1}$$

Figure 12.2 illustrates this for three different scales. The lengths of the approximated curves or perimeters, in each case, are given as $L_1 = N_1 r_1$ and $L_0 = N_0 r_0$ and from the assumptions implied in equation (12.1), it is easy to show that $L_1 > L_0$. This provides the formal justification that the length of the line increases without bound, as the chord size (or scale) r converges towards zero.

The relationship in (12.1) can be formally equated if it is assumed that the ratio of the number of chord sizes at any two scales is always in constant relation to the ratio of the lengths of the chords. Then

$$\frac{N_1}{N_0} = \left(\frac{r_0}{r_1}\right)^D \tag{12.2}$$

where D is defined as the fractal dimension. If halving the scale gives exactly twice the number of chords, then equation (12.2) implies that $D = 1$, and that the line would be straight. If halving the scale gives four times the number of chords, the line would enclose the space and the fractal dimension would be 2. Equation (12.2) can be rearranged as

$$N_1 = (N_0 r_0^D) r_1^{-D} = \alpha \, r_1^{-D} \tag{12.3}$$

where the term in brackets $(N_0 r_0^D)$ acts as the base constant α in predicting the number of chords N_1 from any interval of size r_1 relative to this base.

From equations (12.2) and (12.3), a number of methods for determining D emerge. Equation (12.2) suggests that D can be calculated if only two scales are available (Goodchild, 1980). Rearranging equation (12.2) gives

$$D = \log \frac{N_1}{N_0} \Big/ \log \frac{r_0}{r_1} \tag{12.4}$$

However, most analyses not only involve a determination of the value of D but also of whether or not the phenomenon in question actually is fractal (the assumption made in equation (12.2) above), and more than two scales are required to ascertain whether this is or is not the case. Generalizing equation (12.3) gives

$$N(r) = \alpha \, r^{-D} \tag{12.5}$$

where $N(r)$ is the number of chords associated with any r. Using logarithms, we can linearize equation (12.5) as

$$\log N(r) = \log \alpha - D \log r \qquad (12.6)$$

Equation (12.6) can be used as a basis for regression by using estimates of N and r across a range of scales. The related formula involving the length of the curve or perimeter L is derived from equation (12.5) as

$$L = N r = \alpha\, r^{(1-D)} \qquad (12.7)$$

Equation (12.7) can in turn be linearized by taking logarithms,

$$\log L = \log \alpha + \beta \log r \qquad (12.8)$$

where $\beta = (1-D)$. It is clear that the intercepts α in equations (12.6) and (12.8) are identical and the slopes are related to the fractal dimension D in the manner shown.

This, in essence, was the method originally used by Richardson (1961) to measure the length of coastlines and frontiers by manually walking a pair of dividers along the boundaries at different scales and then determining D from equation (12.8) by applying ordinary least squares regression to the logged measurements. Our own experiments were automated generalizations, which investigated the appropriate range of scales over which measurements of urban edges might be made, as well as the sensitivity of recorded measurements to the point on a digitized base curve at which measurement was initiated. Two such generalizations are illustrated in Figure 12.3.

Our measurements of urban edges, and of individual land parcels within urban areas suggested that in general, scale effects vary with scale itself, and that this was likely to be the result of multiple processes being of different relative importance across the range of scales. We also made some tentative steps towards re-establishing links between the spatial form of urban systems and the processes that governed their evolution, based upon the observation that as cities grow, they come to fill their space more efficiently and homogeneously. For example, it has been suggested (Batty and Longley, 1994) that changes in the fractal dimensions of urban edges over time reflect improved coordination of development and increased control over physical form (perhaps reflecting improvements in technology; see also Frankhauser, 1993). Different, though consistent and equivalent, measures of fractal dimension were also developed and used in subsequent work.

The development and innovation of GIS during the 1980s was not accompanied by the development of significant digital data infrastructure or, indeed, even by wide availability of terribly robust techniques of data capture. The process of linking the form of the city to its functioning required analysis of

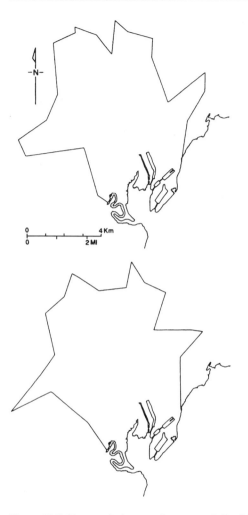

Figure 12.3 Two scaled generalizations of Cardiff's urban edge (Batty and Longley, 1994).

the growth dynamics of city structures and, even today, there is very little information on the system histories of urban settlements. Data input was painfully slow and so, working with Stewart Fotheringham, the emphasis of the 'fractal cities' project shifted towards a different strand to the locational analysis tradition of human geography, namely the development of analogue urban models. In the past, such models had typically emphasized analogies between urban forms and living organisms. In the case of our own work, analogies were forged with the growth and evolution of far-from-equilibrium structures, such as those based upon diffusion-limited aggregation

Figure 12.4 The 'classic' diffusion-limited aggregation structure.

(DLA). A 'classic' DLA structure is illustrated in Figure 12.4 and whilst this particular form is not obviously 'city-like' (though some of those in Fotheringham *et al.*, 1989 are), there are a number of ways in which such structures encourage insights into urban structure and form. The key to the usefulness of such analogies is governed by the principle of model parsimony, i.e. the degree to which the model usefully represents aspects of the system of interest without the confounding impact of extraneous factors. With regard to fractal models of urban structure, both the DLA (Batty *et al.*, 1989) and dielectric breakdown models (Batty, 1991) have been used to model urban growth patterns, and have generated insights by virtue of the system histories that have been created through the medium of the simulations. For example, Batty *et al.* (1989) use analogue models based upon DLA in order to suggest that conventional wisdom about the decline in urban density gradients over time may, in fact, be an artefact of the measurement method. Such work has also advanced the use of fractal geometry as a means of identifying space-filling 'norms' with particular fractal dimensions.

The common thread through much of this research has been to instate the geographical as the most important medium through which model forecasts may be visualized. Many of the most important and enduring theories in geography are also the most simple, but (most) academics and students alike become disaffected with them as they try to move from idealized lattices and other networks towards messy inelegant empirical structures which are not dominated by 'pure' geometry. Fractal geometry provides a robust means of simulating the jaggedly irregular in a visually acceptable manner, and it is GC through the graphical medium that has provided the most important stimulus to this work. Also important has been the interactive computer environment that fosters experimentation with a range of graphical scenarios for urban analysis. Yet it is with improvements in the availability of digital data coupled with improved data handling techniques that exploratory graphical analysis is becoming relevant to the world of empirical application, and it is to this that we now turn.

12.4 Towards data-rich models of urban morphologies

The previous section has alluded to the view that, despite much of hype and boosterist claims surrounding the development of GIS during the 1980s (see Aangeenbrug, 1991), and the view of GIS as an important technology and data-led innovation (Martin, 1996), its adaptation to urban modelling practice was not as straightforward as might be supposed, largely because there were no general-purpose digital datasets suitable for urban modelling. This situation is now changing, however. First, developments that are taking place within remote sensing are leading to the creation of high-resolution, spatially comprehensive and frequently updateable images of urban morphologies. In particular, continuing improvements in the spatial resolution of optical satellite sensors mean that it is becoming possible to derive accurate information both on the location and extent of 'urban' areas and on their internal morphology. This has been accompanied by well-known parallel developments in computer technology, notably increases in processing capability, reductions in the processing costs, increases in the total capacity of mass data storage devices and new methods of processing information (e.g. see Chapter 8 for a review which emphasizes developments in neuro-computing). These have brought sophisticated digital image processing and spatial data analysis to the desktop, and offer the potential to develop new, more advanced ways of analysing remotely-sensed images of urban areas in near-real-time.

In order to identify land use, rather than land cover, however, ancillary digital data sources are required. Fortunately the development of digital socio-economic data and digital mapping products (such as Ordnance Survey (GB)'s ADDRESS-POINT and LAND-LINE, and the UK Department of the Environment's Land Use Change Statistics) has now reached the point where these can usefully be used to estimate and improve existing data and to inform comparative urban analysis.

Taken together there exists the possibility that such integrated datasets may be created for monitoring the function of urban areas as well as their outline physical forms. Such hybrid datasets will doubtless continue to be vulnerable to the sources and operations of the errors and uncertainties that result from combining different datasets with different data structures and standards (Fisher, 1999; Martin *et al.*, 1994). Moreover, the land use classes within such tailor made 'RS-GIS' (Mesev *et al.*, 1995) datasets will likely be predicated upon human activity patterns, and these will remain inherently subjective. Nevertheless, there is much that can be achieved here in what Martin (1996) terms the 'data transformation' stage of GIS operations, and which can lay a reasonably firm foundation for analytical models of urban structure.

As an extension of the 'fractal cities' project, we have begun to develop hybrid 'RS-GIS' data models in order to investigate the ways in which space is filled by different land use categories, and to begin empirically to diagnose scaling relations between different aspects of urban morphology (Mesev *et al.*, 1995). Specifically we have begun to investigate the ways in which census small area statistics (SAS) may be used as ancillary information in order to modify the prior probabilities in the classification of Landsat satellite images. Very few satellite image classifiers are able to statistically accommodate external non-spectral information, but Mesev *et al.* (1995) show how a population surface model (Bracken and Martin, 1989) of SAS may be used to vary the *a priori* probabilities of class membership within a standard (ERDAS) maximum likelihood classifier. This is achieved by developing a Bayesian decision rule which is also used empirically to improve the detailed classification of urban dwelling types. The population surface model is also used as an input to the selection of training samples for the supervised image classification, and in post-classification sorting. The objective of this classification exercise is to generate a fine-scale model of the distribution of domestic properties of all types, and to use the census surface model as a simple, spatially approximate device for 'weeding out' pixels potentially misclassified as 'residential' from other 'non-residential' categories.

Experience to date has suggested that hybrid 'RS-GIS' data models provide a suitable means of identifying the spatial distribution of housing attributes, consistent with satellite representations of the physical layout of urban structures. Variants of this approach might utilize other datasets (e.g. postcodes, mail delivery points and road centrelines) in similar ways to further different classification objectives. Figure 12.5 illustrates four of the different data models (based upon postal, census small area statistics, population surface model and Landsat TM geographies) which can be used to provide the requisite information for deriving fractal measures of space-filling, and Figure 12.6 illustrates the sorts of profiles of cumulative and incremental density that may be developed consistent with established urban density gradient as well as fractal theory.

Empirical analysis (e.g. Mesev *et al.*, 1999) suggests that fractal dimensions generated from hybrid 'RS-GIS' are reliable and very much consistent

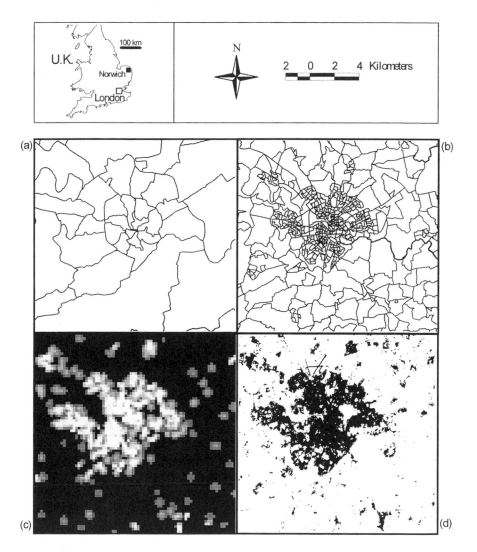

Figure 12.5 Four possible components of a hybrid 'RS-GIS' data model of Norwich, UK: (a) postal geography; (b) census enumeration districts; (c) population surface model; and (d) Landsat TM image (Longley and Mesev, 1999).

with fractal theory of urban form. Moreover, the profiles of the densities of different urban land uses are being used to validate the sorts of hypotheses that could only be investigated using analogue models in earlier stages of the 'fractal cities' project. For example, one of the principal findings advanced on the basis of analogue DLA simulations by Fotheringham *et al.* (1989) was how density profiles of growing city structures exhibited different

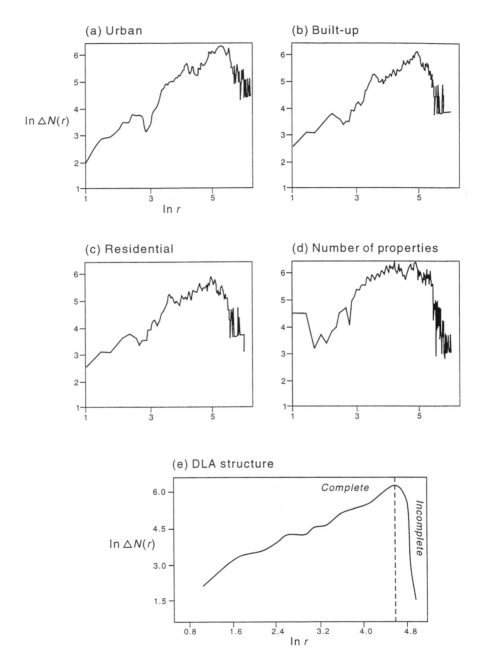

Figure 12.6 Four density profiles (a)–(d) of urban land-uses in Norwich, UK, and density profile of DLA structure (e) for comparison. N(r) denotes the number of occupied ('urban') points in each radial distance band r from the historic centre of the city (Longley and Mesev, 1997).

space-filling norms between that part of the urban development that could be considered to be 'complete', versus that portion in which urban growth is still actively taking place. Figure 12.6(e) shows the density profile of the diffusion-limited aggregation structure of Figure 12.4, for comparison with the 'real-world' distributions of Norwich. Recent research is suggesting that this and other findings derived from analogue models are more or less directly applicable to real-world urban systems.

12.5 Discussion and evaluation

Openshaw (1991, 1996) has observed that the spatial analysis agenda attained only a low priority in the major GIS initiatives of the late 1980s, and the view advanced here has been that the urban modelling community in particular has also failed to fully grasp the opportunities and prospects that GIS-based analysis and computation offers. Indeed, a qualitative impression is that the 'fractal cities' project has generated rather more interest in biological and computer science than in geography. Within geography, by contrast, much of urban morphological research (as reviewed by Whitehand, 1992) has remained largely confined to historiographies of particular urban areas, with quantitative perspectives being regarded with some suspicion. Together, this has created a partial vacuum in the quest for morphological generalization, which the rash of recent research council 'cities' initiatives seems likely, at best, only partially to fill. Arguably, many of the broader questions about urban form and function are not being addressed, and there is too little emphasis upon the explicitly spatial consequences of urban policies – ranging from the effectiveness of green belts (Longley et al., 1992) to prescriptions about residential zoning and local planning.

There is a need for theory-led research to build upon the vastly improved data models that can now be created within GIS. A physicalist goal of such research should be success in linking the forms of urban settlements to their functions across all significant scale ranges of urban development. This might in turn form the basis to comparison of the detailed morphologies of a range of settlements at different levels of settlement hierarchies, yet would move beyond conventional, physicalist, conceptions of urban morphology. Such data models can be made much more representative of the daily activity patterns of urban residents and can develop more meaningful conceptions and analyses of density of activity and land use. The empirical data models of Mesev et al. (1999) represent some first steps towards such data-rich models of geographical reality, and are, in turn, amenable to further spatial analysis either within or adjacent to GIS. The exciting prospect, then, is of our increasing ability to devise much improved empirical 'models of models' which will permit generalization consistent with theory across space and time, and will reinvigorate urban modelling as an empirically-based intellectual activity.

In the same spirit as earlier research in urban modelling, contemporary GIS research is concerned with seeking generalizations from quantitative analysis. The broader question that this raises is whether these developments represent substantial improvements in our abilities to represent geographical reality in any meaningful senses. There is a danger that GC becomes overwhelmingly identified with a 'data trawling' approach (Openshaw, 1989, 1996), a kind of unbridled empiricism in which computation overwhelms human intervention. It is boosterist claims for such approaches that promote some detractors of GIS to deride the notion that having 'massive amounts of information... provides one with a better understanding of the world' (Curry, 1995, p. 78), and can even blinker geographers to the more measured view that digital computation opens up a wide range of avenues within geography, beyond as well as within empiricism as traditionally conceived (Johnston, 1999). The challenge is, then, to make clear how even the most data-rich models are guided by human intervention and interpretation, since any 'model' of reality necessarily entails simplifications which must be bridged through human interpretation rather than mechanistic or statistical assumption. The implications of the digital world of the late 1990s for urban modelling as traditionally conceived are at least as profound, for regional science and much of spatial econometrics has thus far proved extraordinarily resilient towards modelling any physical land uses and patterns, never mind more abstract human activities. Such considerations simply cannot be discarded or assumed away, now that precisely geo-referenced data are so readily available. Response to this challenge offers the incentive of moving urban modelling back towards the mainstream of geography.

It is in this spirit that the 'fractal cities' project has begun to model cities as systems of organized complexity, whose geometry betrays regularities of scale and form of which we have hitherto been largely unaware. Fractal geometry allows us to use mathematics more realistically to portray this underlying order and regularity, and this in turn is beginning to suggest a deeper sense of how the morphology of cities might be understood in terms of form and process, scale and shape, statics and dynamics. The fractal simulation of urban structure represents a break with the tradition that sees cities as simple, ordered structures, in which at worst space requires no explicit representation at all, and in which at best over-all physical forms are represented only by schematized smooth lines and shapes. The former shares characteristics with much of everything from contemporary social theory to regional science, while the latter is consistent with the ways in which planners have conceived the physical and social patterning of urban areas. Neither entails the kind of abstraction that reflects and ultimately may explain the structured irregularity of the real world; it may be simple and effective in its reductionism, but in explicitly spatial terms it is simply wrong. There is a pressing need to demonstrate the ways in which the organization of the city is reflected in its form or morphology, and prevailing data-led technologies provide a means of achieving this.

Hitherto, there have been few if any models which take the level of analysis down to the physical form of the city or to the relationship between urban activities, land uses and their physical forms. In visual terms, fractal models and their picturescapes provide a way of making our theories more real and of communicating more meaning through our analyses; in a 'black box' sense the fractal model can also provide realistic rendering for high theory and inductive generalization alike. Use of fractal geometry does not presume intricate theory (Saupe, 1991), although its intrinsic characteristics of hierarchy and recursion in measurement and simulation of urban morphologies are demonstrably consistent with the central tenets of locational theory (Arlinghaus, 1985; Wong and Fotheringham, 1990). Some of the 'fractal cities' simulation work, however, inclines more towards current fashions in biology, in that cities are viewed as complex organisms, evolving and changing at microscales according to local rules and conditions, but also manifesting more global order across many scales and time periods.

An initial stimulus to fractal analogue modelling was that simulated structures were a substitute for detailed morphological measures of the real world, and an enduring use is that it is possible to identify the system histories of evolving structures. A logical sequel to this work has been to begin to analyse the size-density relations of real world settlements, and we have begun to reformulate urban density analysis in terms of allometric relations (Longley and Mesev, 1997, 1999). We have only touched upon the multitude of urban theories in the most cursory terms, yet it is clear that fractal geometry has appealing properties with respect to cities in ideas concerning space-filling, self-similarity, and density. It remains for future empirical research to substantiate our informed yet subjective understanding of urban morphology consistent with spatial theory.

Further reading

Batty, M. and Longley, P. (1994) *Fractal Cities: a Geometry of Form and Function*, London and San Diego: Academic Press.

Mandelbrot, B. B. (1967) 'How long is the coast of Britain? Statistical self-similarity and fractal dimension', *Science*, 155, pp, 636–638.

Peitgen, H.-O., Jurgens, H. and Saupe, D. (1992) *Fractals for the Classroom: Part 1: Introduction to Fractals and Chaos*, New York: Springer Verlag.

References

Aangeenbrug, A. (1991) 'A critique of GIS', in Maguire, D., Goodchild, M. and Rhind, D. (eds) *Geographical Information Systems: Principles and Applications*, London: Longman, pp. 101–107.

Arlinghaus, S. L. (1985) 'Fractals take a central place', *Geografiska Annaler*, 67B, pp. 83–88.

Barnsley, M. and Donnay, J.-P. (eds) (1999) *Remote Sensing and Urban Analysis*, London: Taylor & Francis.

Batty, M. (1991) 'Cities as fractals: simulating growth and form', in Crilly, T., Earnshaw, R. A. and Jones, H. (eds) *Fractals and Chaos*, New York: Springer Verlag, pp. 41–69.

Batty, M. and Longley, P. (1986) 'The fractal simulation of urban structure', *Environment and Planning A*, 18, pp. 1143–1179.

Batty, M. and Longley, P. (1987a) 'A fractal-based description of urban form', *Environment and Planning B*, 14, pp. 123–134.

Batty, M. and Longley, P. (1987b) 'Fractal dimensions of urban shapes', *Area*, 19, pp. 215–221.

Batty, M. and Longley, P. (1988) 'The morphology of urban land use', *Environment and Planning B*, 15, pp. 461–488.

Batty, M. and Longley, P. (1994) *Fractal Cities: A Geometry of Form and Function*. London, Academic Press.

Batty, M., Longley, P. and Fotheringham, A. S. (1989) 'Urban growth and form: scaling, fractal geometry and diffusion-limited aggregation', *Environment and Planning A*, 21, pp. 1447–1472.

Bertuglia, C. S., Clarke, G. P. and Wilson, A. G. (1994) *Modelling the City: Performance, Policy and Planning*. London: Routledge.

Bracken, I. and Martin, D. (1989) 'The generation of spatial population distributions from census centroid data', *Environment and Planning A*, 21, pp. 537–543.

Can, A. (1992) 'Specification and estimation of hedonic housing price models', *Regional Science and Urban Economics*, 22, pp. 453–474.

Curry, M. R. (1995) 'GIS and the inevitability of ethical inconsistency', in Pickles, J. (ed.) *Ground Truth: The Social Implications of Geographic Information Systems*, New York: Guilford Press, pp. 68–87.

Fisher, P. (1999) 'Models of uncertainty in spatial data', in Longley, P., Goodchild, M., Maguire, D. and Rhind, D. (eds) *Geographical Information Systems: Principles, Techniques, Management, Applications*, New York: Wiley, Vol. 1, pp. 191–205.

Fotheringham, A. S., Batty, M. and Longley, P. (1989) 'Diffusion-limited aggregation and the fractal nature of urban growth', *Papers of the Regional Science Association*, 67, pp. 55–69.

Frankhauser, P. (1993) 'La fractalité des structures urbaines', Thèse de Doctorat. Paris: UFR de Geographie, Université de Paris I.

Getis, A. and Ord, J. K. (1996) 'Local spatial statistics: an overview', in Longley, P. and Batty, M. (eds) *Spatial Analysis: Modelling in a GIS Environment*, Cambridge: GeoInformation International, pp. 261–277.

Goodchild, M. F. (1980) 'Fractals and the accuracy of geographical measures', *Mathematical Geology*, 12, pp. 85–98.

Haggett, P., Cliff, A. D. and Frey, A. (1977) *Locational Analysis in Human Geography*, London: Arnold and New York: Wiley.

Johnston, R. (1999) 'GIS and Geography', in Longley, P., Goodchild, M., Maguire, D. and Rhind, D. (eds) *Geographical Information Systems: Principles, Techniques, Management, Applications*, New York: Wiley, Vol. 1, pp. 39–47.

Longley, P. and Dunn, R. (1988) 'Graphical assessment of housing market models', *Urban Studies*, 25, 1, 21–34.

Longley, P. and Mesev, V. (1997) 'Beyond analogue models: space filling and density measurement of an urban settlement', *Papers in Regional Science*, 76, 4, pp. 409–427.

Longley, P. and Mesev, V. (1999) 'Measuring urban morphology using remotely-sensed imagery', in Barnsley, M. and Donnay, J.-P. (eds) *Remote Sensing and Urban Analysis*, Taylor & Francis, London, in press.

Longley, P., Batty, M., Shepherd, J. and Sadler, G. (1992) 'Do green belts change the shape of urban areas? A preliminary analysis of the settlement geography of South East England', *Regional Studies*, 26, 5, pp. 437–452.

Mandelbrot, B. B. (1967) 'How Long is the Coast of Britain? Statistical self-similarity and fractal dimension', *Science*, 155, pp. 636–638.

Mandelbrot, B. B. (1982) 'Comment of computer rendering of fractal stochastic models', *Communications of ACM*, 25, pp. 581–583.

Mark, D. M. (1984) 'Fractal dimension of a coral reef at ecological scales: a discussion', *Marine Ecology Progress Series*, 14, pp. 293–294.

Martin, D. (1996) *Geographic Information Systems and Their Socioeconomic Applications*, 2nd edn, London: Routledge.

Martin, D., Longley, P. and Higgs, G. (1994) 'The use of GIS in the analysis of diverse urban databases', *Computers, Environment and Urban Systems*, 18, pp. 55–66.

Mesev, T. V., Batty, M., Longley, P. and Xie, Y. (1995) 'Morphology from imagery: detecting and measuring the density of urban land use', *Environment and Planning A*, 27, pp. 759–780.

Mesev, T. V., Gorte, B. and Longley, P. (1999) 'Modified maximum likelihood classifications and their application to urban remote sensing', in Barnsley, M. and Donnay, J.-P. (eds) *Remote Sensing and Urban Analysis*, London: Taylor & Francis, in press.

Openshaw, S. (1989) 'Computer modelling in human geography', In Macmillan, B. (ed.) *Remodelling Geography*, Oxford: Blackwell Publishers, pp. 70–88.

Openshaw, S. (1991) 'A spatial analysis research agenda', in Masser, I. and Blakemore, M. J. (eds) *Handling Geographic Information: Methodology and Potential Applications*, London: Longman, pp. 18–37.

Openshaw, S. (1996) 'Developing GIS-relevant zone based spatial analysis methods', in Longley, P. and Batty, M. (eds) *Spatial Analysis: Modelling in a GIS Environment*, Cambridge: GeoInformation International, pp. 55–73.

Orford, S. (1999) *Valuing the Built Environment: GIS and House Price Analysis*, Aldershot: Ashgate.

Peitgen, H.-O., Jurgens, H. and Saupe, D. (1992) *Fractals for the Classroom: Part 1, Introduction to Fractals and Chaos*. New York: Springer Verlag.

Pumain, D., Sanders, L. and Saint-Julien, T. (1989) *Villes et auto-organisation*. Paris: Economica.

Richardson, L. F. (1961) 'The problem of contiguity: an appendix of "statistics of deadly quarrels" ', *General Systems Yearbook*, 6, pp. 139–187.

Saupe, D. (1991) 'Random fractals in image synthesis', in Crilly, A. J., Earnshaw, R. A. and Jones, H. (eds) *Fractals and Chaos*, New York: Springer Verlag, pp. 89–118.

Whitehand, J. (1992.) 'Recent advances in urban morphology', *Urban Studies*, 29, pp. 619–636.

Wong, D. W. S. and Fotheringham, A. S. (1990) 'Urban systems as examples of bounded chaos: exploring the relationship between fractal dimension, rank size, and rural-to-urban migration', *Geografiska Annaler*, 72B, pp. 89–99.

Wrigley, N. (1985) *Categorical Data Analysis for Geographers and Environmental Scientists*, Harlow: Longman.

Wrigley, N. and Longley, P. A. (1984) 'Discrete choice modelling in urban analysis', in Herbert, D. T. and Johnston, R. J. (eds) *Geography and the Urban Environment: Volume 6: Progress in Research and Applications*, Chichester: Wiley, pp. 45–94.

Wrigley, N., Longley, P. and Dunn, R. (1988) 'Some developments in the specification, estimation and testing of discrete choice models', in Golledge, R. G. and Timmermans, H. (eds) *Behavioural Modelling in Geography and Planning*, London: Croom Helm, pp. 96–123.

Chapter 13

Cyberspatial analysis: appropriate methods and metrics for a new geography

Shane Murnion

13.1 Introduction

13.1.1 The importance of the Internet and WWW information flows

Information is a commodity. The role of information businesses (creators, collectors, vendors and distributors) will become increasingly important in the coming years. This will be particularly true for advanced economies which are dependent on the service rather than manufacturing industries. The UK economy, for example, is already heavily dependent on this type of commerce. It is likely that the World Wide Web (WWW) protocol and the Internet infrastructure will play a central role in information transactions. A recent study predicts that the on-line travel market alone will be worth $9 billion within the next five years (Gardner, 1997) and it is suggested that approximately $2 billion worth of research and development on new Internet access services will be carried out by the private sector this year alone (Hertzberg, 1997). That the Internet should be so important is not surprising, when it is considered that the number of Internet users is predicted to grow to over 100 000 000 within the next 10 years (Cerf, 1993).

The importance of this sector has not gone unrecognized at government level and a number of strategic initiatives designed to support an information economy infrastructure have recently been implemented in the UK (UKERNA, 1993), the USA (Clinton and Gore, 1997), Malaysia (Durham, 1997) and other states. The UK SuperJanet initiative, for example, aims to support and develop prototype applications which require and utilise high-speed broadband communication networks. The Malaysian initiative involves the provision of a superb networking infrastructure for one region, in the hope of attracting information businesses to that area. Although the various initiatives are diverse in implementation, they all have the same basic aims, to improve the national communication infrastructure and to develop the skills within the work-force that will allow them to compete for information business. Information business is attractive for a number of reasons, requiring no natural resources and involving minimal transportation costs for product delivery.

The growth of the information economy will also have major consequences on the way we work. If we have a truly effective global communication network there is no reason that a company's work-force should even be located in the same hemisphere. Businesses will be able to employ staff from any region without the complications of work visas or immigration legislation. British Airways recently moved some of their software development from the UK to India (Leung, 1996).

13.1.2 The need for quantitative analysis

Although no one doubts that the WWW will provide fantastic opportunities for business, it is equally true that there are very few examples of financially successful web-based ventures and numerous well-publicised failures, e.g. The Spot (Tullis, 1997). A successful business plan for any new commercial venture will usually include some estimation of the numbers of customers expected and the likely catchment areas that these customers will come from. Using this information, probable turnover and profits can be calculated and furthermore, services could be tailored to match customer requirements. In the case of WWW or Internet-based businesses, this type of calculation is impossible to make, since the whole commercial sector is still in its infancy. A number of basic questions remain unanswered.

If a WWW based service is set up at one location on the Internet, it should in theory, be accessible from any other Internet location. If the services offered are suitably global in nature, would we then expect customers using this service to be equally distributed around the Internet? Alternatively, we might expect more customers from Internet regions local to the server. Unfortunately, there is little evidence to support either view. Another way of viewing this issue is to ask if distance will have an influence on Internet transactions? If distance is important, then what metric can be used to measure it?

There are other important commercial issues that remain unexplored. Do cultural factors such as language influence which WWW based services a customer will use? Is it important where an on-line service is located, either geographically or in Internet terms. Is there any relationship between geographical and Internet space (cyberspace)?

Part of the interest in potential Internet commerce is in high-volume businesses, such as pay-per-view TV, global radio-broadcasting, Internet telephone services and video-conferencing. A single video tape, for example, can store approximately 4 Gb of movie data. To view a film across the Internet would thus require a continuous information transfer of at least 500 Kb of data per second over a two hour period. This type of information transaction will require a sophisticated networking infrastructure to operate effectively. The topology of the existing infrastructure will therefore define the possible footprint or catchment area of the service. Analysis is needed to define the required improvements to make such services feasible.

While there is currently little quantitative spatial analysis in this area, there are indicators that the situation may soon change. In recent times, there has been a growing interest in mapping the Internet, from a number of different perspectives, for visualization purposes (Dodge, 1996; MIDS, 1997; Quarterman *et al.*, 1994) and it seems likely, based on the experience of GIS development, that initial attempts at quantitative cyberspatial analysis will follow.

13.1.3 What role can geography play?

At first glance it is not obvious that the discipline of geography has much relevance to areas of research such as the WWW, computer-networking, multi-media services, etc. However, in this instance appearances are deceptive. Traditionally geographers have studied flows of people from a source location to a destination location in tourism and migration analysis, but the process of a person viewing information from a remote WWW site could be considered in terms of a virtual visit from the viewer's source location to a destination at that site. Alternatively, reversing our view, it could be considered that the information from the WWW site travels to the viewer's computer. If we know the number of visits made from each source to each destination and the number of persons who could make such a journey then we could consider applying spatial interaction methods to model such a process. Traffic flows have also been analysed by geographers using network models within GIS for the purposes of urban and regional planning. Information flows along the Internet can also be considered as traffic flowing along a physical network. In both types of analysis, the broad experience of geographers and the rich diversity of analytical methods that have been developed on previous problems could be re-applied to this new context to provide meaningful answers to some important questions.

The application of spatial analysis methods to cyberspace is not a straight-forward one however. Spatial analysis clearly depends on some definition of space. For cyberspatial analysis, we therefore need a suitable metric for space and distance, that is appropriate for this new geography. Customers of on-line services are people, and in estimating numbers of customers it would be useful to have sources of cyberdemographic information. Thus before we can attempt to apply traditional spatial analysis methods to cyberspace we must overcome two obstacles: firstly, we must define space in a useful and relevant way, preferably allowing us to map it to geographical space and, secondly, we must identify sources of information that will allow us to carry out some useful research.

13.2 Defining the space in cyberspace

The backbone of the Internet consists of fibre-optic connections allowing transmission of information at the speed of light. Thus it might be expected that any site on the Internet is essentially co-located with any other site and that space or time has no meaning in terms of information transfer. However, there are an abundance of indicators which show that there is a spatial extent to cyberspace. One obvious example of this is the use of 'mirror' sites of popular Internet resource archives (e.g. SUNSITE). The mirrors are identical copies of the archives which are available from multiple servers distributed around the Internet. The reason these mirrors exist is that it is usually faster to download any resource from a local archive than a centrally placed server. This suggests two important points. Firstly, that there is some relationship between geographical space and cyberspace distance. Secondly, that Internet users perceive distances in cyberspace in terms of time, i.e. the time required to download or transfer information. It seems likely that time will play an important part in any metric used in analysing information flows in cyberspace.

The reasons for the disparity between the theoretical situation of light-speed information transfers and the real-world situation of slow web-page downloads lies in the hardware and software protocols used to transfer information across the Internet. Transmission control protocol/information protocol (TCP/IP) controls how information is passed from one site to another (Hunt, 1996). When information (a text document, image or piece of software) is sent, the information is broken down into packets. Each packet contains the destination address to which the packet is to be sent. The information is sent from the source machine to the destination machine through a sequence of routers. Each router is simply a piece of equipment (e.g. a computer running a router system) that receives packets of information, examines the address in each packet and works out which router the packet should be sent to next. How this route determination is carried out depends on the router in question. Some simple routers use predefined paths, while others calculate the best direction to send material based on current traffic flows in the vicinity of the router. Whatever the method used, the router will decide where to send the information next and will pass it on to the appropriate location. The process is repeated until all the packets arrive at their destination, at which point the packets are reassembled into the original information. Even when information traffic is low there will be a small delay, as each router in the chain receives and processes each packet. Furthermore, when information traffic levels are high, the router may not be able to immediately receive the packet for processing. When this type of congestion occurs the previous router may have to attempt to send the packet a number of times before it is successfully received. Delays occurring at each stage of the journey from server to client accumulate, increasing the actual time it takes for the information to proceed from source to destination. Since we can see that cyberspace has some spatial extent, or more precisely some spatio-temporal extent, we need to find some metric that can be usefully employed to measure it.

13.2.1 Option 1: true geographical space

One possible metric would be to use true geographical space, since on the basis of our software archive mirrors example, there is apparently some relationship between geographical space and cyberspace. If this metric is used then we could simply define the location of our clients in equation terms using geographical co-ordinates. Unfortunately while such an approach might be partially successful, it cannot provide a complete answer. Routers will send information packets on to a further router based on the quickest rather than the physically shortest path (e.g. it is not uncommon for information sent to France from the UK to travel via the USA).

13.2.2 Option 2: the WWW as a network using traceroute

A second possibility would be to examine the route taken by the information. The information is travelling along a network, with routers as the nodes at the junctions of the information network. Since there are readily available network analysis methods developed for road and rail networks within GIS it seems plausible that these methodologies could be applied to Internet traffic.

Using a network model, there are two possible ways we could define locations. Firstly we could once again use the geographical locations of the source, destination and intermediate router machines to locate the nodes of the network in geographical space. The connections between these nodes could have attached attribute information representing the time required for information to travel along that segment of the network. Secondly we could consider the locations in terms of internet protocol (IP) addresses, which are in the format XXX.XXX.XXX.XXX, where X is a number between 0 and 9. Since every node in the network will have both an Internet and a geographical address, it is possible to map cyberspace as represented by the network to real geographical space on a map. However, as we will see, it is often easier to obtain the Internet address of a particular Internet location than it is to obtain the geographical co-ordinates of that node.

Another factor that supports the utility of the network model is that computational tools already exist which allow us to map the Internet network. A useful application, originating on UNIX machines known as *traceroute* (Rickard, 1996), can be used to generate information on the Internet network routes. Figure 13.1 illustrates information generated by *traceroute* about the current information path between a UNIX workstation

1	148.197.254.6 (148.197.254.6)	2 ms 2 ms 2 ms
2	smds-gw.ulcc.ja.net (193.63.203.33)	9 ms 13 ms 18 ms
3	smds-gw.mcc.ja.net (193.63.203.97)	53 ms 22 ms 20 ms
4	sj-gw.qub.ja.net (193.63.91.34)	31 ms 33 ms 35 ms
5	ntp0.qub.ac.uk (143.117.41.6)	45 ms 33 ms 32 ms
6	boris.qub.ac.uk (143.117.9.39)	32 ms 32 ms 37 ms

Figure 13.1 Results from *traceroute* between alpha4.iso.port.ac.uk and boris.qub.ac.uk. Each entry shows one router through which the information passed. The time for the packet to travel from the source to that router is measured three times.

at the University of Portsmouth and a WWW server at Queen's University of Belfast.

Figure 13.1 is created by the *traceroute* command in six stages. In the first stage, it measures how long it takes to send a packet of information to the machine at IP address 148.197.254.6 and back. In the second stage, it measures how long the round trip takes to send a packet from the source machine to smds-gw.ulcc.ja.net and back. This process is repeated until information about a complete path between the source machine and the destination machine is built up. Using the example above one might assume that the 148.197.254.6 to smds-gw.ulcc.ja.net segment of the network takes on average 13.33 ms to traverse, by subtracting the average of the three timings taken during the first stage of the *traceroute* execution (2, 2 and 2 ms) from the average of the three times measured in the second stage (9, 13 and 18 ms).

The results from the *traceroute* command offer two possible metrics. Firstly, as a rather crude metric, the number of routers the information passes through could be used as an indicator of the cyberspace distance. Alternatively it would be possible to map the network branching from a server, with each segment of the network coded to represent the time taken for a single packet to traverse that segment. Figure 13.2 illustrates a geo-referenced version of a *traceroute* between a computer at Portsmouth and one in Japan. The visualisation was carried out using the *GeoBoy* software package.

There are a number of tools available in GIS that would allow users to calculate the areas that could be reached within a certain time using such a network. The resulting map could be used to define the cyberspace distance between client and server. There are however problems with both the suggested approaches.

If the first approach is considered, using the example above, the WWW server in Belfast is six steps away from the local machine in Portsmouth. However, it is clear from Figure 13.1 that each step does not take equal time

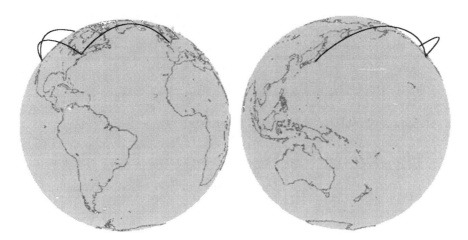

Figure 13.2 Results of a *traceroute* between a computer in Portsmouth and one in Japan displayed on a global map.

1 148.197.254.6 (148.197.254.6)	2 ms 2 ms 2 ms	
2 smds-gw.thouse.ja.net (193.63.203.46)	25 ms 20 ms 21 ms	
3 eu-gw.ja.net (193.62.157.245)	20 ms 12 ms 11 ms	
4 stockholm-gw.triangle.ja.net (193.62.157.21)	18 ms 20 ms 19 ms	
5 Stockholm1.triangle.net (192.12.54.26)	64 ms 59 ms *	
6 syd-gw.nordu.net (192.36.148.205)	51 ms 53 ms 50 ms	
7 fi-gw.nordu.net (192.36.148.54)	67 ms 65 ms 89 ms	
8 nic.funet.fi (128.214.248.6)	81 ms 69 ms *	

Figure 13.3 Results from a *traceroute* command between a Portsmouth computer and *www.funet.fi* (*indicates no value returned).

to traverse and thus each segment is not of equal importance. It would also be impossible to compare steps that arise from different routes taken through the Internet.

There are also problems with the second approach. One of these problems is clearly indicated in Figure 13.3, which shows the results of a *traceroute* from a local machine at Portsmouth to a WWW server in Finland.

From the results of this *traceroute* execution the smds-gw.thouse.ja.net to eu-gw.ja.net segment of the network takes on average −7.7 seconds to traverse! The reason for this unusual result is that the Internet is in a constant state of congestion and the time taken to traverse a particular segment of the network is more dependent on traffic levels than the physical nature of the

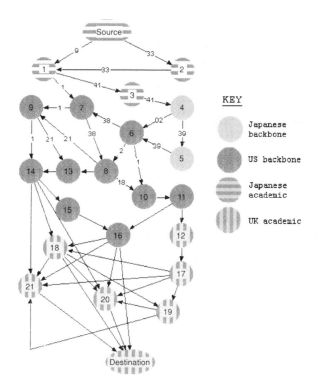

Figure 13.4 Routes taken during 40 *traceroute* runs by information travelling along the 'shortest path' from a computer in Japan to a computer in the UK. The network flow diagram is an accumulation over the 41 runs. Each circle represents a unique router. The numbers on some of the arrows indicate the number of times that segment of the route was used.

segment. In the example above, there was probably a traffic surge during the second stage of building the table. It would conceivably be possible to overcome this difficulty by carrying out multiple runs of the *traceroute* command and averaging the results over time. However as traffic levels on the segment fluctuate, a further complication arises. When a router receives a packet, it opens the packet, examines the final destination address and calculates where to send it next. As traffic levels fluctuate, the next router it sends the information to may change, i.e. the actual route taken between any source and destination, will change over time. Figure 13.4 shows the results of 41 *traceroute* commands, following the paths of information flow via various routers from a computer in Japan to one in the UK.

As is obvious from Figure 13.4, the actual path taken by information moving from one location to another varies dramatically even over a short space of time. Thus in order to use a network model, the cost values would

Sun Jul 1995 03:17:01 1995				
1	133.74.8.254	3 ms	2 ms	3 ms
2	isas-gw.isas.ac.jp	5 ms	5 ms	4 ms
3	sinet-gw.isas.ac.jp	24 ms	23 ms	23 ms
4	nishi-chiba.sinet.ad.jp	25 ms	23 ms	24 ms
5	nacsis-gate3.sinet.ad.jp	26 ms	24 ms	24 ms
6	sl-stk-7-S4/0-5M.sprintlink.net	137 ms	135 ms	135 ms
7	sl-stk-5-F0/0.sprintlink.net	137 ms	136 ms	135 ms
8	sl-dc-6-H1/0-T3.sprintlink.net	199 ms	198 ms	200 ms
9	icm-dc-2b-F1/0.icp.net	201 ms	201 ms	199 ms
10	icm-dc-1-F0/0.icp.net	203 ms	200 ms	200 ms
11	icm-london-1-S1-1984k.icp.net	338 ms	320 ms	347 ms
13	smds-gw.ulcc.ja.net	292 ms	*	*
14	* * *			
15	* * *			
16	* * *			
17	mssly1.mssl.ucl.ac.uk	341 ms	336 ms	*

Figure 13.5 Incomplete *traceroute* result (* indicates no value returned).

have to be calculated in real time. Alternatively some sort of averaging would have to be applied.

A third problem with using *traceroute* and a network model is that information about each segment of the route is not always supplied. An example of an incomplete return from one of the *traceroute* runs is shown below in Figure 13.5.

There are many reasons why incomplete data is returned. Some systems have a commonly found bug which may result in a null return, other routers may be too busy to return a response within the *traceroute* time-out period and others, such as MIT's C gateway, are simply not compatible at all with the *traceroute* command.

In conclusion, it seems that whilst network analysis of information flows is feasible, it is complicated by rapid changes in the fastest route between two locations and hampered by potential gaps in the collected data. A simpler approach may produce better results and will be considered next.

13.2.3 Option 3: simple source to destination distances using ping

Ping is another common utility that originated in UNIX systems. The *ping* command sends a single packet of information to a remote host. A response

```
>ping -c3 www.yahoo.com
PING www12.yahoo.com (204.71.177.77): 56 data bytes
64 bytes from 204.71.177.77: icmp_seq=1 ttl=240 time=291 ms
64 bytes from 204.71.177.77: icmp_seq=2 ttl=240 time=290 ms
----www12.yahoo.com PING Statistics----
3 packets transmitted, 2 packets received, 33% packet loss
round-trip (ms)  min/avg/max = 290/290/291 ms
```

Figure 13.6 Example output from *ping* command.

from the server is echoed back to the original machine and the time for the round trip is calculated. As with the previous method, a suitable definition of the client location would be either geographical or the client's IP address. The packet transfer is repeated a number of times and an average is calculated. An example output from a *ping* command is illustrated in Figure 13.6.

In the example shown in Figure 13.6, it took on average 290 ms to send a packet of 56 bytes on a round trip from a UK computer to the Yahoo server in the USA and back. Not only does this method allow easy calculation of the time taken to transfer information to a remote machine, it is also much more consistent with the practice of web browsing than the previously discussed network model. Users of the WWW are unaware of the routes taken by information, but the time taken to do so is apparent. The journey time returned by the *ping* command is usually referred to as latency and is already widely used in mapping of the Internet information flows for visualization purposes, creating what has been termed as 'Internet Weather Maps' (MIDS, 1997), although latency has yet to be applied in analysis of information flows on the WWW. Latency provides a useful metric for defining the cyberspace distance.

Although it is possible to measure latencies between the UK and other globally distributed domains now, there are unfortunately no records available of latency measurements over the last couple of years. This would suggest that the analysis could only include current WWW server access statistics, since the latency measurements might not be relevant to the latencies that were extant one or more years ago.

13.2.4 Temporal variation and global data collection problems

The time taken by information travelling between client and server on the Internet is heavily dependent on traffic conditions at the time of the transaction. On a larger time scale, the Internet network infrastructure is constantly growing and changing. This suggests that any measurement of distances should be made at or close to the time of the information transfer to be valid. Although some form of temporally averaged values could be used for large-scale or long-term studies, some interesting variations in information browsing habits may be lost in the process. It might be the case that WWW users in Europe may access information differently in the morning, before most users in the US come on-line, than they do in the afternoon when Internet traffic congestion increases.

Finally there remains the problem of global cyberspace distance data collection. If in an attempted analysis we are interested in measuring the cyberspace distance between a local WWW server and globally distributed clients it would be a straightforward matter to use the *ping* command from our WWW server computer to measure the distance between it and any clients accessing the WWW server. However, if we are attempting to model transactions between globally distributed WWW servers and globally distributed clients, e.g. to build a spatial interaction model, then problems arise. It is unlikely that we will have access to user accounts on those WWW server computers and thus we have no way of using *ping* to measure the distance between the remote WWW servers and remote clients. It is possible to direct, to a certain degree, the route travelled in a *traceroute* command and it might be possible to use this feature to collect data on the cyberspace distance between two remote Internet locations. Such a method would have to be used with caution, however, since it would not be possible to carry out latency measurements from each Internet location to every other Internet location on a continuous basis without putting stress on what is already a congested system. There is strong need for the development of some simple method of determining the latency between two points anywhere in cyberspace utilising as few measurements as possible.

13.3 Counting visits in cyberspace

The question of how to count visits to a WWW site is the subject of ongoing research. The commercial advertising sector was one of the pioneers in using the WWW commercially and is understandably interested in tracking individual visitors to WWW sites which display advertisements (Stehle, 1996). This issue is a difficult and complex one for many reasons. The main problem in tracking individuals who visit WWW sites is that the method of delivering web pages is a stateless protocol. Each web page transfer (and transfers of images or other objects embedded in the web page) involves a completely separate information transfer. If, for example, two web pages from the same WWW site are downloaded shortly after each other, the server has no way of automatically connecting the first transfer to the second. Indeed it is impossible to tell if the recipient of the web pages is the same person or two separate individuals sharing the same IP address. Some WWW servers also keep track of the client software used. In theory, this information might be used to obtain the number of individuals using the WWW site. For example

Table 13.1 Common variants of access statistics available from WWW sites

Software	Total transfers by time period	Total transfers by client domain		Total transfers by client sub-domain		Total transfers by object
	Requests	Requests	Bytes	Requests	Bytes	Requests
Getstats	✓	✓		✓		✓
Wwwstat	✓	✓	✓	✓	✓	✓
Analog	✓	✓	✓	✓	✓	✓
Wusage	✓	Top ten only	Top ten only	Top ten only	Top ten only	✓

if there are 40 information transfers from an individual IP address and the client used was a Microsoft Windows 3.1 version of Netscape, it might be possible to assume that since Windows 3.1 is not a multiuser operating system, then a single IP address indicates a single user. If the information was transferred to a single IP address using a Digital OSF client, then one might assume that multiple users would share that IP address. In this case, it would be necessary to examine the temporal pattern of the information transfers to determine the number of individuals accessing the site. The issue is further complicated by the fact that some Internet providers maintain a pool of IP addresses that are only issued when needed. Thus a single IP address assigned to a Windows 3.1 client might in fact represent two users that were sequentially assigned that particular IP address.

It is clear that it is difficult to count the number of individual users accessing a WWW site. It is even more difficult to track individual users visiting the same site at different times, although there are methods for doing this (Thompson, 1997). For the purposes of simple analysis, it should be sufficient simply to track the total number of transfers to each client and assume that this relates to the number of individuals using the clients. This information is usually stored in the access logs of WWW servers. The detail available will depend on the level of access to the information contained in the logs. Generally access to the full details of the server logs for WWW servers will only be available to individuals at that institution and so there will probably be a reliance on aggregated and processed information sometimes made public in the form of server access statistics. There are now a wide variety of programs or scripts available to WWW server administrators that allow them to create summaries of WWW server access statistics automatically. Unfortunately, there is wide variation in the quality of information made available in this way. Table 13.1 covers some of the main types of published statistics found. Many of these packages allow the creation of highly customized reports, the features listed here are simply the commonly found variants of reports produced.

If all that was needed was the number of information transfers from a server then any of the above report types would be suitable. Unfortunately, it is also necessary to obtain the number of information transfers to a client at individual locations. Of the server statistics that are available, it is obvious that any statistics provided by the Wusage package are incomplete. They are also commonly displayed in a graphical image, e.g. a pie-chart, which renders the information almost unusable. All of the other report formats give data on information transfers to clients, but usually only down to the subdomain level. To get down to individual client transfers would usually require access to the logs themselves.

Wwwstat and Analog also provide information about the number of bytes transferred and this may provide important data for information vendors that deal with large volume information objects such as movies or multimedia objects.

In summary, two levels of analysis can be carried out. At the detailed level, where the actual server logs are available, individual transactions can be analysed and the latencies can be calculated to the individual client. At the less detailed level, server statistics at a subdomain, national or international level can be gathered and applied with appropriately averaged latency values.

13.3.1 Problems with collecting global data

While it is true that many WWW sites do provide summary WWW site access statistics as part of their content delivery, some do not, as the information may be commercially sensitive, thus we will only be able to obtain a partial picture at any point in time. Furthermore, to collect and process all the publicly available WWW server access statistics would be an enormous task, and since many WWW sites only publish current WWW server access statistics, this data collection would have to be carried out on a continuous basis.

13.4 Determining Internet populations

As part of a cyberspatial analysis of information transactions between two Internet locations it would be useful to know the total number of Internet users at each location that could potentially initiate such a transaction. Unfortunately, Internet population statistics is still an area in its infancy and as a result there is currently no accurate measure of the total number of Internet users, never mind spatially dis-aggregated counts. However, all is not lost. Almost every computer that has access to the Internet has an IP address. These IP addresses are organized into domains. Within each domains are subdomains; for example, all the computers at the University of Portsmouth have an IP address of the form 148.197.XXX.XXX (where X is a number between 0 and 9) which corresponds to the domain .port.ac.uk, i.e. all the computers belong to the UK top level domain (.uk), the academic

> >nslookup www.yahoo.com
>
> >Non-authoritative answer:
>
> >Name: www7.yahoo.com
>
> >Address: 204.71.177.72
>
> >Aliases: www.yahoo.com

Figure 13.7 Results from an *nslookup* command.

sub-domain (.ac.uk) and the University of Portsmouth subsubdomain (.port.ac.uk). Even within the University of Portsmouth subdomains, there will be further division into smaller groups, e.g., all the computers in the Geography Department are grouped into the geog.port.ac.uk subdomain, which corresponds to the IP addresses of 148.197.55.XXX. Each of these domains and subdomains must be registered with a central domain-name database. It is likely that there is a strong correlation between the number of registered subdomains and the number of users. If this is true then these counts could be used to get a spatially dis-aggregated measure of the number of computer users on the Internet. If we are carrying out a cyberspatial analysis of information flows at national level, then it is fairly simple to obtain counts of registered subdomains, since this information is published on a regular basis (Network Wizards, 1997). Most top-level domains correspond to particular countries (e.g. .uk represents the United Kingdom, .nl the Netherlands, etc.). However, there is a problem with this system. The .com top-level domain is commonly used by commercial organisations all over the globe and a significant proportion of the total number of Internet users have access through .com domains (e.g. Compuserve users at compuserve.com and America On-line users at aol.com). At the current time, it could probably be assumed that the majority of .com users originate from the USA, however this may not hold true indefinitely.

13.4.1 Problems with detailed analysis

If we wish to carry out a cyberspatial analysis at a smaller than national scale then counting subdomains and geographically mapping these domains becomes a little more problematical. Fortunately a computational software tool known as *nslookup* exists that could provide part of the solution. In its most basic mode of use, *nslookup* will provide an IP address for a domain name or vice-versa. It does this by accessing the central domain name database and returning the relevant information. Figure 13.7 shows an example of *nslookup* in use.

But *nslookup* can provide much more information if used properly. We can use this utility to list all the subdomains in the database that exist within a domain. It is possible therefore, to produce custom dis-aggregated counts of each subdomain even down to, for example, the number of domains registered within the University of Portsmouth. A secondary problem at this level of analysis would be the difficulty of mapping these subdomains to a geographical area. An experimental extension to the domain registration that includes geographical location information (Davis, 1997) is growing in popularity and when in widespread use this will allow automatic querying of the location of computers on the Internet. New software tools such as *GeoBoy* can use this information to produce geo-referenced plots of information obtained from *traceroute* or *ping* (an example of this is shown in Figure 13.2). An alternative approach could involve the development of an Internet Gazetteer and there is already growing interest in this area (Harrington, 1996).

In conclusion, therefore, it seems that the information we need is either currently available, or will soon be available, to allow the automatic collection of cyberspatial demographic information.

13.5 Potential areas for further research

13.5.1 Spatial analysis of browsing habits

It would be useful to know if there is a spatial aspect to use of the WWW, i.e. do people use Internet access services that are close to them in cyberspace, or is cyberspatial distance irrelevant. Using the access logs of WWW servers and the *ping* utility, it should be possible to measure the latency or cyberspace distance between the server and the clients who are accessing information from the server. It is plausible to imagine that more users will access a server if they can download that server's web pages quickly, i.e. if the latency between client and server is small. If the latency is large then the user may look for a similar information service closer to them on the Internet. A general distance decay function may exist which represents how the number of users falls off with cyberspace distance, and if it does exist then it could be used to determine the catchment area of that WWW server.

13.5.2 WWW site location studies

Different domains have different numbers of Internet users, e.g. there are many more Internet users in the UK (.uk) domain, than in the South African (.za) domain. If a distance decay relationship exists as discussed above, then it may be preferable for South African online services to place their information content on a WWW server that is closer to the .uk domain than the .za domain. Location of a businesses WWW site on the Internet may be the most important factor controlling the amount of business that the WWW server generates, yet is hardly considered at the current time when setting up an online information service. It would be useful to set up parallel information services at different locations on the Internet and analyse the amount of accesses each server receives from different domains in combination with latency measurements and cyberspace demographic information to see if any variation occurs and if so if it can be explained.

13.5.3 Quantitative analysis of cultural barriers

One of the main barriers in accessing information on the WWW for many non-English speaking users is language. It would be useful to set up parallel information services, one in English only and another with multilingual content and examine the spatial differences in accesses to each service. It may be possible to quantify the language barrier and determine the importance of multilingual content for online services.

13.5.4 Information warfare and the effects of natural disasters

It was mentioned previously that it is not uncommon for information travelling across the Internet between two countries to be routed via a third country. This comes about because both countries may have high-speed connections to the third state but not directly between them. It is thus faster to route the information through the third country. If one country creates a superb information infrastructure and is strategically placed on the Internet, then it would not be surprising for a situation to arise where most of the world's information transfers would travel through that infrastructure. In this situation, it would be feasible for that country to engage in information warfare. It would be relatively simple to block any information that originates from or is destined for a particular state. The Internet was designed to protect information flows against such a possibility and the information will almost always be rerouted successfully through an alternative path. However, the alternative route may be much slower than the original and as a result of this services which require high-speed connections, such as video-conferencing or video-on-demand, may fail. It is conceivable therefore that one nation with a strategic advantage in terms of infrastructure may be able to destroy or damage the Internet-based export businesses of another state. It would be useful to carry out large-scale latency measurements and conduct simulations of the possible effects of such an occurrence. A similar situation might also arise as a result of a large-scale natural disaster such as an earthquake. At a national level, it may be important to plan infrastructure development to guard against such an occurrence.

13.6 Conclusions

The current situation of WWW analysis is mirroring the early development of digital mapping and GIS, where initial mapping and visualisation tools progressed to become powerful and practical spatial analysis tool-kits and spatial decision support systems. It is likely that a similar scenario will occur with Internet information flows and cyberspatial analysis. In this chapter, we have illustrated some possible metrics and information sources that could support such an analysis and outlined some of the many potential areas of study in this new geography. Three main obstacles exist to carrying out useful global analysis at the current time.

The first obstacle involves the problem of measuring the distance between two Internet locations in cyberspace. As mentioned previously *ping* and *traceroute* can both be used. However, it becomes difficult to use *traceroute* and impossible to use *ping* if both the Internet locations are remote sites. If cyberspatial analysis is to provide useful global information, it is essential that some method be developed that will allow the automatic and regular mapping of all Internet locations, and latencies between locations, in cyberspace. Such a method should also not add to the burden of an already congested network.

The second obstacle is the problem of counting the number of visits made to web sites located all over the Internet. Many web sites provide publicly accessible access statistics. There is a requirement for the development of some type of WWW spider or robot, similar to those used by WWW search engines, that can continuously discover and collect access statistics for use in tandem with latency measurements.

The third obstacle is the limited availability of cyber-demographic information. Spatially dis-aggregated counts are needed of Internet users and WWW sites. It may be possible to produce a partial census head count if enough WWW server access logs could be gathered together in a single archive. It should certainly be feasible for universities to contribute their WWW server access logs to a central archive for the purposes of Internet research.

Although these problems exist I would like to finish on one optimistic note. Cyberspatial research is possibly the only field of geography where almost all the necessary data can be gathered without leaving our desk.

References

Cerf, V. (1993) Testimony before the US House of Representatives, Committee on Space, *Science and Technology*, March 23.

Clinton W. J. and Gore A. (1997) 'A Framework For Global Electronic Commerce', *http://www.iitf.nist.gov/eleccomm/ecomm.htm*

Davis, C. (1997) 'RFC 1876 Resources Putting Locations into the DNS: An Overview', *http://www.kei.com/homepages/ckd/dns-loc/*

Dodge, M. (1996) 'Mapping the world wide web', *GIS Europe*, 5, 9, pp. 22–24.

Durham, T. (1997) 'Malaysian PM appeals for cyber city partners', *The Times Higher Education Supplement*, June 13.

Gardner, E. (1997) 'Study says online travel market will approach $9B within five years', *Web Week*. 3, 11.

Harrington, P. J. J. (1996) 'W7-95 an internet gazetteer', *http://warp.dcs.st-and.ac.uk/warp/reports/gazetteer/gazetteer.html*

Hertzberg, R. (1997) 'Clinton's net initiative', *Web Week*, 3, 4.

Hunt, C. (1996) *Networking Personal Computers with TCP/IP*. Sevastopol, CA: O'Reilley & Associates.

Leung, J. (1996) 'Winging their way to global might British Airways . . .', *TimesNet Asia*, *http://web3.asia1.com.sg/timesnet/data/ab/docs/ab1163.html*

MIDS (1997) 'MIDS Internet Weather Report (IWR)', *http://www3.mids.org/weather/index.html*

Network Wizards (1997) *Internet Domain Survey*, *http://www.nw.com/zone/WWW/top.html*

Quarterman, J. S., Smoot, C. M. and Gretchen, P. (1994) 'Internet Interaction Pinged and Mapped', in *Proceedings of INET'94 Conference*, Prague: Internet Society, *http://info.isoc.org/isoc/whatis/conferences/inet/94/papers/522.ps.gz*

Rickard, J. (1996) 'Mapping the internet with *Traceroute*', *BoardWatch Magazine*, 38, December, *http://www.boardwatch.com/mag/96/dec/bwm38.htm*

Stehle, T. (1996) 'Getting real about usage statistics', *http://www.wprc.com/wpl/stats.html*

Thompson, B. (1997) 'How to create and use cookies', *Internet Magazine*, 27, February 1997, pp. 113–115.

Tullis, P. (1997) 'Staff take hit for web financial failure', *Now Magazine*, May, *http://www.now.com*

UKERNA (1993) 'The SuperJANET Project', *http://www.ja.net/SuperJANET/SuperJANET.html*

Chapter 14

Integrating models and geographical information systems

Roger Bivand and Anne Lucas

14.1 Introduction

Since geographic information systems (GIS) currently dominate our perception of how computing and geography should interface, and since GeoComputation (GC) is providing analysts of spatial phenomena with ever more powerful computing tools, it may be helpful to examine the experience that has accrued concerning links between them. Our examination is both empirical and normative, and the reader may find it useful to repeat at least some of our literature surveys, since new papers and articles are accumulating rapidly. Searching on the key words 'GIS' and 'model*' or 'integrat*', where '*' is the wild card, led to a wide range of hits both in ISI Science and Social Science Citations Indices, and in OCLC-FirstSearch. These sources primarily contain journal articles, while conference proceedings may be searched at the Ohio State University GIS Master Bibliography Project, and more recently through the web-sites of conference organizers, such as NCGIA and GISDATA in Europe. Adding these resources to what we already knew about the issues involved, we were able to scan the field for interesting regularities, trends, and citation clustering.

In our initial hypotheses, we believed that there existed a common understanding of what integrating models and GIS entailed. In fact, our search illustrated that views on GIS-model integration were disparate, and no consensus has yet emerged through either practice or theory in the GIS or modelling communities. This early disappointment proved to be a useful pointer towards a more fruitful way forward, by problematizing both the contents and contexts of the concepts involved, together with the practices in which they are entrained. Views that we had expected to resonate through the literature can be quoted from presentations at the Boulder, Colorado, conference on environmental modelling and GIS held in 1991. Something big, interesting, and synergistic was to be had by combining GIS and environmental models. Both use maps (or map-like display) as an expression of reality, both have lots of data to deal with, and both require analysis of these data in some map-like form. GIS is presented as a general purpose technology for satisfying the following specific needs (Goodchild, 1993):

- preprocessing data from large stores into a form suitable for analysis including reformatting, change of projection, resampling, and generalization;
- supporting analysis and modelling, such that analysis, calibration of models, forecasting and prediction are all handled through instructions to the GIS;
- postprocessing of results including reformatting, tabulation, report generation, and mapping (these operations are expected to be available under a graphical user interface or GUI).

Dangermond (1993) saw the future as bringing more user-friendly GIS, with less need for GIS specialist knowledge, because both software and hardware would be more powerful, more graphical, less expensive, and easier to use. In addition, artificial intelligence would support users behind the scenes, there would be more input/output formats, the natural environment and its variability would be easier to represent as hardware prices fell, animation would become feasible for displaying results, and in ESRI, the trend towards integration into a single system: raster with vector, CAD with GIS, image processing with GIS, would continue. All of these should enhance the integration of models and GIS and be offered as the rational solution.

Maybe this is a solution looking for a problem, but it is not necessarily a viable solution to problems faced by many modellers. The fact that there is still a need to discuss integration of models and GIS in such 'general' terms suggests that either this expectation has not been fulfilled to the anticipated extent, and/or that there is something inhibiting 'easy' integration. Enough experience with models should have been accumulated after 30 years; GIS has been around just as long in one form or another. A review of the literature reveals a lack of clarity in the definition of 'integration', which can take place at many levels and in many forms. This issue needs to be re-examined. Integration has been promoted as a 'good thing', and that with better integration will come better, faster, easier systems and enhanced analysis. One would then expect to be able to follow the trend in the literature, culminating in some 'well integrated' system. We need to examine the literature, to determine how integration has proceeded over the years.

We have found that the majority of the models discussed in the literature are environmental, rather than social or economic. It seems that social and economic topics are mostly covered in the integration of spatial and exploratory data analysis and statistical tools in GIS. Consequently, we will by and large restrict this presentation to environmental models, based on theory drawn from physics, chemistry, or biology. We begin with a survey of the background concerning GIS, what integration in GIS has meant, and modelling. Continuing to develop the theme of integrating models and GIS, we examine the reasons for such steps, the problems encountered, how integration may be defined, and what alternative system architectures may be employed. We find that the issues involved in data modelling are of considerable importance for integration projects, and discuss a range of requirements for the viability of such projects. To illustrate these points, we describe an integration project with data from the Baltic Sea. In conclusion, we discuss some consequences not only for integrating environmental models and GIS, but possibly also for GC in more general terms, addressing data modelling in particular.

14.2 Integrating models and GIS

In this section, we will consider the background of geographic information systems, in order to understand integration in such systems, and the background for models, chiefly of environmental systems. We will be concerned both with the formalizations proposed for, and the practices entailed by, these technologies and/or this science; GIS seems to be more like technology, while modelling is seen more as a scientific activity (Wright *et al.*, 1997). Of course, there is a continuum between technology and science, or engineering and science, but this does not map directly onto the applied–pure continuum. There are differing reasons for the assumptions made, and differing motivations for the choice of topics reflected in the results achieved and the methods employed to achieve them.

14.2.1 Geographic information systems

Adopting Goodchild's description of a geographic information system given above, and illustrated in Figure 14.1, we quickly find that there are information systems fulfilling his criteria which perhaps do not term themselves GIS, and on the other hand there are software products accepted as and called GIS which only partly satisfy the same criteria. If a GIS is composed of hardware, software, liveware, and organizational procedures regulating its running (modified after Federal Interagency Coordinating Committee, 1988), GIS can hardly be used as a valid description of statistics packages or spreadsheets with map functions, visualization programs with layers, or indeed many cartographic or mapping products. The introduction of client/ server architectures to GIS has, for example, changed our understanding of what is a GIS; databases may now be distributed across platforms and user-friendly 'front ends' provide limited access and functionality to some of the larger, more complex systems.

Some of the confusion has doubtless come from the dramatic reduction in the cost of most hardware components needed for a GIS. Desktop hardware now regarded as standard for office automation is far more powerful than the minicomputers and workstations on which most GIS were initially

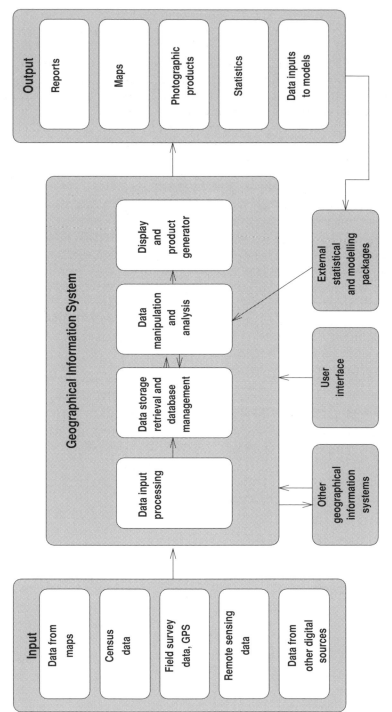

Figure 14.1 Component subsystems of geographic information systems (Fischer and Nijkamp, 1993, p. 5).

developed. Software has not fallen in cost to anything like the same extent, and most scientists would see commercial GIS as being expensive, especially in the context of the need for considerable expertise in their installation, running, and customizing. Modellers frequently dwell in the domain of community (shared) models and public domain software. In addition, data capture and purchase remain very expensive, again requiring skill, training and experience, most often the experience of learning from one's own mistakes.

For a mainstream GIS shop, a utility, public authority, or private company using the system for automated mapping and facilities management, the point of the GIS is to render more efficient and flexible operations already undertaken, related to data the organization largely generates itself. The 'geographic' adds value to existing and evolving corporate information systems, keying and relating disparate records by location in space, and breaking down artificial barriers between information stakeholders. In such a setting, the cost of establishing and maintaining knowledge of chosen GIS software is minimal compared to the benefits that the organization may reap, also compared to the costs of data acquisition. This is, of course, because the running of the GIS becomes very repetitive once established, calling for no substantial new learning after launch. In fact, the value-added for GIS vendors may lie more in customizing systems under long-term service contracts than in pushing shrink-wrapped packages.

Such customers do not need the standardization of GIS procedures, at most agreement on standards for digital data exchange to avoid unnecessary repetition of costly data capture. They are vitally interested in, and willing to purchase, integration in GIS where this leads to results of significance for their own goals and operations; this should surprise nobody, and is indeed one of the forces for progress in developing GIS. Before examining the effects of this view of GIS on modellers, who by and large have neither repetitive tasks to accomplish, nor budgets permitting major investments of time in learning how to set up and run information systems, we will survey the meaning given to 'integration' in the GIS context.

14.2.2 Integration in GIS

Integration has been proposed as one of the characteristics which distinguish GIS from other information systems. It has, for instance, been held by Shepherd (1991) that users can take a unified view of their data by integrating information in a GIS. He goes on to define information integration as 'the synthesis of geographical information in a computer system which depends for its effectiveness on information linkage (i.e. of spatial and attribute data) within a coherent data model. This brings together diverse information from a variety of sources, requires the effective matching of supposedly similar entities in these sources and demands information consistency across the data sets.'

This requires, Shepherd continues, the bringing together of spatial data from a variety of sources, the creation of geometrical descriptions of the earth's surface within consistent topological frames, the inter-conversion of raster and vector models of the world within a single software system, the provision of a comprehensive set of geographic information handling functions within a unified software environment, and the interlinking of both spatial and attribute data within a single coherent representation or model. These normative postulates are rather strong, and it can be argued that, to date, no GIS, production or prototype, satisfies them.

Rather more realistically, Flowerdew (1991) defines spatial data integration as the process of making different datasets compatible with each other so they can reasonably be displayed on the same maps and so that their relationships can be sensibly analysed. This reduces integration to the coregistration of data such that positioning errors of one data set or map layer with reference to others, or to the underlying phenomena, are minimized. It also implies prior decisions on the data models employed in representing the data sets. An issue which deserves attention in this connection, and which has bedevilled integration in GIS, is the raster/vector cleavage, associated with but not the same as the object/field distinction. It is to these that we now turn in a brief presentation of the modelling background.

14.2.3 Models

According to Steyaert (1993), models for our purposes may be defined as: 'computer based mathematical models that realistically simulate spatially distributed time dependent environmental processes in nature'. Environmental processes in the real world are three-dimensional, time-dependent and complex, frequently involving non-linearity, stochastic components, and feedback loops over multiple space-time scales. A range of the types of topics modelled are shown in Table 14.1. Models are still only simplifications of real-world processes despite their complexity; reality is only ever approximated in them. Among the different kinds of model, we could mention scale or analogue models, conceptual or symbolic models, and mathematical, non-deterministic (stochastic) or deterministic, steady-state or dynamic. A chosen scientific problem, topic and empirical arena may validly be modelled using a variety of approaches, which may or may not be in themselves well-integrated, but which may each shed light on the issues under study. While we are concerned here with computer-based mathematical models, other models will doubtless feed into them, in the same way that field or laboratory trials are used to tease out interesting relationships.

Models are constructed to give us representations of reality we can manipulate, allowing us both to demonstrate how real 1:1 scale environmental processes function, and how sensitive output is to changes in input and system parameters. However, there is often 'too much' reality to measure,

Table 14.1 A classification of environmental modelling domains and attempts to integrate with GIS (modified after Kemp (1993) and Sklar and Constanza (1991))

Discipline	Model objective	Typical modelling variables	Data model requirements[1]	Typical GIS-related problems encountered	Cross-section of references
Hydrological models	Predict flow of water and constitutents over land and through upper layer	Momentum, acceleration, depth, friction	2.5D plus time	Requires quality digital elevation model Input data (precipitation, evaporation) and stream gauges usually have limited coverage	Bian et al., 1996; Gorokhovich and Janus, 1996; Krysanova et al., 1996; MacMillan et al., 1993; Maidment, 1996a, b; Moore, 1996; Mueller-Wohlfeil et al., 1996; Ramanarayanan et al., 1996; Streit and Wiesmann, 1996
Land surface and subsurface models	Predict flow of materials (soil, surficial deposits, groundwater) where flow is contrained by the medium through which it flows	Momentum, acceleration, depth, friction	2.5 or 3D plus time	Little substantive theory; models based on a few empirically-based equations High demand for accurate spatial data Poorly defined spatial variables (e.g. slope, surface roughness)	Charnock et al., 1996; Ellis, 1996; Fisher, 1993; Harris et al., 1993; Laffan, 1996; Raper and Livingstone, 1996; Rundquist et al., 1991; Schell et al., 1996; Slawecki et al., 1996; Wilson, 1996
Ecological models	Predict resource distribution and number or size of population	Migration/ diffusion, density, birth/death/ growth, resources (nutrients, light, etc.)	2D plus time	Entities difficult to bound Fundamental theories often qualitative Non-linearities involved and complex interaction between factors	Akcakaya, 1996; Hart et al., 1996; Harvey, 1996; Horne et al., 1996; Hunsaker et al., 1993; Johnston et al., 1996; Liff et al., 1994; Maas and Doraiswamy, 1996; Mackey, 1996; Michener and Houhoulis, 1996

Table 14.1 (cont'd)

Discipline	Model objective	Typical modelling variables	Data model requirements[1]	Typical GIS-related problems encountered	Cross-section of references
Landscape ecological models	Predict flows between points and mechanisms of change in a spatial pattern	As for ecological models plus, density, transition rates, habitat, mass, turbulence	2D plus time	Uses complex, non-linear mathematics Hierarchic processes need to be linked at multiple scales	Aspinall, 1993, 1994; Bogs et al., 1996; Bowser, 1996; Dodson and Turner, 1996; Dungan and Coughlan, 1996; Gustafson and Gardner, 1996; Haines-Young et al., 1993; Logsdon, 1996; McKeown et al., 1996; Ojima and Parton, 1996; Rupp, 1996; van Horssen, 1996; Yarie, 1996
Atmosphere/ ocean models	Predict velocity, mass and direction of flows in atmosphere and/ or ocean system	Momentum, turbulence, temperature, moisture (air), salinity (oceans), density, pressure	3D plus time	Very dynamic: high rate of change for conditions Fully 4D entities Sparse observations relative to volume of model-produced data Complex models requiring supercomputer servers	Dragosits et al., 1996; Galagan and Howlett, 1994; Lee and Pielke, 1996; Lucas, 1996; Walker et al., 1996; Wu, 1996

[1] 2.5 dimensional settings involve the draping of two dimensional data over a relief model.

and the same applies to shoe-horning observed data representing reality into data models, needed for digital modelling. If reality could be entirely captured as objects, or entities, whose position and relationships could be described accurately, then integrating models and GIS would be trivial, because GIS presuppose both entitation and known position and topology. As Burrough (1992a, 1996) shows, however, reality by no means consists of such fully defined objects (Figure 14.2). Most natural phenomena are rather fields displaying smooth, continuous spatial variation, or spatial entities which are either incompletely defined or incompletely definable (Peuquet, 1984). Csillag (1996) seems to oversimplify the issue by mapping the raster/vector dichotomy onto the object/field; the problem is perhaps more oblique, and challenges Shepherd's desire to be able to shift between raster and vector representations at will or as needed.

A further issue brought up in the definition of models above is that reality is not two-dimensional. Maps are two-dimensional artifacts, but even they are arbitrary representations of regions of a lumpy spheroid. Lurkers in GIS and GPS newsgroups know that questions about projections are among the most frequently posted. Even modern introductions to GIS (Worboys, 1996) often simply choose to assume that data have been projected onto the plane, although awareness that this is artificial is present (Willmott et al., 1996). Elevation or depth, taken as the third dimension, is not integrated into GIS, but is a fundamental requirement of most environmental models; the same applics to time, the fourth dimension, for very many models (Peuquet, 1994). As we will see in the next section, the mismatch between GIS, with a representation of reality founded on $(x, y, \text{attribute}, \ldots)$, and models $(x, y, z, t, \text{attribute}, \ldots)$ can be overcome (Goodchild, 1992), but not necessarily in general.

14.2.4 Integration of models and GIS

Albrecht (1996a), justifying the need to interpose a virtual layer between GIS and the modeller, writes that: 'current GIS have little to offer to the scientist who is interested in modelling spatial phenomena. They are so difficult to use that it takes some expertise to handle them, and it is not unusual to work a whole year before an operator masters a GIS. This is especially cumbersome for cursory users (such as environmental modellers) that employ GIS as one tool among many others'. Despite his pessimism, in fact a good deal has been done where the requirements of models and GIS are found to be commensurable. Indeed, models are but one of a number of modules for integration with GIS, others being decision support tools, statistical procedures, and visualization programs. A characteristic typology of modes of integration is shown in Figure 14.3; it seems that loose coupling is the mode that has achieved the widest acceptance so far, since it can be implemented by sharing file formats. Tight coupling needs mechanisms in

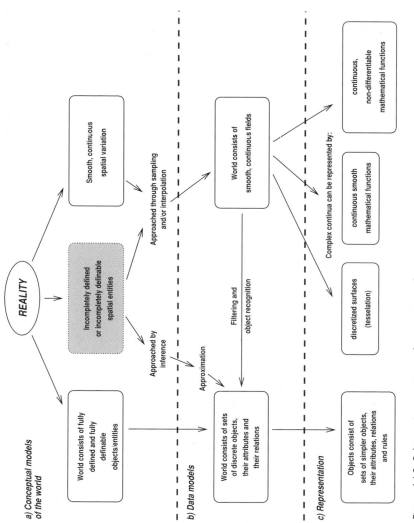

a) Conceptual models of the world

REALITY

World consists of fully defined and fully definable objects/entities

Incompletely defined or incompletely definable spatial entities

Smooth, continuous spatial variation

Approached by inference

Approached through sampling and/or interpolation

b) Data models

Approximation

World consists of sets of discrete objects, their attributes and their relations

Filtering and object recognition

World consists of smooth, continuous fields

Complex continua can be represented by:

c) Representation

Objects consist of sets of simpler objects, their attributes, relations and rules

discretized surfaces (tesselation)

continuous smooth mathematical functions

continuous, non-differentiable mathematical functions

Figure 14.2 Schematic overview of progression from conceptual models of the world to their representation (Burrough, 1996, p. 8).

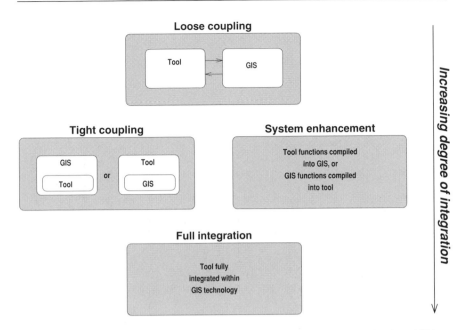

Figure 14.3 Four ways of linking GIS and modelling technologies (after Fischer, 1994, p. 237).

the running programs for data exchange, for instance remote procedure calls, while both system enhancement and full integration require access to source code.

In this section, we will consider some of the reasons for integrating models and GIS, examine the functional requirements for such integration, and survey chosen system architectures. In the next section, we move to the issues identified above as being crucial for progress with integration, data and data modelling, so they will receive less attention here.

14.2.5 To integrate, or not to integrate?

Parks (1993) gives three reasons to integrate. Firstly, spatial representation is critical to environmental problem solving, but GIS lack the predictive and analytic capabilities to examine complex problems (GIS need to be able to operate outside their planar framework). Secondly, modelling tools lack sufficiently flexible GIS-like spatial analytical components and are often inaccessible to the non-specialist. Finally, modelling and GIS can both be made more robust by their linkage and co-evolution. Integration is clearly understood as loose or tight coupling, rather than the subsumption of models within GIS, a much weaker use of the term than by Sadler and Hamid (1992). They differentiate three modes of linkage: connected systems (with

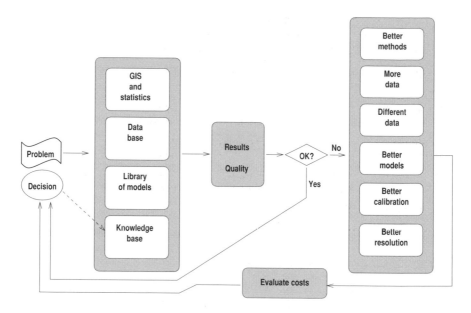

Figure 14.4 The components of an intelligent geographic information system with feed-back on the quality of results and how to improve them (Burrough, 1992b, p. 9).

standard file formats for file transfer); interfaced systems (data exchange between systems is completed without exiting either system); and integrated systems (a change in one data set is immediately reflected in all the other data sets). Our use of 'integration' includes all three of their types, including loosely coupled or connected systems, through more advanced and more demanding modes.

In Figure 14.4, Burrough (1992b) illustrates how integration can be problem-driven, when both models and GIS are subject to the need for the scientist to solve pure or applied research problems. The criteria guiding the choice of appropriate tools to be used together is the researcher's evaluation of the quality of the obtained results in the context of the problem, and the costs involved in improving result quality. Changes in the right-hand components can affect any or all of the left-hand boxes. For instance, better methods can signify the need to refine the GIS tools employed, more or better data – the database, better models – the library of models, and better calibration and resolution – the knowledge base. If the GIS tools employed enhance result quality, they will be drawn into the centre of the research effort, because they contribute to a better solution of the problem. If, on the other hand, they are costly or do not contribute as much, say, as alternative data marshalling or presentation tools, they may not deserve to be considered at all.

Table 14.2 The 20 universal GIS operations (Albrecht, 1996b)

Search	*Interpolation*	*Thematic search*	*Spatial search*	*(Re-)classification*
Locational analysis	Buffer	Corridor	Overlay	Thiessen/ Voronoi
Terrain analysis	Slope, aspect	Catchment, basins	Drainage, network	Viewshed analysis
Distribution/ neighborhood	Cost, diffusion, spread	Proximity	Nearest neighbor	
Spatial analysis	Multivariate analysis	Pattern, dispersion	Centrality, connectedness	Shape
Measurements	Measurements			

14.2.6 Functional specifications

Accepting that most environmental research is problem-driven, integration can be simplified to establishing the functions that GIS can perform for models, either exclusively, or better than alternative systems. Table 14.2 shows a listing of universal GIS functions that Albrecht (1996b) has compiled; they are similar to and build on the set defined by Tomlin (1990) as a cartographic modelling language.

These analytical functions slot neatly into a wider range given by Goodchild (1993):

- efficient methods of data input;
- alternative data models, especially for continuous data;
- an ability to compute relationships between objects based on geometry and handle attributes of pairs of objects;
- a range of geometric operations (area, distance);
- an ability to generate new objects on request based on geometric rules (e.g. Thiessen polygons);
- assigning new attributes to objects based on existing attributes and complex numeric and logical rules;
- transfer of data to and from other packages (simulation models and statistics).

Four of his central points are covered by Albrecht's more detailed specification. Goodchild (1993) concludes: 'as with all GIS applications, the needs of environmental modelling are best handled not by integrating all forms of geographic analysis into one GIS package, but by providing appropriate linkages and hooks to allow software components to act in a federation'. All of these functions, it may be argued, may most efficiently be carried out

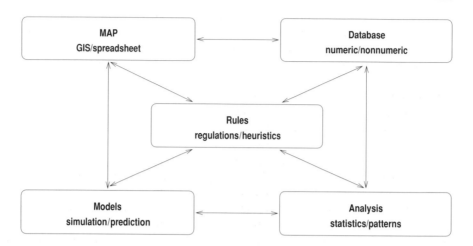

Figure 14.5 The RAISON paradigm: an integrated system with five primary components linked to each other (Lam, 1993, p. 271).

within a GIS. It follows that modelling that involves calling on sufficiently many of these functions to cover entry costs will find integration with GIS functionally rational.

14.2.7 Achieving integration

The RAISON system shown in conceptual form in Figure 14.5 is an example of model/GIS integration, achieved by stepping outside both (Lam, 1993; Lam and Swayne, 1993; Lam *et al.*, 1994a, b; Lam *et al.*, 1996). RAISON is problem-driven, centered around a conceptual knowledge-base kernel, which may represent formalized rules in expert system mode, or a looser set of heuristics driven by the practice of environmental research and management. Five phases are distinguished in constructing such an information system:

- the exploration phase examining where the data are located and what the sample sites intersect with, for instance soil units. This phase uses database queries on spatial units, but requires functions beyond the standard GIS repertoire to get background data for modelling, say from satellite images.
- the tool assembly phase often involves much work to assemble suitable database and program libraries. Considerable flexibility is required here.
- the trial and verification phase is connected to tool assembly by iteration, but also entails overlaying data layers to see which data sets or parameters will be most useful.

- the production and verification phase should institute and run repetitive and demanding test procedures before the system moves into production.
- the report and presentation phase needs flexibility to show different variables, and combinations of variables and scenario parameters.

The RAISON team stepped beyond standard GIS because of difficulties in the systems integration of modelling, statistical data analysis, databases, and GIS, all of which tend to be delivered as large and sophisticated systems, often also having more functionality than that required by the tasks at hand. One requirement found to be important for environmental modelling and management was conducting simple queries for time series of attribute values at chosen sites. It is clear that GIS, viewing data as $(x, y,$ attribute, $\dots)$, will lose control of the time dimension if the time series are packed into the site attributes, and will be grossly inefficient if all the separate time-slice map layers have to be searched to reconstruct the series for a queried site. Since RAISON has been implemented for watershed management and planning, timely recovery of series of instrument values for chosen sites is crucial in this application.

RAISON can be seen as a toolkit with smoothly interconnecting components, where the user only need select those components required for a particular application, and where external applications, e.g. models, can be joined in orderly fashion. It may be, however, that the modules are so tightly integrated that the potential universality of the design is seriously reduced; this is full integration in terms of Figure 14.3. Production is tied to the use of information systems more in management than research, and brings RAISON closer to the customization and decision support paradigm we see in facilities management, albeit for agencies charged with environmental protection. Dimmestøl and Lucas (1992) compared three approaches to less-tight system integration, which they termed the model-centered approach, the GIS-centered approach, both system enhancements in terms of Figure 14.3, and the shell or tool-box approach, also known as loose or tight coupling in Figure 14.3. The model-centered approach assumes either that the user is a highly qualified scientist or that the model is constrained such that the user needs to change few parameters. Building in the constraints and a suitable interface can only be afforded in terms of development effort if the system is to be used repetitively. Because GIS functionality is limited, there is little leeway for exploratory analysis, subjecting the data to visual interrogation. Knowledge resides *a priori* in the model, with little opportunity for the user to contribute. The GIS-centered approach may be feasible, but means that four-dimensional data for input to model modules, and their output, have to be coerced into hostile data models, leading to grossly inefficient database use. Finally, the shell or tool-box approach seemed at that stage to solve many of the difficulties of the model and GIS-centered approaches. It provides the best selection of tools and demands little in user

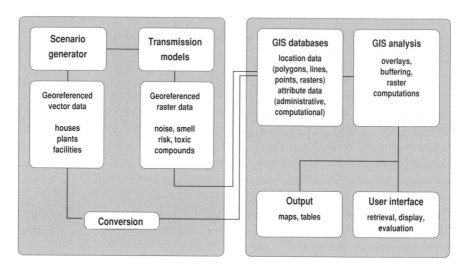

Figure 14.6 The structure of an environmental zoning application (ten Velden and Kreuwel, 1990, p. 123).

interface or systems programming, but requires consistent conversion routines between formats, assuming no loss of information, imposing metadata requirements. On the other hand, as with all toolboxes, the user is made responsible for the choice of tools to apply to tasks, getting at best a warning message about the correct usage of the tool, at worst meaningless results, if an inappropriate tool is chosen. This issue can be tackled at the shell level, by providing some often used combinations of underlying tools ready-packaged, for a variety of specific modeller/user domains.

Figure 14.6 shows an example of tool-box-style integration for modelling environmental impact for an environmental zoning application (ten Velden and Kreuwel, 1990). GIS functions are used where they are most efficiently executed, while model components are linked through data conversion procedures, letting each part of the hybrid do what it does best. However, as Engel *et al.* (1993) point out in integrating the AgNPS (Agricultural Non-Point Source) pollution model and the GRASS GIS, this style of integration is difficult to implement without access to the source code of the component tools. It not infrequently occurs that, even when applications programming interfaces and data formats are well documented, the demands of integration press the software beyond its presumed use, and into the realms of 'undocumented' features. Having access to source code permits such features to be instrumented and debugged. In other cases reading source code can indicate that the underlying design of the candidate module will prove unsatisfactory, prompting the choice of an alternative.

Djokic *et al.* (1996) have summarized experience with integration to date.

They found that the integration of GIS and modelling software has often involved the construction of one-off systems with much time spent creating and maintaining specialized interfaces and data exchange formats. Most models seem only to concern a portion of some larger cycle, with few attempts to develop comprehensive solutions because of the complexity of the problem, or the solution, or finally the lack of data to support the degree of complexity required. While more computational power can help us with the second problem, and remote sensing data with the third, the scientific issues in the first remain central to modelling. They see two viable approaches to integrating models and GIS: either fund and build a whole new comprehensive model, or take pieces from existing models and string them together, reusing code that has been tested and approved by the scientific community. In this spirit, they identify data-level integration as a least-resistance path forward, and propose a data exchange approach to common data formats, combining descriptions of file structures with drivers for data conversion to and from common formats from ranges of specific or proprietary formats.

14.2.8 Data and data modelling

Djokic *et al.* (1996) are not alone in concluding that the key to integration of models and GIS is at the moment to be found in data and data exchange, rather than in loftier system architectures. Given the significance of the role played by data conversion in the tool-box approach to integration, this is perhaps not so much a new beginning as the accumulation of experience of what can be achieved in practice. Hunsaker *et al.* (1993) report that work in modelling ecological systems and processes in space is moving in a similar direction. Maidment (1996a) indicates that hydrological modelling is being rethought. In the past, the tendency was to build the best model to represent the physical processes in reality, and then worry about finding data that would fit into that model. The GIS database was then linked to the existing model. Now there is a trend to identify the data first, then build a spatial hydrological model that uses the data that are actually available. This new approach can then take advantage of the spatial data organizing capabilities of the GIS, a reversal of the traditional priorities. Much the same is reported by Wilson (1996) for the many soil erosion and non-point source pollution models developed with GIS. Integration here has typically addressed input data requirements and the role of GIS for some specific models. Emphasis has been placed on data resolution and input data estimation and interpolation, also touching on the modifiable areal unit problem, one of the key symptoms of ill-bounded entities.

  In this section, we will move from observed research practice up to normative descriptions and conceptualizations of geographical data models, not because the latter are unimportant, but because researchers doing integration are in general problem-driven, and have arrived at the position sketched above having experienced both GIS that did not suit models or data, and theory-based models that were far too demanding with regard both to empirical data, and to the digital representation of the concepts in the models. Both these extremes led to an inability to deliver usable results, or even to calibrate models satisfactorily. Here we will point to ongoing work on data, before rounding off with a summary of issues connected to geographical data models. Readers interested in following the ongoing discussions in spatial data modelling in detail are referred to the following: Burrough and Frank (1996), Goodchild (1993), Kemp (1993, 1996a, b), and Peuquet (1984, 1994).

14.2.9 Data

Kirchner (1994) identifies three groups of data required for simulation modelling: data for constructing the model, data for running the model, and data for analysing the model (including uncertainty analysis, sensitivity analysis, validation, and verification). Many researchers are not sufficiently aware that data management is needed because of the many different types of data required; in addition, model runs generate large volumes of derived data. Most GIS attempt to encourage the researcher to document the origin of data, and metadata can accommodate indications of data accuracy (Aspinall and Pearson, 1996; Burrough *et al.*, 1996; Hunter and Goodchild, 1993, 1994), in addition to the basic information. While the research context is not one in which much use tends to be made of metadata, it remains good practice to adapt the fields of metadata records associated with data to document by whom the data set was created, when, for what reason, and how. The latter ought to cover the input data used, and model version and parameters if related to a model run. Handling such metadata adequately is an important challenge for data conversion utilities. For a more comprehensive treatment of metadata questions, see Chrisman (1994) and Lucas *et al.* (1994).

Communication about data, and the sharing of documented data sources is increasing in significance. Driel and Loveland (1996) report on steps being taken by the USGS to produce data specifically for environmental modelling and monitoring, inventory and management, with the goal of making a multiscale, multipurpose, database at scales from 1:24 000 to 1:2 million. One practical reason for modellers to use GIS is to extract data from such public repositories for further use as background information for the processes of interest. While many GIS are not the tools of choice for preprocessing remotely sensed data, these also add richly to the range of information sources at hand, albeit at a price. The advent of frequently repeated coverage of much of the earth adds a major challenge to using these kinds of volumes of raw data in GIS: one can anticipate that the registration and collation of repeated observations may take longer than the repeat period itself, calling for performance and automation enhancements orders of magnitude better than present practice.

Lees (1996) draws attention to the difficulties that may arise when data are derived from paper map sources, particularly when the apparently new digital database is derived from choropleth maps, for instance of vegetation or soils. The original document may have been best practice when it was drawn, but is strictly inappropriate for such continuous phenomena. As Burrough (1996) puts it: 'This should worry managers of natural environment resource databases and their clients because (it implies) that the fundamental conceptual models used to represent reality are insufficient to capture the complexities of multiscale, polythetic phenomena, and that therefore the information they store or use is incomplete.'

14.2.10 Data modelling

As Couclelis (1996) puts it: 'Few outside the community of GIS researchers (and perhaps also cartographers) would think that the representation of geographic objects with undetermined boundaries is a problem at all, or that the distinction of geographic entities into well-bounded and ill-bounded recognizes an important division of things in nature'. We are returning here to themes introduced in Figure 14.2, to one of the major reasons for the lack of success, by and large, of model/GIS integration projects, and to the current trend for data-driven integration.

Table 14.3 constitutes a distillation of the issues engaged in considering data models. We have followed Albrecht's (1996b) ordering here, rather than Frank's (1992), going from concepts through data models to data structures. It is clear that if large-scale model/GIS integration projects are to succeed, they must achieve coherence at all three levels, unless some procedure for encapsulation or scoping is put in place. This is precisely the solution proposed by Albrecht, who would like to permit researchers with varying knowledge, traditions, and understandings access to common data structures on their own premises, internal to their own disciplines, and then allow multidisciplinary results to grow bottom-up. In this way, divergence and dissonance in concepts, and even in data models, would be scoped out of discipline clusters, which would then be free to use the data structures and the universal operations adhering to them to solve problems seen as relevant. A further important reason to shy away from coherence at the conceptual level is that, as Albrecht points out, it is not possible to design at this level to accommodate concepts that are, as yet, unknown.

Finally, it is worth mentioning attempts to use hierarchical data structures (quadtrees) by Csillag (1996), and object-oriented design by Raper and Livingstone (1995) as alternative approaches to spatial modelling, in addition to widespread interest in fuzzy representations of geographic objects (see, for instance, Burrough, 1996, and Fisher, 1996). All hold promise,

Table 14.3 A comparison of definitions of conceptual, data model, and data structure levels given by Frank (1992) and Albrecht (1996b)

Frank (1992), p. 412	Concepts (specifically spatial concepts and geometry)	Ideas, notions, and relations between them which are used by humans to organize and structure their perception of reality. They differ depending on the task at hand, the circumstances, and the experience of the persons. They are either formally defined but cannot be implemented, because of fundamental restrictions of computer systems (e.g. limitations of finite machines), or informal, that is not formally defined or currently definable.
	Data models (specifically geometric data models)	A comprehensive set of conceptual tools to be used to structure data. They are defined formally and are constructed such that they can be implemented.
	Data structures (specifically geometric and spatial data structures)	Detailed and low-level descriptions of storage structures (traditional data structures) and the pertinent operations, with details of how the desired effects are achieved. They will not only provide a specific function (i.e. fulfill the conditions of an operation) but also are fixed in terms of performance, storage utilization, etc. They are a specific solution to a generic problem.
Albrecht (1996b), p. 327	Geographical models	Are the conceptual models used by environmental modellers as they develop an understanding of the phenomenon being studied and extract its salient features from the background of infinite complexity in nature. Models at this level cannot be completely specified.
	Spatial data models	Are formally defined sets of entities and relationships that are used to discretize the complexity of geographic reality (Goodchild, 1992). The entities in these models can be measured and the models completely specified.
	Data structures	Describe details of specific implementations of spatial data models.

not least as spurs to thinking more carefully about the digital rendering of reality in models.

14.3 Case study: integration of ocean model and GIS for the Baltic

This application was initially designed as the basis for impact assessment in the event of a chemical and/or oil spill (Lucas, 1996). The goal was to integrate the results from a high resolution, multilayer simulation model with a spatial database containing habitat, wildlife and economic information. The simulation model used is the three-dimensional, sigma-coordinate,

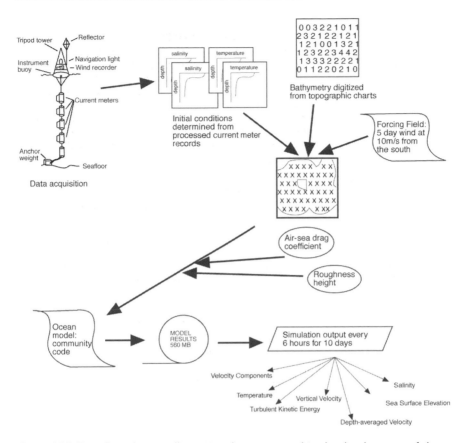

Figure 14.7 Data flow diagram illustrating data types used in the development of the numerical model and output variables (after Lucas, 1996).

coastal ocean circulation model of Blumberg and Mellor (1987), also known as the Princeton Ocean Model (POM). The specifics of this Baltic implementation of the model are described in detail elsewhere (Aukrust, 1992). In order to save time in the event of an environmental emergency, model simulations corresponding to combinations of representative winds and initial conditions were to be run on a mainframe, compiled and stored in advance in a GIS-workstation.

The model set-up is shown schematically in Figure 14.7. A 5×5 minute model grid was established for the entire Baltic (containing more than 9000 data points); this grid is approximately equal to a 5 km resolution in the longitudinal direction and 10 km in the latitudinal direction. Bathymetry was obtained on the same 5×5 minute grid (personal communication, F. Wulff, University of Stockholm). The model was initialized with temperature and salinity values derived from averaged summer 1988 measurements stored in the HELCOM database. The first scenario was driven by an idealized wind forcing of velocity 10 m/s over a 10-day period.

Results were viewed periodically during the model run and output every 6 hours during the 10-day simulation. The model resolves 14 levels in the vertical and at each grid point the following conditions were estimated for each level: temperature, salinity, turbulent kinetic energy, turbulent macro length scale, velocity components (u, v), and vertical velocity. Sea surface elevation and depth-averaged velocities were also produced.

14.3.1 Data modelling

In order to fit the obviously four-dimensional data into an essentially 2.5-dimensional GIS, the data modelling of the model output had to be handled in three stages. Firstly, the quantity of data needed to be selectively reduced for practical reasons. Secondly, the data needed to be structured according to the constraints imposed by the GIS software. Thirdly, GIS data needed to be transformed for ease of display and analysis.

14.3.2 Stage 1

For a single simulation, more than 1 gigabyte of output was produced. This was reduced by almost half by generalizing the data to produce sets of daily conditions. Although the fine-scale temporal variations were lost, this loss was determined to be acceptable since it was felt that users could visually interpolate between the daily views; there were no practical alternatives. In the vertical dimension, the 14 levels were reduced to 4 for input into the GIS database. These four were chosen in consultation with both modellers and marine biologists in order to minimize loss of potentially significant data. The basic structure is shown in Figure 14.8. It must be emphasized that three of these model 'levels' are, in fact, density surfaces, and are not associated with a consistent depth. This is at odds with the traditional 'layer' concept in GIS. A fourth level was selected at a constant depth of 60 m as a reference plane and permitted comparisons between sites at the same depth.

14.3.3 Stage 2

The attributes associated with each point were stored in the database internal to the commercial GIS. A separate database table was established for each day in the model run; 10 tables in all. In each table, the 15 attributes plus the grid point identifier were encoded for each point resulting in a table of size 16×9340. These attributes included bathymetry, temperature, salinity, turbulent kinetic energy and sea surface elevation. Owing to software limitations, each table was linked to a separate copy of the model

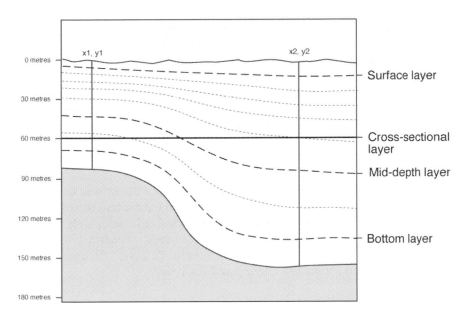

Figure 14.8 Cross-sectional view through the water column at two locations: (xl, yl) and (x2, y2). Only four of the fourteen levels produced by the numerical models are used in the GIS implementation (after Lucas, 1996).

grid. Velocity components (horizontal) were stored separately in a similar set-up.

This structure was a compromise between many small tables (40 tables of 4 attributes × 9340 model grid points) and one large (unmanageable) table (150 attributes × 9340): However, the cost of maintaining 10 versions of the spatial data (model grid) was high in terms of disc storage and potential version update. There was provided easy 'point and click' querying to invest-igate conditions for each variable or combinations of variables on a given day with this structure. It should be noted that this was less than optimal for querying and displaying temporal trends, since each of the 10 tables needed to be accessed in turn.

14.3.4 Stage 3

The model grid and its attributes are best spatially represented by points, as this is closest to the original form of the data (which are not, in fact, data in the traditional sense, but rather model output). The point-attribute combina-tion is suitable for searching at each grid location, but this form does not lend itself to either query at random locations between grid points (solving the 'what's here?' query) or useful visualization of the data.

Since the data were to be viewed at at least two scales, two different representations (or data model transformations) were used. Thiessen polygons were constructed around each model grid point and the linked attributes were assigned to the polygon. This produced a map suitable for viewing large areas, e.g. basin scale, with an appropriate level of detail.

The same model data were interpolated into a gridded raster image with a spatial resolution of 500×500 m for viewing small areas (large scale maps). Maps with this level of spatial detail were not suitable for displaying geographic areas as they took too long to plot to screen.

In both cases, the images resulting from the data model transformation from discrete point to complete raster appeared to have 'gained' information in the interpolation procedure. This can be especially critical depending upon the classification scale used to colour code the pixel values. Uninformed users, seeing only the finished product and not the original source, can be left with an inappropriate impression of data accuracy.

14.3.5 Discussion

The data modelling for the simulation results was difficult. Conventional database strategies were not of much help in constructing a four-dimensional database using two-dimensional tables; much of the work was done by trial and error and involved compromises between query capability, efficient data storage, and appropriate display. While the point spatial representation was suitable for linking to the attribute tables, in practice they were difficult to use. Point maps were mostly used to generate new coverages and to convert from one spatial data representation to another. In addition, they served as explicit reminders of the spatial distribution of the original data (i.e. the model grid), but needed to be deliberately plotted on top of secondary layers.

Multiple representations of the data were needed. It was easiest to integrate point maps with other data having different spatial structures, e.g. comparison with satellite images for evaluation of sea surface temperatures. Thematic maps generated from the point data and created as continuous rasters were optimal for displaying individual fields and comparing fields. Classifications could easily be modified through an internal scripting language. However, it was neither efficient to generate these raster maps on the fly, nor create raster maps for every attribute at every time step, both owing to the quantity of data involved. As a result, only the surface characteristics were produced as thematic maps. Once produced, these raster maps took on a planar appearance, when in fact, the map surface followed a density surface of varying depth. A more appropriate visualization would have been as a draped surface over the density surface height. Contour maps were not used, but would have been useful for the modellers.

Multiple scales were also required. For the detailed view, a raster cell resolution of 500×500 m was used to produce the series of thematic maps.

These maps were intended for visual analysis at the subbasin scale. While this resolution produced aesthetically-pleasing maps, especially when combined with the shoreline mask at the same resolution, this level of detail was much better than the original model grid (approximately 5 km × 10 km), and implied a high degree of data accuracy. As a tool for the oceanographic and biological modellers, this presumed map detail was not an issue. However, the intent to combine these compiled maps at a later date with site-specific habitat data for impact assessment, and the known persistence and replication of digital map files, created a concern for potential inappropriate use of the data.

The much coarser resolution afforded by the Thiessen polygons is suitable for investigation of patterns and processes at the basin scale. These polygons are a better representation of the actual data content since they were at the same resolution as the model grid. However, when combined with a shoreline in either raster or vector structure, the apparent mismatch in detail between the two coverages caused viewers to question the reliability of the data themselves. This was more pronounced when users 'zoomed-in' to smaller geographic areas.

14.4 Conclusion

The case study demonstrates how four-dimensional numerical models (or more precisely model results) can be incorporated into a commercial 2.5-dimensional GIS. It also identifies the cost of doing so, in terms of requirements for multiple representations (e.g. data structures), at multiple scales (e.g. spatial resolutions), in a multitude of two-dimensional tables (e.g. data storage). Finding the optimal solution means balancing user needs and system capabilities, and then settling for a tool that only marginally fulfills the task.

In this examination of the integration of environmental models and GIS, we have been concerned to balance normative views with the experiences of the research communities involved. Much of this experience can help us to make better use both of models and concepts, and of geographic information in the future, even when some of the projects from which experience is derived have not achieved all of their goals. We would argue that three conclusions may be drawn about relationships between these models and GIS, and by analogy, between GC models in more general forms and GIS.

• Integration is of greater benefit when it is problem-driven by a process relating available data, whether from observations or model runs, and the quality and applicability of results achieved, subject to cost constraints on result improvement. Integration appears to suffer from being driven by normative goals linked to system architecture design as such, and experience seems to show that integration for its own sake is usually both costly and yields little.

- Shell and tool-box approaches as modes of implementing integration accord better with current software development practice and theory than alternatives. Attention to scoping in designing both information systems and data structures used in models, together with adequate and quality-controlled documentation of file formats, interface library functions, etc., can enable the researcher to undertake the one-off customization necessary to achieve integration without compromising the integrity or quality of output.
- As Kemp (1996b) argues, without working on data modelling, integration projects are likely to founder. We would argue that it is not only integration projects that place the scientific value of their results at risk by ignoring data modelling, but that this conclusion applies with full force right across the range of GC. Data modelling, although it might seem to be dry and abstract computer science, far from 'muddy boots' environmental work, is actually at the heart of being able to use digital computers on 'muddy' conceptual problems. It implies thinking very carefully through entitation, and is essential for the expression of object-oriented design (Raper and Livingstone, 1995).

Finally, as Burrough and Frank (1995) indicate, current GIS are less than truly generic. They can in turn benefit from more comprehensive work on data modelling, enriching the simple repertoire to which they have traditionally, and understandably, been constrained. Until recent years, GIS proponents have tended to see integration as beneficial in itself. Integration is maturing, and with it our understanding of the challenges it involves.

References

Akcakaya, H. R. (1996) 'Linking GIS with models of ecological risk assessment for endangered species', in *Proceedings of the 3rd International Conference on Integrating GIS and Environmental Modelling*, CD-ROM, UC Santa Barbara: NCGIA.

Albrecht, J. (1996a) 'Geographic objects, and how to avoid them', in Burrough, P. A. and Frank, A. U. (eds) *Geographic Objects with Indeterminate Boundaries*, London: Taylor & Francis, pp. 325–331.

Albrecht, J. (1996b) 'Universal GIS operations for environmental modelling', in *Proceedings of the 3rd International Conference on Integrating GIS and Environmental Modelling*, CD-ROM, UC Santa Barbara: NCGIA.

Aspinall, R. (1993) 'Use of geographic information systems for interpreting land-use policy and modelling effects of land-use change', in Haines-Young, R., Green, D., and Cousins, S. (eds) *Landscape Ecology and Geographic Information Systems*, New York: Taylor & Francis, pp. 223–236.

Aspinall, R. (1994) 'GIS and spatial analysis for ecological modelling', in Michener, W. K., Brunt, J. W. and Stafford, S. G. (eds) *Environmental Information Management and Analysis: Ecosystem to Global Scales*, New York: Taylor & Francis, pp. 377–396.

Aspinall, R. and Pearson, D. M. (1996) 'Data quality and error analysis: analytical use of GIS for ecological modelling', in Goodchild, M. F., Steyaert, L., Parks, B., Johnson, C., Maidment, D. Crane, M. and Glenndinning, S. (eds) *GIS and Environmental Modelling: Progress and Research Issues*, Fort Collins CO: GIS World Books, pp. 35–38.

Aukrust, T. (1992) 'Adapting the three-dimensional sigma-co-ordinate model by Blumberg and Mellor to the Baltic Sea', *IBM Bergen Scientific Center Paper 92/1*, Bergen.

Bian, L., Sun, H., Blodgett, C., Egbert, S., Li, W., Ran, L. and Koussis, A. (1996) 'An integrated interface system to couple the SWAT model and ARC/INFO', in *Proceedings of the 3rd International Conference On Integrating GIS and Environmental Modelling*, CD-ROM, UC Santa Barbara: NCGIA.

Blumberg, A. F. and Mellor, G. L. (1987) 'A description of a three-dimensional coastal ocean circulation model', in Heaps, N. (ed.) *Three-dimensional Coastal Ocean Models*, Vol. 4, Washington: American Geophysical Union Monograph.

Bogs, F. C., Newell, J. and Fitzgerald, J. W. (1996) 'Modelling spatial effects of landscape pattern on the spread of airborne fungal disease in simulated agricultural landscapes', in *Proceedings of the 3rd International Conference On Integrating GIS and Environmental Modelling*, CD-ROM, UC Santa Barbara: NCGIA.

Bowser, G. (1996) 'Integrating ecological tools with remotely sensed data: modelling animal dispersal on complex landscapes', in *Proceedings of the 3rd International Conference On Integrating GIS and Environmental Modelling*, CD-ROM, UC Santa Barbara: NCGIA.

Burrough, P. A. (1992a) 'Are GIS data structures too simple minded?' *Computers & Geosciences*, 18, pp. 395–400.

Burrough, P. A. (1992b) 'Development of intelligent geographical information systems', *IJGIS*, 6, pp. 1–11.

Burrough, P. A. (1996) 'Natural objects with indeterminate boundaries', in Burrough, P. A. and Frank, A. U. (eds) *Geographic Objects with Indeterminate Boundaries*, London: Taylor & Francis, pp. 3–28.

Burrough, P. A. and Frank, A. U. (1995) 'Concepts and paradigms in spatial information: are current geographical information systems truly generic?', *IJGIS*, 9, pp. 101–116.

Burrough, P. A. and Frank, A. U. (eds) (1996) *Geographic Objects with Indeterminate Boundaries*, London: Taylor & Francis.

Burrough, P. A., van Rijn, R. and Rikken, M. (1996) 'Spatial data quality and error analysis issues: GIS functions and environmental modelling', in Goodchild, M. F., Steyaert, L., Parks, B., Johnson, C., Maidment, D., Crane, M. and Glenndinning, S. (eds) *GIS and Environmental Modelling: Progress and Research Issues*, Fort Collins CO: GIS World Books, pp. 29–34.

Charnock, T. W., Elgy, J. and Hedges, P. (1996) 'Application of GIS linked environment models over a large area', in *Proceedings of the 3rd International Conference on Integrating GIS and Environmental Modelling*, CD-ROM, UC Santa Barbara: NCGIA.

Chrisman, N. R. (1994) 'Metadata required to determine the fitness of spatial data for use in environmental analysis', in Michener, W., Brunt, J. and Stafford, S. (eds) *Environmental Information Management and Analysis: Ecosystem to Global Scales*, London: Taylor & Francis, pp. 177–190.

Couclelis, H. (1996) 'Towards an operational typology of geographic entities with ill-defined boundaries', in Burrough, P. A. and Frank, A. U. (eds) *Geographic Objects with Indeterminate Boundaries*, London: Taylor & Francis, pp. 45–55.

Csillag, F. (1996) 'Variations on hierarchies: toward linking and integrating structures', in Goodchild, M. F., Steyaert, L., Parks, B., Johnson, C., Maidment, D. Crane, M. and Glenndinning, S. (eds) *GIS and Environmental Modelling: Progress and Research Issues*, Fort Collins CO: GIS World Books, pp. 423–437.

Dangermond, J. (1993) 'Software vendors in GIS and environmental modelling', in Goodchild, M. J., Parks, B. O. and Steyaert, L. T. (eds) *Environmental Modelling with GIS*, Oxford: Oxford University Press, pp. 51–56.

Dimmestøl, T. and Lucas, A. (1992) 'Integrating GIS with ocean models to simulate and visualize spills', in Artime, K. (ed.) *Proc. Scan GIS 92*, Helsinki: Helsinki University of Technology.

Djokic, D., Coates, A. and Ball, J. (1996) 'Generic data exchange – integrating models and data providers', in *Proceedings of the 3rd International Conference On Integrating GIS and Environmental Modelling*, CD-ROM, UC Santa Barbara: NCGIA.

Dodson, R. and Turner, D. P. (1996) 'Using GIS to enable diagnostic interaction with a spatially distributed biogeochemistry model', in *Proceedings of the 3rd International Conference On Integrating GIS and Environmental Modelling*, CD-ROM, UC Santa Barbara: NCGIA.

Dragosits, U., Place, C. J. and Smith, R. I. (1996) 'Potential of GIS and coupled gis/conventional systems to model acid deposition of sulphur dioxide', in *Proceedings of the 3rd International Conference On Integrating GIS and Environmental Modelling*, CD-ROM, UC Santa Barbara: NCGIA.

Driel, N. and Loveland, T. (1996) 'The U.S. Geological Survey's land cover characterization program', in *Proceedings of the 3rd International Conference On Integrating GIS and Environmental Modelling*, CD-ROM, UC Santa Barbara: NCGIA.

Dungan, J. L. and Coughlan, J. C. (1996) 'Redefining the spatial support of environmental data in the regional hydroecological simulation system', in *Proceedings of the 3rd International Conference On Integrating GIS and Environmental Modelling*, CD-ROM, UC Santa Barbara: NCGIA.

Ellis, F. (1996) 'The application of machine learning techniques to erosion modelling', in *Proceedings of the 3rd International Conference On Integrating GIS and Environmental Modelling*, CD-ROM, UC Santa Barbara: NCGIA.

Engel, B. A., Srinivasan, R. and C. Rewerts (1993) 'Modelling agricultural non-point-source pollution', in Goodchild, M. J., Parks, B. O. and Steyaert, L. T. (eds) *Environmental Modelling with GIS*, Oxford: Oxford University Press, pp. 231–237.

Federal Interagency Coordinating Committee Technology Working Group (FICC) (1988) *A Process for Evaluating GIS*. Technical Report 1. USGS Open File Report 88–105(1200).

Fischer, M. M. (1994) 'From conventional to knowledge-based geographic information systems', *Computing, Environment and Urban Systems*, 18, pp. 233–242.

Fischer, M. M. and Nijkamp, P. (1993) 'Design and use of geographic information systems', in Fischer, M. M. and Nijkamp, P. (eds) *Geographic Information Systems, Spatial Modelling and Policy Evaluation*, Berlin: Springer Verlag, pp. 3–13.

Fisher, P. F. (1996) 'Boolean and fuzzy regions', in Burrough, P. A. and Frank, A. U. (eds) *Geographic Objects with Indeterminate Boundaries*, London: Taylor & Francis, pp. 87–94.

Fisher, T. R. (1993) 'Use of 3D geographic information systems in hazardous waste site investigations', in Goodchild, M. J., Parks, B. O. and Steyaert, L. T. (eds) *Environmental Modelling with GIS*, Oxford: Oxford University Press, pp. 238–247.

Flowerdew, R. (1991) 'Spatial data integration', in Maguire, D., Goodchild, M. F. and Rhind, D. (eds) *Geographical Information Systems*, London: Longman, pp. 375–387.

Frank, A. U. (1992) 'Spatial concepts, geometric data models and geometric data structures', *Computers & Geosciences*, 18, pp. 409–417.

Frank A. U. (1996) 'The prevalence of objects with sharp boundaries in GIS', in Burrough, P. A. and Frank, A. U. (eds) *Geographic Objects with Indeterminate Boundaries*, London: Taylor & Francis, pp. 29–40.

Galagan, C. and Howlett, E. (1994) 'Integrating GIS and numerical models', in *Proceedings of Workshop on Requirements for Integrated GIS*, Ann Arbor, MI: ERIM, pp. 147–153.

Goodchild, M. F. (1992) 'Geographical data modelling', *Computers & Geosciences*, 18, pp. 401–408.

Goodchild, M. F. (1993) 'The state of GIS for environmental problem solving', in Goodchild, M. J., Parks, B. O. and Steyaert, L. T. (eds) *Environmental Modelling with GIS*, Oxford: Oxford University Press, pp. 8–15.

Gorokhovich, Y. and Janus, L. (1996) 'GIS applications for watershed management', in *Proceedings of the 3rd International Conference on Integrating GIS and Environmental Modelling*, CD-ROM, UC Santa Barbara: NCGIA.

Gustafson, E. and Gardner, R. H. (1996) 'Dispersal and mortality in a heterogenous landscape matrix', in *Proceedings of the 3rd International Conference on Integrating GIS and Environmental Modelling*, CD-ROM, UC Santa Barbara: NCGIA.

Haines-Young, R., Green, D. and Cousins, S. (1993) *Landscape Ecology and Geographic Information Systems*, New York: Taylor & Francis.

Harris, J., Gupta, S., Woodside, G. and Ziemba, N. (1993) 'Integrated use of a GIS and a 3-dimensional, finite element model: San Gabriel Basin groundwater flow analysis', in Goodchild, M. F., Parks, B. O. and Steyaert, L. T. (eds) *Environmental Modelling with GIS*, New York: University of Oxford Press, pp. 168–172.

Hart, T., Greene, S. and Afonin, A. (1996) 'Mapping for germplasm collections: site selection and attribution', in *Proceedings of the 3rd International Conference on Integrating GIS and Environmental Modelling*, CD-ROM, UC Santa Barbara: NCGIA.

Harvey, L. E. (1996) 'Macroecological studies of species composition, habitat and biodiversity with a GIS and an integrated optimization method', in *Proceedings of the 3rd International Conference on Integrating GIS and Environmental Modelling*, CD-ROM, UC Santa Barbara: NCGIA.

Horne, J. K., Jech, J. M. and Brand, S. B. (1996) 'Spatial modelling of aquatic habitat from a fish's perspective', in *Proceedings of the 3rd International Conference on Integrating GIS and Environmental Modelling*, CD-ROM, UC Santa Barbara: NCGIA.

Hunsaker, C. T., Nisbet, R., Lam, D. C. L., Browder, J. A., Baker, W., Turner, M. G. and Bodkin, D. B. (1993) 'Spatial models of ecological systems and processes: the role of GIS', in Goodchild, M. J., Parks, B. O. and Steyaert, L. T. (eds) *Environmental Modelling with GIS*, Oxford: Oxford University Press, pp. 248–264.

360 Roger Bivand and Anne Lucas

Hunter, G. J. and Goodchild, M. F. (1993) 'Managing uncertainty in spatial databases: putting theory into practice', *URISA Journal*, 5, 2, pp. 55–62.

Hunter, G. J. and Goodchild, M. F. (1994) 'Design and application of a methodology for reporting uncertainty in spatial databases', *Proceedings of the URISA '94 Conference*, Milwaukee, 1, pp. 771–785.

Johnston, C., Sersland, C., Bonde, J., Pomroy-Petry, D. and Meysembourg, P. (1996) 'Constructing detailed vegetation databases from field data and airborne videography', in *Proceedings of the 3rd International Conference On Integrating GIS and Environmental Modelling*, CD-ROM, UC Santa Barbara: NCGIA.

Kemp, K. (1993) 'Environmental modelling with GIS: a strategy for dealing with spatial continuity', *NCGIA Technical Report 93-3*, UC Santa Barbara, CA.

Kemp, K. (1996a) 'Managing spatial continuity for integrating environmental models with GIS', in Goodchild, M. F., Steyaert, L., Parks, B., Johnson, C., Maidment, D. Crane, M. and Glenndinning, S. (eds) *GIS and Environmental Modelling: Progress and Research Issues*, Fort Collins CO: GIS World Books, pp. 339–343.

Kemp, K. (1996b) 'Easing traditional environmental models into GIS', in *Proceedings of the 3rd International Conference On Integrating GIS and Environmental Modelling*, CD-ROM, UC Santa Barbara: NCGIA.

Kirchner, T. B. (1994) 'Data management and simulation modelling', in Michener, W., Brunt, J. and Stafford, S. (eds) *Environmental Information Management and Analysis: Ecosystem to Global Scales*, London: Taylor & Francis, pp. 357–376.

Krysanova, V., Müller-Wohlfeil, D.-I. and Becker, A. (1996) 'Mesoscale integrated modelling of hydrology', in *Proceedings of the 3rd International Conference On Integrating GIS and Environmental Modelling*, CD-ROM, UC Santa Barbara: NCGIA.

Laffan, S. (1996) 'Rapid appraisal of groundwater discharge using fuzzy logic and topography, in *Proceedings of the 3rd International Conference on Integrating GIS and Environmental Modelling*, CD-ROM, UC Santa Barbara: NCGIA.

Lam, D. C. L. (1993) 'Combining ecological modelling, GIS and expert systems: a case study of regional fish richness model', in Goodchild, M. J., Parks, B. O. and Steyaert, L. T. (eds) *Environmental Modelling with GIS*, Oxford: Oxford University Press, pp. 270–275.

Lam, D. C. L. and Swayne, D. A. (1993) 'An expert system approach of integrating hydrological database, models and GIS: application of the RAISON system', in *Hydro GIS 93: Application of Geographic Information Systems in Hydrology and Water Resources*, pp. 23–33.

Lam, D. C. L., Wong, I., Swayne, D. A. and Fong, P. (1994a) 'Data knowledge and visualization in an environmental information systems', *Journal of Biological Systems*, 2, pp. 481–497.

Lam, D. C. L., Mayfield, C. I., Swayne, D. A. and Hopkins, K. (1994b) 'A prototype information system for watershed management and planning', *Journal of Biological Systems*, 2, pp. 499–517.

Lam, D. C. L., Swayne, D. A., Mayfield, C. and Cowan, D. (1996) 'Integration of GIS with other software systems: Integration vs. interconnection', in *Proceedings of the 3rd International Conference on Integrating GIS and Environmental Modelling*, CD-ROM, UC Santa Barbara: NCGIA.

Lee, T. J. and Pielke, R. A. (1996) 'GIS and atmospheric modelling: a case study', in Goodchild, M. F., Steyaert, L., Parks, B., Johnson, C., Maidment, D. Crane, M.

and Glenndinning, S. (eds) *GIS and Environmental Modelling: Progress and Research Issues*, Fort Collins CO: GIS World Books, pp. 231–234.

Lees, B. (1996) 'Improving the spatial extension of point data by changing the data model', in *Proceedings of the 3rd International Conference on Integrating GIS and Environmental Modelling*, CD-ROM, UC Santa Barbara: NCGIA.

Liff, C. I., Riitters, K. II. and Hermann, K. A. (1994) 'Forest health monitoring case study', in Michener, W. K., Brunt, J. W. and Stafford, S. G. (eds) *Environmental Information Management and Analysis: Ecosystem to Global Scales*, New York: Taylor & Francis, pp. 101–112.

Logsdon, M. (1996) 'Modelling land-cover change from measures of spatial landscape structure', in *Proceedings of the 3rd International Conference on Integrating GIS and Environmental Modelling*, CD-ROM, UC Santa Barbara: NCGIA.

Lucas, A. E. (1996) 'Data for coastal GIS: issues and implications for management', *GeoJournal*, 39, 2, pp. 133–142.

Lucas, A. E., Abbedissen, M. B. and Budgell, W. P. (1994) 'A spatial metadata management system for ocean applications: requirements analysis', in *Proceedings of Workshop on Requirements for Integrated GIS*, Ann Arbor, MI: ERIM, pp. 63–74.

Maas, S. and Doraiswamy, P. C. (1996) 'Integration of satellite data and model simulations in a GIS for monitoring regional evaporation and biomass production, in *Proceedings of the 3rd International Conference on Integrating GIS and Environmental Modelling*, CD-ROM, UC Santa Barbara: NCGIA.

Mackey, B. (1996) 'The role of GIS and environmental modelling in the conservation of biodiversity', in *Proceedings of the 3rd International Conference on Integrating GIS and Environmental Modelling*, CD-ROM, UC Santa Barbara: NCGIA.

MacMillan, R. A., Furley, P. A. and Healey, R. G. (1993) 'Using hydrological models and geographic information systems to assist with the management of surface water in agricultural landscapes', in Haines-Young, R., Green, D. and Cousins, S. (eds) *Landscape Ecology and Geographic Information Systems*, New York: Taylor & Francis, pp. 181–209.

Maidment, D. R. (1996a) 'GIS and hydrologic modelling – an assessment of progress', in *Proceedings of the 3rd International Conference on Integrating GIS and Environmental Modelling*, CD-ROM, UC Santa Barbara: NCGIA.

Maidment, D. R. (1996b) 'Environmental modelling within GIS', in Goodchild, M. F., Steyaert, L., Parks, B., Johnson, C., Maidment, D. Crane, M. and Glenndinning, S. (eds) *GIS and Environmental Modelling: Progress and Research Issues*, Fort Collins CO: GIS World Books, pp. 315–323.

McKeown, R., Ojima, D. S., Kittel, T. G. F., Schimel, D. S., Parton, W. J. *et al.* (1996) 'Ecosystem modelling of spatially explicit land surface changes for climate and global change analysis', in *Proceedings of the 3rd International Conference on Integrating GIS and Environmental Modelling*, CD-ROM, UC Santa Barbara: NCGIA.

Michener, W. and Houhoulis, P. (1996) 'Identification and assessment of natural disturbaces in forested ecosystems: the role of GIS and remote sensing', in *Proceeding of the 3rd International Conference on Integrating GIS and Environmental Modelling*, CD-ROM, UC Santa Barbara: NCGIA.

Moore, I. D. (1996) 'Hydrologic modelling and GIS', in Goodchild, M. F., Steyaert, L., Parks, B., Johnson, C., Maidment, D., Crane, M. and Glenndinning, S. (eds)

GIS and Environmental Modelling: Progress and Research Issues, Fort Collins CO: GIS World Books, pp. 143–148.

Mueller-Wohlfeil, D., Lahmer, W., Krysanova, V. and Becker, A. (1996) 'Topography-based hydrological modelling in the Elbe drainage basin', in *Proceedings of the 3rd International Conference on Integrating GIS and Environmental Modelling*, CD-ROM, UC Santa Barbara: NCGIA.

Ojima, D. S. and Parton, W. J. (1996) 'Integrated approach to land use analysis', in *Proceedings of the 3rd International Conference on Integrating GIS and Environmental Modelling*, CD-ROM, UC Santa Barbara: NCGIA.

Parks, B. O. (1993) 'The need for integration', in Goodchild, M. J., Parks, B. O. and Steyaert, L. T. (eds) *Environmental Modelling with GIS*, Oxford: Oxford University Press, pp. 31–34.

Peuquet, D. (1984) 'A conceptual framework and comparison of spatial data models', *Cartographica*, 21, pp. 66–113.

Peuquet, D. (1994) 'It's about time: a conceptual framework for the representation of temporal dynamics in geographic information systems', *Annals of AAG*, 84, 3, pp. 441–461.

Ramanarayanan, T. S., Srinivasan, R. and Arnold, J. G. (1996) 'Modelling Wister Lake watershed using a GIS-linked basin-scale hydrologic/water quality model', in *Proceedings of the 3rd International Conference on Integrating GIS and Environmental Modelling*, CD-ROM, UC Santa Barbara: NCGIA.

Raper, J. and Livingstone, D. (1995) 'Development of a geomorphological spatial model using object-oriented design, *IJGIS*, 9, 4, pp. 359–383.

Raper, J. and Livingstone, D. (1996) 'Spatio-temporal interpolation in four dimensional coastal process models', in *Proceedings of the 3rd International Conference on Integrating GIS and Environmental Modelling*, CD-ROM, UC Santa Barbara: NCGIA.

Rundquist, D. C., Peters, A. J., Di, L., Rodekohr, D. A., Ehrman, R. L. and Murray, G. (1991) 'Statewide groundwater-vulnerability assessment in Nebraska using the DRASTIC/GIS model', *GeoCarto International*, 2, pp. 51–58.

Rupp, S. (1996) 'Landscape-level modelling of spruce seedfall using a geographic information system', in *Proceedings of the 3rd International Conference on Integrating GIS and Environmental Modelling*, CD-ROM, UC Santa Barbara: NCGIA.

Sadler, G. and Hamid, A. (1992) 'Integrated analytical tools to monitor the urban environment', in Harts, J., Ottens, H. and Scholten, H. (eds), *Proceedings of EGIS '92*, Utrecht: EGIS Foundation, pp. 772–781.

Schell, T. T., Avecedo, M. F., Bogs, F., Newell, J., Dickson, K. L. and Mayer, F. (1996) 'Assessing pollutant loading to Bayou Chico, Florida by integrating an urban stormwater runoff and fate model with GIS', in *Proceedings of the 3rd International Conference on Integrating GIS and Environmental Modelling*, CD-ROM, UC Santa Barbara: NCGIA.

Shepherd, I. D. H. (1991) 'Information integration and GIS', in Maguire, D., Goodchild, M. F. and Rhind, D. (eds) *Geographical Information Systems*, London: Longman, pp. 337–360.

Sklar, F. H. and Constanza, R. (1991) 'The development of dynamic spatial models for landscape ecology: a review and prognosis', in Turner, M. G. and Gardner, R. H. (eds) *Quantitative Methods in Landscape Ecology*, Berlin: Springer Verlag, pp. 239–288.

Slawecki, T., Raghunathan, R. K., Bierman, V. J. and Rodgers, P. W. (1996) 'Modelling resuspension of river sediments using ARC/INFO', in *Proceedings of the 3rd International Conference on Integrating GIS and Environmental Modelling*, CD-ROM, UC Santa Barbara: NCGIA.

Steyaert, L. (1993) 'A perspective on the state of environmental simulation modelling', in Goodchild, M. J., Parks, B. O. and Steyaert, L. T. (eds) *Environmental Modelling with GIS*, Oxford: Oxford University Press, pp. 16–30.

Streit, U. and Weismann, K. (1996) 'Problems of integrating GIS and hydrological models', in Fischer, M. M., Scholten, H. J. and Unwin, D. (eds) *Spatial Analytical Perspectives on GIS*, London: Taylor & Francis, pp. 161–174.

ten Velden, H. E. and Kreuwel, G. (1990) 'A geographical information system based on decision support system for environmental zoning', in Scholten, H. J. and Stillwell, J. C. H. (eds) *Geographical Information Systems for Urban and Regional Planning*, Dordrecht: Kluwer Academic, pp. 119–128.

Tomlin, C. D. (1990) *Geographic Information Systems and Cartographic Modelling*, Englewood Cliffs, NJ: Prentice Hall.

van Horssen, P. (1996) 'Ecological modelling in GIS', in *Proceedings of the 3rd International Conference on Integrating GIS and Environmental Modelling*, CD-ROM, UC Santa Barbara: NCGIA.

Walker, H., Leone, J. M. and Kim, J. (1996) 'The effects of elevation data representation on mesoscale atmospheric model simulations', in *Proceedings of the 3rd International Conference on Integrating GIS and Environmental Modelling*, CD-ROM, UC Santa Barbara: NCGIA.

Willmott, C., Raskin, R., Funk, C., Webber, S. and Goodchild, M. F. (1996) Spherekit: the spatial interpolation toolkit. http://www.ncgia.ucsb.edu/pubs/spherekit/main.html

Wilson, J. (1996) 'GIS-based land surface/subsurface modelling: new potential for new models?', in *Proceedings of the 3rd International Conference on Integrating GIS and Environmental Modelling*, CD-ROM, UC Santa Barbara: NCGIA.

Worboys, M. (1996) *GIS: A Computing Perspective*, London: Taylor & Francis.

Wright, D. J., Goodchild, M. F. and Proctor, J. D. (1997) 'Demystifying the persistent ambiguity of GIS as tool vs science', *Annals of Association of Professional Geographers*, 87, 2, pp. 346–363.

Wu, L. (1996) 'An integration of a surface energy balance climate model with TIN and GRID in GIS', in *Proceedings of the 3rd International Conference on Integrating GIS and Environmental Modelling*, CD-ROM, UC Santa Barbara: NCGIA.

Yarie, J. (1996) 'A forest ecosystem dynamics model integrated within a GIS', in *Proceedings of the 3rd International Conference on Integrating GIS and Environmental Modelling*, CD-ROM, UC Santa Barbara: NCGIA.

Chapter 15

Limits to modelling in the Earth and environmental sciences

Mike Kirkby

15.1 Introduction

Even the most 'physically based' models of environmental systems rely on gross simplifications of the detailed process mechanics, and substantial advances in understanding are achieved mainly through this paradigm of scientific simplification rather than through developing more complex model structures. Instead, the priorities for computationally intensive research are thought to lie primarily in two other areas: reconciling models at different scales, and improved calibration and validation.

Relevant models exist at many spatial scales, which range from the soil crumb to continental or global scales, a range of 10^{10} times. Similarly relevant time scales span from fractions of a second up to the age of the Earth, an even greater relative range. Most models have time or space scopes of at most 10^3 times, so that fundamentally different structures have to be used to cover the full range of relevant scales. Reconciliation between models at different scales is by formal integration or, increasingly, by heavy computation. The greatest challenge is then to understand the structural changes which are associated with non-linearity.

To date, only some hydrological models with 4–6 parameters have been simple enough to allow thorough investigation of their parameter space, often requiring the use of parallel processing facilities to achieve the large number of replications required. There is a need to extend this process to a wider class of models, and to explore more computationally efficient ways of understanding the range of uncertainty where models have a substantial number of parameters and/or significant run times. There is also a need to consider the use of qualitative data for calibration, rather than rely exclusively on hydrological outputs, which suppress much of the richness of potential change in the landscape.

One of the central goals of environmental science is to increase our understanding of the processes and forms around us. A mathematical or computational model can, in principle, be established from a self-consistent and comprehensive quantitative understanding of the mechanisms involved, and should then allow results to be transferred from place to place and to different time and space scales. Because data is generally concentrated on spatial distributions (e.g. from remote sensing), stratigraphic sequences or short period process studies, models have an essential role to fulfil in allowing extrapolation to other combinations of time and space scales, and this generally requires models with an explicit physical basis (Kirkby, 1990, 1996).

In practice, this physical basis is most generally provided by application of the continuity or storage equation for sediment or water discharge:

$$\frac{\partial z}{\partial t} + \nabla.S = a \tag{15.1}$$

where z is elevation, S is vector sediment transport, a is areal accumulation (e.g. dust), and t is elapsed time.

Since S generally depends more or less linearly on gradient ∇z, among other terms, it may be noted that there are substantial diffusive terms in the equation. Diffusion terms reasonably represent soil creep, rainsplash and gelifluction near divides, and wash or fluvial sediment transport elsewhere. For evolution in the long term ($>>10$ years), these processes can be generalized within the family of empirical Musgrave (1947) expressions:

$$S \propto A^m \Lambda^n \tag{15.2}$$

where S = sediment transport rate, A = catchment area, Λ = slope gradient, and m,n are empirical exponents (normally $0 \le m \le 2$; $1 \le n \le 2$).

Other processes, such as rapid mass movements, need not have an appreciable diffusive component, but often act in combination with diffusive processes. For the shorter term (seconds to hours), more explicit process models are available in some areas of hydraulics and sediment transport, but there have been few attempts to scale these up to the longer term, either computationally or through formal integration.

To date, the most successful models have been those which make full use of the scientific paradigm of simplification, usually neglecting all but the dominant two, or at most three, processes. For example, an important part of fluvial hydraulics is built on the Reynolds' and Froude numbers, which are concerned with deciding which of the two sets of viscous, inertial and gravity forces need to be considered in any situation. Similarly, it may be argued that landscape form is controlled primarily by the processes which are dominant in landscape denudation. We may therefore analogously consider the ratio of mass movement rates to wash rates, or of wash rates to solution rates, as similar dimensionless ratios which determine the hillslope 'regime'.

These methods have had moderate success in creating a consensus on some aspects of landform development, but major difficulties remain. Models are built on the assumption that the landscape systems of interest are essentially deterministic, even if they are driven by climatic events which can often be treated as stochastic by the geomorphologist.

There are, however, clear limits to our ability to make useful forecasts. They arise from the inherent unpredictability of non-linear systems, from the great breadth of uncertainty bands where these can even be defined, and from the difficulties in moving between time and space scales. These limitations have implications for where our ever-expanding computing power can most effectively be applied.

15.1.1 Inherent predictability

It cannot always be assumed that effective forecasting models exist. There are limits, both to our level of understanding and to the inherent predictability of our systems of study. Inherent predictability is closely tied to the concept of chaos, or extreme sensitivity to initial conditions, which limits the effective range of forecasts. Chaos can occur within most non-linear systems, but its impact may be limited. Where systems are predominantly linear, the non-linearities may only be important under certain restricted conditions.

The response to differences in initial conditions z can generally be described approximately as an exponential growth over time and space, of the form $z \exp(t/t_0)$. Loss of effective predictability occurs beyond the so-called Lyapunov horizon, t_0, where the differences between almost identical initial conditions becomes macroscopically significant. In chaotic systems, the horizon becomes very short in certain regions, and all effective predictability is lost. This occurs, for example, in areas of cyclogenesis, or in the classical Newtonian three-body problem wherever orbits come close to one another. In geomorphology, the spatial location of stream heads or the transitions between braided and meandering stream patterns are thought to be chaotic.

Newtonian astronomical systems are characterized by gravitational attraction between bodies, and more generalized gravity models have also been used in geographical modelling. In general, the attractive force between two bodies distance d apart, and of attractiveness m_1 and m_2, takes the form $m_1 m_2 / d^n$, where $n = 2$ for Newtonian gravitation. Such dynamical systems with more than two bodies show strong tendencies to chaos, which increases with the exponent n. In contrast, dispersive systems, in which particles repel one another with the same form show convergent behaviour, in which the final forms have much in common despite diverse initial states. Diffusion corresponds roughly to a dispersive system with $n = 1$. Where systems are strongly diffusive, their Lyapunov horizon becomes negative and the growth of initially small non-linear terms is generally damped out. Thus, under common process conditions, initial differences in slope form are progressively damped out, and hillslopes become increasingly 'characteristic' (Kirkby, 1971) of the processes which formed them. Another kind of attenuation

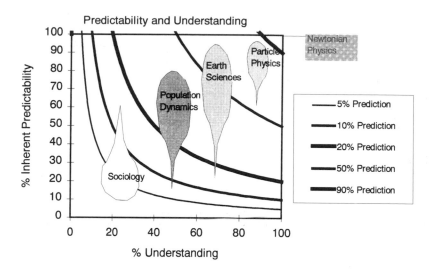

Figure 15.1 Conceptual dependence of actual predictive power on inherent predictability and completeness of understanding.

occurs in hydrological modelling. At the end of each storm, the recession period is one where the inputs (rainfall less evapotranspiration) become spatially uniform, and the discharges decay towards zero, so that the system approaches a steady state condition. The effect of these kinds of convergent behaviour is to make forecasting easier, but at the expense of making it impossible to reconstruct the past beyond its (negative) Lyapunov horizon.

These kinds of near-linear, and therefore forgiving, behaviours are most common in the physical sciences, and least in the social sciences, so that there is a general decrease in inherent predictability in this physical to social direction. The general level of understanding of the processes involved also tends to fall off in the same direction, partly as a result of the difficulty of directly linking cause and effect in unpredictable systems. Figure 15.1 sketches the way in which these two components interact to produce levels of practical forecasting potential. The potential of modelling can only be realized within these practical limits.

Another source of unpredictability in environmental systems lies in the distribution of the climatic and other external inputs. It is generally assumed that the distribution of external events or their geomorphological impacts is well behaved, as sketched in Figure 15.2, with a well-defined modal value and an exponential or similarly well-behaved extreme value distribution (Wolman and Miller, 1960). Thus, for large event magnitude, M, the corresponding frequency, $f \sim \exp(-M/M_0)$. A dominant event size can then be defined from the cumulative frequency, Mf, which is the event size which is cumulatively most effective. For the exponential asymptotic form, the

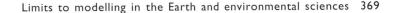

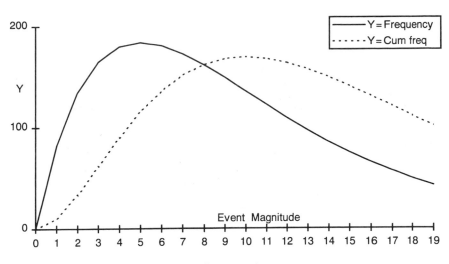

Figure 15.2 Magnitude and frequency of geomorphic events.

dominant event is associated with magnitude M_0, and in general the dominant event is larger and rarer than the modal event. If however, the frequency curve falls off more slowly than the exponential form, it may be impossible to define the dominant event, or the mean event size or its variance. The variance exists only if $\int_0^\infty M^2 f\, dM$ is finite, and the mean only if $\int_0^\infty Mf\, dM$ is finite. If the frequency declines as a power law, $\sim M^{-n}$, rather than an exponential, then the variance only exists if $n > 3$, the mean only if $n > 2$ and the dominant event only if $n > 1$. For impacts which themselves depend on powers of the magnitude (for instance, sediment transport proportional to flow squared), still higher powers of n are required for good behaviour, and in general we may be suspicious of any power law.

It is fair to observe that most short-term statistical studies indicate an approximately exponential form, but that for longer periods there are good reasons to question whether the distribution remains well behaved. Evidence for this is based on long records, such as those for the Nile (Hurst, 1950; Kirkby, 1987), in which it is shown that the re-scaled range statistic increases not as (length of record)$^{0.5}$ as expected, but with a higher exponent (0.7–0.9), indicating persistence of autocorrelation over very long time periods. Related fractals can also be used to describe these time series. Such long-term persistence, associated with power laws, provides the kind of series which no longer has well-behaved means and/or variances. In this case, the long-term behaviour of the system can no longer be described by any meaningful average and/or confidence band, but is specific to the historic sequence of events. Where there are non-linear positive feedbacks resulting from the individual event, new sources of chaos are also being continually added to the system.

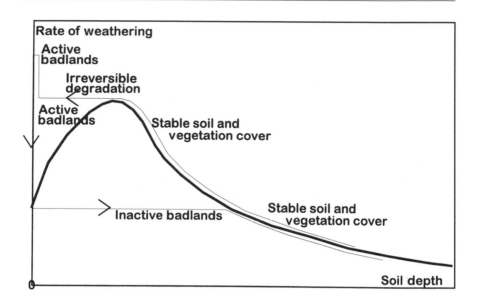

Figure 15.3 Hysteresis in badland erosion and re-stabilization.

Philosophically, many environmental scientists are unwilling to treat history as a poorly behaved statistical distribution, and prefer to seek causes in the external driving system which structure the distributions. One very simple cause can lie in the non-stationarity of some systems, associated with, for example, climate change. If, as now begins to seem possible with externally validated regional climate proxy records from ice cores (e.g. Dansgaard and Oeschger, 1989), process rates can be explicitly linked to climate, then this source of non-stationarity may be reduced in the future. Other sources of variability may be linked to unwitting mixing of event distributions which are modified by such varied causes as different circulation weather types, meteorite impacts, volcanic eruptions and ice ages.

15.1.2 Memory, hysteresis and bifurcations

In a non-linear environmental system, the order in which any sequence of events takes place will have some effect on the final outcome, and this effect is strongest where major events cross significant thresholds. When or where a threshold is exceeded, the outcomes are qualitatively different, and influence the course of subsequent events for a significant period.

A simple example occurs in the evolution of slopes in fine grained materials in semi-arid climates (Kirkby, 1995). The heavy curve in Figure 15.3 shows the relationship between the rate of weathering (conversion of rock to soil) and the soil depth. The form of the curve shows a single maximum, which may be explained as follows (Carson and Kirkby, 1972, p. 105). At shallow depths, runoff is rapid and water does not remain in contact with the rock for long enough to reach chemical equilibrium. Consequently rates of weathering are low and increase with depth. At large soil depths, greater depth progressively prevents circulation and replacement of well-equilibrated water so that weathering rates decrease with depth. The maximum occurs where replacement is effective, and contact times are long enough for equilibration, and probably occurs at a depth of a few centimetres.

The weathering process is combined with processes of mechanical denudation, by landslides and soil erosion which reduce the soil thickness. In general, the rate, for a given gradient, etc., increases modestly with soil depth because of the greater clay fraction and associated loss of cohesion. The curves for erosion and weathering intersect twice or not at all. With two intersections, the left-hand intersection is an unstable equilibrium, since deviations are magnified. For example, a local increase in soil depth leads to greater weathering, and hence further increases in soil depth, and vice-versa. The right-hand intersection is stable, with local differences leading to a return to the equilibrium.

If such a system is subjected to a changing environment where, say, climate change or uplift produces steadily increasing erosion rates, then the stable equilibrium will move gradually up and left along the right hand limb of the heavy curve in Figure 15.3. This phase is characterized by a maintenance of a stable soil and vegetation cover, under which erosion is dominated by mass movements in the clay. Continued increases in erosion can only be accommodated by a complete stripping of the soil, indicated by the 'irreversible degradation' in Figure 15.3, and any further increases in erosion can only maintain this state, which is characterized by the development of steep badlands on which water erosion is the dominant process.

When the environment changes to a condition of reducing erosion rates, then the badlands will survive until they drop back to the heavy curve, matching the weathering rate on bare parent material. Thereafter the soil will begin to thicken, and the badlands become inactive, maintaining their gross morphology of closely dissected channels, but with a stabilizing and deepening layer of soil and vegetation. Eventually the landscape will return to the stable limb of the weathering curve, and the landscape will again evolve slowly by mass movements under a stable soil and vegetation cover.

This evolution not only shows a simple bifurcation catastrophe, but also a considerable asymmetry in formation and recovery. The irreversible degradation into badlands can occur within a period of decades, while the recovery process is controlled by the rate of weathering, and may take 10^3–10^5 years. Such an evolution is very sensitive to whether the badland

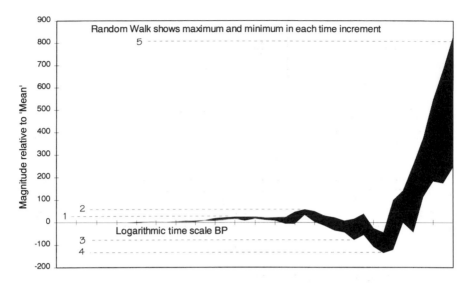

Figure 15.4 Simulated random walk showing events (1–5) which are large enough to leave clear evidence which will not be obliterated.

threshold has been reached, and landscapes may show very different evolutionary paths on either side of the threshold.

Except in such extreme cases, the same set of events realized in different sequential orders has a relatively minor influence on the final outcome of most geomorphic events, but different random sequences drawn from the same distribution may differ dramatically. This is readily seen by comparing different realizations of a random walk with sequences constrained so that there are an equal number of forward and backward steps.

15.1.3 Uniqueness of historical time

Another common, indeed inevitable, perspective on the landscape is through the filter of the present and the actual history leading up to it. Typically, our interpretation is heavily biased by the imprint of the more recent major events. We can perhaps see the last major event, and, before that the last prior event which was significantly larger, and so on into the past, so that our landscape history may be dominated by a description of the effects of the last big flood, the largest Holocene Flood, the last Pleistocene glaciation, and the Tertiary geological history responsible for the regional relief. This type of sequence is sketched in Figure 15.4, in which the minimum and maximum values of a simple random walk with equally likely steps of magnitude 1 are shown with logarithmic time increments. In the sketch, a stratigraphic reconstruction might clearly identify events 1 to 5, but all others could be at least partially obliterated.

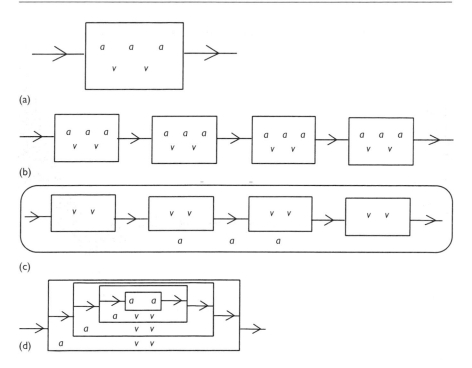

Figure 15.5 Alternative ways of linking models together: (a) simple model; (b) inefficent serial chaining of simple models; (c) efficient serial chaining of simple models; (d) efficient nesting with parameters replaced by state variables from finer scales. a = parameter; v = state variable.

15.1.4 Linking models together and up-scaling

Simple physical models are generally described by a set of parameter values and a set of state variables, together with an appropriate set of functional relationships (Figure 15.5(a)). Such simple models can be combined by chaining them together in series (Figure 15.5(b)) or in parallel. If, however, each model retains its own set of parameters, the number of parameters soon becomes intractable. Both greater physical realism and greater applicability are usually achieved if the submodels share some or all of their parameters, which hopefully describe the common universe occupied by all the submodels (Figure 15.5(c)). Both time or space scales and levels of understanding can often be structured as a series of models nested within one another (Figure 15.5(d)), and this can provide a valuable method for providing parameters for an outer model from the state variables of an inner model. Thus, for example, the state of a very detailed model which predicts the evolution of microtopography during soil erosion may give a roughness parameter which can be more relevant to use in a coarser-scale model.

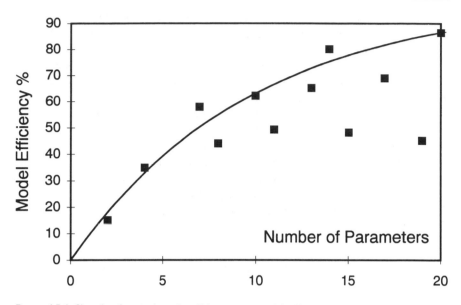

Figure 15.6 Sketch of typical trade-off between model efficiency and number of parameters for 'good' models.

The clear premise of this discussion is that large numbers of parameters are undesirable. To be more precise, there is no difficulty in incorporating parameters which have a physical significance which is wider than the model, and for which values can be independently derived. Many model parameters, however, are closely bound up with the model itself, and can, in many cases, only be derived by a process of optimization against a validation data set. Such parameters add to the complexity of the model, and their proliferation makes the model difficult to calibrate or to transfer between areas. However, additional parameters generally add to the goodness of fit of a model and, among state of the art models, more parameters tend to give improved efficiency (however evaluated) as sketched in Figure 15.6. The choice of model complexity is generally a trade-off between greater comprehension and transferability with few parameters against greater forecasting precision with more parameters. The choice of methods for linking submodels together is one means of eliminating unnecessary parameters, and is closely related to an understanding of all the significant interactions both within and between the systems represented by their submodels.

One aspect of economy in parameters is related to the use of spatially distributed models. Ideally perhaps, each grid cell should be represented by its own parameter values, but this provides a daunting exercise in either measurement or calibration. In some contexts, for example a distributed model for hydrograph forecasting, it also requires a very large amount of input parameter data to generate a rather modest output of catchment discharge. There is thus a very high ratio of input to output information, which minimizes the value of the modelling exercise, particularly if the spatially distributed parameter may vary over time.

It is argued that, although growing computer power allows models to be linked together in more complex chains or other networks, the limits of effectiveness are not generally set by computer power, but by our understanding of the increasing numbers of interactions between the various subsystems included. With n subsystems, the number of interactions increases with $n!$, and we generally have to recognize that the complexity of a suitable model may approach this limit, and is rarely confined to the n interactions of the serial chain. Even in areal distributed models, where the number of interactions $\sim n^2$, the complexity of the parameter space is prohibitive unless it can be assumed that all grid cell parameters fall into a small number of mappable classes.

15.1.5 Alternative uses for additional computing power

Because it is not only computing power which limits our modelling capability, we must be discriminating in how to use additional resources. There seem to be two main areas in which these resources can be used, first in exploring more complex models, and second in understanding the uncertainties in model forecasts.

Greater complexity, it has been argued above, must be linked to a greater understanding of the relevant interactions between component subsystems as we either create or combine models. Issues of time and space scales are intimately involved in this understanding. When we bring together models of interacting systems, or models of the same system at different time or space resolution, then our crucial task is to reconcile the models. In some cases, this may be done analytically, but this approach is often too simple, particularly where the dynamics are non-linear. For example, we may attempt to reconcile scales by a process of formal integration which reveals the way in which system states at the detailed scale are linked to parameters at the coarser scale. In straightforward cases, variances at the detailed scale become significant parameters at the next coarser scale, as for example with Reynolds' shear stresses (over time) and microtopographic roughness (over space). An important part of this reconciliation is with our scientific understanding of the processes, since this comparison may help to provide a better physical basis for model parameters, hopefully releasing them from the need for optimization though a model-specific calibration.

The hydrology and environment science communities are much concerned with problems of up-scaling at present, partly due to the need to bridge from their traditional plot and small catchment scales up to the GCM scale of the climate modellers. This has accelerated the search for an understanding of processes and interactions between processes. A great deal of computer time has been devoted to increasing the complexity of models and to making distributed models of ever larger areas, but there is still only a limited understanding of the additional interactions, over longer times and over larger areas, which this requires. One of the lessons of chaos theory and systems analysis is that large-scale organization is not simply a replication of many small-scale organizations, but that new processes and structures emerge at larger scales. This re-emphasizes the need to combine and reconcile models native to different scales, and not only to build steadily up or steadily down in scale from one extreme.

The second major area where there is a need for additional computing power is to improve the process of calibration, validation and uncertainty estimation. In the environmental sciences, the formal process of calibration and validation has rarely been attempted outside catchment hydrology. For TOPMODEL (Beven and Binley, 1992), the Generalized Likelihood Uncertainty Estimation (GLUE) methodology has proved highly effective for a compact model with about five relevant parameters, although still requiring large numbers (about 10 000) of replicate runs to estimate uncertainty bounds. For larger models, and those with larger numbers of parameters, even large computers have difficulty in running sufficient replicates to define the parameter sets which provide acceptable output forecasts.

The traditional solution to this problem is, firstly, to define a unique optimal parameter set and assign error bands to these parameter values and, secondly, to draw parameter sets at random from these distributions and, thirdly, to use the set of forecasts as an envelope of probable outcomes. The weakness of this approach is that the global optima are often poorly defined and that non-contiguous points in the parameter space lead to acceptable forecasts. One alternative approach is to use a genetic algorithm to 'breed' an acceptable population of parameter sets, which provides an efficient method for converging on good solutions, while still allowing disparate parameter sets to contribute to the set of outcomes, corresponding to a number of genotypes in the population. In common with GLUE and some other methods, this genetic approach allows new calibration data to be incorporated into future forecasts, allowing the model to adapt to changing conditions where appropriate.

Hydrological models traditionally concentrate on forecasting discharge at the outlet, which is much more sensitive to channel conditions than to the detailed response of hillslopes. Erosion models similarly tend to concentrate on water and sediment yields at the outlet. Nevertheless, many of the quantitative impacts of hydrology or erosion are also strongly seen in a distributed form. Thus we may wish models to reflect not only the integrated outputs but also the distribution of effects within the catchment. These data will rarely be available in a systematic and quantitative form, but we may also seek to use methods of fuzzy logic or dummy variables to provide a richer set of forecasts and a richer context for calibration.

Even with the use of these or other new approaches, uncertainty estimation makes heavy demands on computing power, and allows both existing and new models to realize their full potential as forecasting tools.

15.2 Conclusion

Although increasing computer power provides many opportunities for larger, more complex and better calibrated models, there remain many limitations to what can reasonably be achieved. These limitations relate to the inherent predictability of the phenomena, to our understanding of the processes and their interactions, and to the limitations of data gathering, particularly data over time for places and periods without instrumental records. As with other aspects of research, there is a need to use computing tools with discrimination, and not assume that power conquers all.

References

Beven, K. J. and Binley, A. M. (1992) 'The future of distributed models: calibration and predictive uncertainty', *Hydrological Processes*, 6, pp. 279–298.

Carson, M. A. and Kirkby, M. J. (1972) *Hillslope Form and Process*, Cambridge: Cambridge University Press.

Dansgaard, W. and Oeschger, H. (1989) 'Past environmental long-term records from the Arctic', in Oeschger, H. and Langway, C. C. Jr (eds) *The Environmental Record in Glaciers and Ice Sheets*, New York: Wiley, pp. 287–318.

Hurst, H. E. (1950) 'Long-term storage capacity of reservoirs', *Proceedings of the American Society of Civil Engineers*, 76, p. 11.

Kirkby, M. J. (1971) 'Hillslope process-response models based on the continuity equation', *Transaction, Institute of British Geographer, Special Publication*, 3, pp. 15–30.

Kirkby, M. J. (1987) 'The Hurst effect and its implications for extrapolating process rates', *Earth Surface processes and Landforms*, 12, pp. 57–67.

Kirkby, M. J. (1990) 'The landscape viewed through models', *Zeitshcrift fur Geomorphologie*, NF 79, pp. 63–81

Kirkby, M. J. (1995) 'Modelling the links between vegetation and landforms', *Geomorphology*, 13, pp. 319–335.

Kirkby, M. J. (1996) 'A role for theoretical models in geomorphology?', in Rhoads, B. L. and Thorn, C. E. (eds) *The Scientific Nature of Geomorphology*, Chichester: Wiley, pp. 257–272.

Musgrave, G. W. (1947) 'The quantitative evaluation of factors in water erosion, a first approximation', *J. Soil and Water Conservation*, 2, pp. 133–138.

Wolman, M. G. and Miller, J. P. (1960) 'Magnitude and frequency of forces in geomorphic processes', *Journal of Geology*, 68, pp. 54–74.

Chapter 16

GeoComputation research agendas and futures

Stan Openshaw, Manfred M. Fischer,
George Benwell and Bill Macmillan
with an introduction by Stan Openshaw

16.1 Introduction

As a finale to this book it seemed appropriate to try to identify part of a
GeoComputation (GC) research agenda as well as inviting other contributors
to write down their ideas about the future of GC, possible research agendas,
interesting speculations, and to share with us their thinking about the future
of the subject. This is not an easy task because it is so open-ended.

Bearing in mind the nature of GC then it should be possible to devise a
research agenda that at least covers some of the key topics of importance. A
useful starting position is to try and devise some criteria for a research
agenda. Some of the desirable common features are:

1. be proactive rather that watching others at work;
2. be based on an acceptable scientific paradigm;
3. involve theory as well as data;
4. utilize new computational developments in high performance computing
   as and when they occur;
5. use relevant statistical, mathematical, and computational tools and
   methods;
6. exploit artificial intelligence and computational intelligence methods
   where relevant;
7. benefit from GIS but without being restricted by it;
8. focus on a balanced mix of themes spanning practically useful, commercially valuable, and theoretically interesting;
9. be bold and ambitious in deliberately seeking to move forward rather
   than look backwards;
10. try to focus on the major challenges whether or not they can all be
    solved using current technology;
11. be technology driven to the extent of being aware of what is now
    becoming possible;
12. seek to escape the problems and limitations that afflicted scientific geo-
    graphy three decades ago;

13. seek active re-engagement with other sciences;
14. identify long-term grand challenge projects and then gather the research resources needed to tackle them;
15. have an exciting, novel, and contemporary feel.

There are certainly many possible areas of research, which would match these criteria. However, there are probably five principal categories or thematic topics of GC interest. They encompass:

1. theoretical developments;
2. empirical analysis;
3. modelling and simulation;
4. societal aspects;
5. automation of analysis and modelling functions.

Couclelis (1998) also offers some useful advice here. She identifies five major challenges (Table 16.1). The view here is that challenges 1 (development of major demonstrator projects), 2 (develop scientific standards), and 5 (move in directions that parallel the most intellectually and socially exciting computational developments of the information age) are the most important ones. Challenge 3 (lack of epistemological definition) is not a problem because of the linkage with computational science whilst 4 (justify the 'geo' prefix) is perhaps not a major issue.

Table 16.1 GeoComputation's five major challenges, after Couclelis (1998)

1	GeoComputation must develop a few major demonstration projects of obvious practical interest and persuade commercial companies to invest in and market applications-orientated GeoComputation-based products.
2	GeoComputation must develop scientific standards that will make it fully acceptable to the quantitative side of geography and to other geosciences.
3	GeoComputation must overcome the current lack of epistemological definition, which makes it appear like little more than a grab-bag of problem-solving techniques of varying degrees of practical utility.
4	GeoComputation must develop a coherent perspective on geographical space, i.e. justify the 'geo' prefix.
5	GeoComputation must move in directions that parallel the most intellectually and socially exciting computational developments of the information age.

Table 16.2 outlines some of the key components of a GC research agenda as envisaged by the writers. They can be expanded upon as follows.

1. *Human systems modelling* is going to become an unavoidable area of considerable practical importance. People are too important to ignore. Currently, we have no good or even tolerably poor models of the beha-

Table 16.2 A research agenda

1	Human systems modelling
2	Physical systems modelling
3	Geographical data mining systems
4	Theory simulation by computational means
5	New theory development
6	New ways of representing geo-information
7	Space-time and real-time geographical analysis and monitoring systems
8	Modelling and forecasting the socio-economic effects of global climatic change
9	Handling broader societal concerns
10	Automation of routine analysis and modelling functions
11	Process dynamics

viour of people. The subject is hard but is it any harder than comparable areas in physics or chemistry or biology? Probably not, so why not flag this topic as a key grand challenge for the future and then set about trying to establish its location on key research agendas. So why not seek to start a 30-year research programme concerned with understanding, describing, modelling, and predicting the spatial behaviour of people in a proper geographical setting.

2. *Physical systems modelling* is another area where major progress should now be possible. So why not start to develop computationally realistic models of those physical, biological and environmental systems of interest to geographers to complement those that already interest other environmental sciences. This involves large-scale three-dimensional computer modelling with dynamics. A key question to consider is what new computational methodologies might be useful here (e.g. cellular automata) and also how would a computationally minded chemist or physicist tackle these problems?

3. *Geographical data mining systems* to exploit GIS databases in order to obtain a better return on the capital invested in their creation. It is very important that we start to develop a distinctly geographical form of inductive data mining technology of applied importance. Current methods treat spatial information as being the same as any other type of data thereby perpetuating the mistake first made by the grandfathers of quantitative geography.

4. *Theory simulation* is another historically neglected area in the geosciences. It would appear important that we start to develop computational laboratories able to simulate and hence test theories of spatial and space-time phenomena.

5. *New theory development* is long overdue. It is imperative that we develop existing theories and find new and novel ways of creating new theories that represent the behaviour of our systems of interest. Scientific discovery can involve a mix of deductive, inductive, and experimental pathways; previously, we have relied on only one of these pathways!

6. *New ways of representing geo-information* are soon going to be required. The GIS world is stuck with a crisp and old-fashioned cartographer's view of the world. The question here is what can a GC perspective offer? Is it possible to develop a fuzzy GIS? How should data uncertainties be handled? How can a new GIS be created in which fuzzy polygons can be handled in a natural way rather than being forced through a crisp filter just so that existing tools can be used?

7. *Space-time and real-time geographical analysis and monitoring systems* for databases are needed. Current technologies for data capture far outstrip analysis technology. Patterns and predictable processes are not being spotted because no one has yet developed and applied appropriate geographical analysis technologies. How much longer can this neglect of the obvious and necessary be entertained?

8. *Modelling and forecasting the socio-economic effects of global climatic change* is an emerging area of massive potential significance. Global climate change seems to be underway but what are the likely land use and other socio-economic impacts likely to be in 50–100 years time at a fine spatial scale?

9. *Handling broader societal concerns* to ensure that GC does not become blind to the broader issues involved with its development. The various social critiques of GIS raise important issues and these concerns should not be overlooked. In particular, how can GC be used to assist in enhancing privacy and data confidentiality, in retaining rather than removing uncertainty, in developing new spatial representations, and in aiding ethical application of legacy technologies?

10. *Automation of routine analysis and modelling functions* is becoming very important as a means of creating end-user friendly technology. So, why not build smart analysis and modelling machines that take in geographical information, model or analyse it, and produce patterns, models, equations, or theory as standard output.

11. *Process dynamics* are currently very neglected. More and more data exists that describes the dynamics of complex systems yet very few dynamic models exist. This is an area of considerable generic importance.

16.2 What do others think about the future of GeoComputation?

A good starting view is that of Manfred Fischer. He provides below a structured and comprehensive review of GC and its constituent technologies noting the principal trends he expects to occur. It is worthy of closer study. This is followed by a second indepth review by George Benwell this time focusing on GIS and GC. Bill Macmillan's views are reproduced next. He asks the important question 'Where is the real intelligence in GC and where is the understanding going to come from?' He illustrates his arguments by reference to a modelling example. This is an important issue that deserves an answer but it is also a question that will have no obvious single answer. The final contribution is by myself. I offer a more alternative sci-fi scary technology-led view of possible future developments. Maybe this future view is too generalized and too computer centric but probably some of it may one day happen.

16.2.1 Manfred Fischer writes on the nature and emerging trends in GeoComputation

GeoComputation (GC) is an emerging field. It is reasonable to expect some differences in scope and definition, and the various contributions to this discussion certainly meet this expectation. But the main message here is that within the diverse set of technologies and applications, GC researchers are addressing a central thread of an important area: unlocking the information that is buried in the enormous stock of spatial data we have already available in GIS and remote sensing (RS) environments and developing the underpinnings for better ways to handle spatial and non-spatial data and support future decision making in the widest sense.

GC is an interdisciplinary field that brings together researchers and practitioners from a wide variety of fields. The major related fields include statistics, pattern recognition, artificial intelligence, neurocomputing, and geographic information systems. Let me briefly clarify the role of each of the fields and how they fit together naturally when unified under the goals and applications of GC, but no attempt is made at being comprehensive in any sense of the word.

Statistics plays an important role primarily in data selection and sampling, data mining, and the evaluation of GC results. Historically, most work in statistics and its spatial version has focused on evaluation of model fit to data and on hypothesis testing. These are clearly relevant to evaluating the results of GC to filter the good from the bad, as well as within the search process for patterns. On the front end of GC, statistics offers techniques for detecting outliers, smoothing data if necessary, and estimating noise parameters. But, the focus of research has dealt primarily with small data sets and addressing small sample problems. On the limitations front, (spatial) statistics has focused primarily on theoretical aspects of techniques, methods and models.

Still nowadays, most work focuses on linear models, additive Gaussian noise models, parameter estimation and parametric techniques for a fairly restricted class of models. Search for patterns and regularities has received little emphasis, with focus on closed-form analytical solutions whenever possible. While this is desirable, both theoretically and computationally, in many real-world situations a user might not have the necessary statistics knowledge to appropriately use and apply the statistical tools. Moreover, the typical approaches demand an *a priori* model and significant domain knowledge of the (spatial) data and of the underlying mathematics for their proper use and interpretation. Issues related to interfaces to geo-databases and dealing with massive data sets have only recently begun to receive attention in statistics.

In pattern recognition, work has historically focused on practical techniques with an adequate mix of rigour and formalism. The most prominent techniques fall under the category of clustering and classification. Pattern recognition contributions are distinguished from statistics by their focus on computational algorithms, more sophisticated data structures, and more search for patterns, both parametric and non-parametric. Significant work in dimensionality reduction, transformation, classification and clustering (including regionalization) has relevance to various steps in GC.

Techniques and methods originating from artificial intelligence (AI) have focused primarily on dealing with data at the symbolic (categorical) level, with little emphasis on continuous variables. Artificial intelligence techniques for reasoning, in particular graphical models for Bayesian modelling and reasoning, provide a powerful alternative to classical density estimation in statistics. These techniques permit prior knowledge about the domain and data to be included in a relatively easy framework. Also other areas of AI, especially knowledge acquisition and representation, and search are relevant to the various steps in a GC process including data mining, data preprocessing and data transformation.

Neurocomputing is a field concerned with information processing systems that autonomously develop operational capabilities in adaptive response to an information environment. The primary information structures of interest in neurocomputing are computational neural networks, comprised of densely interconnected adaptive processing elements, although other classes of adaptive information structures are considered, such as evolutionary computation. Several features distinguish neurocomputing from the field of artificial intelligence in the strict sense. Firstly, information processing is inherently parallel. Large-scale parallelism provides a way to increase significantly the speed of information processing. Secondly, knowledge is encoded not in symbolic structures, but rather in patterns of numerical strength of the connections between the processing elements of the system (i.e. connectionist type of knowledge representation). The adaptive feature makes computational neural networks – in combination with a wide variety of learning techniques – very appealing for GC especially in application domains where one has little or incomplete understanding of the problem to be solved but where training data is readily available. These networks are neural in the sense that they may have been inspired by neuroscience but not necessarily because they are faithful models of biological neural phenomena. In fact, the majority of the networks are more closely related to traditional mathematical and/or statistical models, such as non-parametric pattern classifiers, clustering algorithms, statistical regression models, non-linear filters, than they are to neurobiological models.

The relevance of geographic information systems (GIS) and remote sensing systems is evident. They provide the necessary infrastructure to store, access, and manipulate the raw data. Supporting operations from the GC perspective might become an emerging research area in both the GIS- and RS-communities. Classical database techniques for query optimization and new object-oriented databases make the task of searching for patterns in spatial databases much more tenable.

The rapidly emerging field of GC has grown significantly in the past few years. This growth is driven by a mix of daunting practical needs and strong research interests. The GIS technology has enabled institutions to collect and store information from a wide range of sources at rates that were, only a few years ago, considered unimaginable. Examples of this phenomenon abound in a wide spectrum of fields, including census data and RS-data. For example, NASA's Earth Observing System is expected to return data at rates of several gigabytes per hour by the end of the century.

Why are today's techniques, methods and models for spatial data not adequate for addressing the new analysis needs? The answer lies in the fact that there exist serious computational and methodological problems attached to performing spatial data modelling in high-dimensional spaces and with very large amounts of data. These challenges are central to GC and need urgent attention. Without strongly emphasizing GC development and research, we run the risk of forfeiting the value of most of the GIS- and RS-data that we collect and store. We would eventually drown in an ocean of massive data sets that are rendered useless because we are unable to distill the essence from the bulk.

To date, one factor has inhibited the broad development of GC: the significant effort necessary to develop each GC application. But there have been a few significant successes in this new field recently. Illustrative examples can be found in this volume and in various special issues of: *Computers, Environment and Urban Systems*; *Computers & Geosciences*; *Environment and Planning A*; *Geographical Systems*; and *Transactions in GIS*. However, work is just beginning in this challenging and exciting field. Some are practical challenges and await proper implementation, while others are fundamentally difficult research problems that are at the heart of many fields such as spatial statistics, modelling, optimization, and pattern recognition.

While the current state of the art still relies on fairly simple approaches and great enthusiasm, and seldom on the necessary methodological rigour, powerful results are being achieved, and the efforts and community are rapidly growing in size. The driving force behind this development is a combination of large amounts of spatial data available due to the GIS- and RS-data revolutions, the advent of high performance supercomputers (especially the new era of parallel supercomputing), and the availability of practical and applicable computational intelligence (CI) technologies with evolutionary computation, cellular automata, the concept of artificial life, adaptive fuzzy systems and computational neural networks as the major components. These CI-technologies provide the basis for developing novel styles of spatial analysis and modelling meeting the new large-scale data processing needs in data rich environments.

16.2.2 George Benwell writes on future research in GIS and GeoComputation

The descriptions and directions of future research in GIS and GC are fraught with difficulties and uncertainties. This is partly because the future is not totally predictable, and partly because it may be unclear from what base the predictions are being made. In addition, it is uncertain who the audience is and how they perceive the past, present and future. As a result, extreme scepticism or confidence will reign, with the latter viewpoint being adopted here. Confidence is grounded on the fact that such a difficult task will be nigh on impossible, in the short term, to prove wrong. The longer outlook may well be considerably different.

The ideas presented in this analysis outline future endeavours in GIS and GC as the year 2000 fast approaches. The thrust for these pursuits will be driven by system vendors, users and practitioners. The two sides, vendors and users, are so intertwined it would be difficult to attribute any changes or endeavours to one particular side. It is their interactions, variable interplay and individual imperatives that create the agents for change.

Geographical information systems and GC are reasonably well understood, particularly the former. In this work the two are coalesced into a single meaning: 'geocomputation and geo-informatics that relate to the collection, storage, management, analysis and presentation of spatially or geographically referenced information'. This scope is not dissimilar to a definition solely of GIS, but the inclusion of computational intensity is seen as a logical extension into the realm of spatial analysis where problems and solutions are large and previously constrained by computing power. In addition, the compartments of collection, storage, management, analysis and presentation are the basis of a structured treatment of the descriptions and directions of future research. From these initial topics are synthesized: information management and spatial analysis. Each of these topics will be dealt with in turn. The level of treatment in each case varies depending on the perceived existing importance and the likelihood of any significant research being conducted in the future.

16.2.2.1 Information management

As the use of the World Wide Web (WWW) has amply demonstrated during the last part of this decade, information, access to information and ownership of information are critical to the success of the modern world. Whether this should be so, or whether it is a construct of the information moguls, is outside the scope of this review. Nonetheless, it is obvious that the information age is here and will continue to expand for some time to come. To predict for how long would possibly be both difficult and foolhardy; expansion for at least for another decade would (regardless) be a safe guess.

The WWW has heralded the age of information and information access. In the case of GIS and GC, it is considered here that the role of the WWW will be of less importance than may now be thought. It will be more important to realize what the WWW stands for (a transparent international network) and that it is only an early exemplar of what is to come. Remember, GIS has experienced a meteoric rise since the early 1980s. Just a mere 15 years! This rapid development has been followed by an even greater rise in the demand for information. These combined effects are considered to be more significant in the context of GIS than in other more established areas, such as finance, entertainment, law or other global endeavours. This has come to be because when masses of information were *available*, they were *required* and able to be *managed* and utilized. This is true for GIS and GC today but is less significant for other areas, such as banking. Banking dealt with the system and information explosion three decades ago.

This temporal coincidence for GIS is placing considerable demands on the development of better and improved systems for information management. It is considered that research will occur in three areas;

1. information dissemination;
2. information storage;
3. computing power.

Information dissemination will become the powerhouse of the next decade. It will not be sufficient to have just information; it will be necessary to make this information available to relevant users, local, national and international. To achieve this, users will demand to have knowledge of where information is, what it is, how to access it and at what charge. The early indication of this is already here, the WWW. Such an implementation is but an example! It will be common place to dispatch search applets to locate and retrieve information from any source. The capture and collection of information for a GIS project at present consumes an inordinate amount of effort and time.

Hand in glove with any database developments is a concept of federated databases. This is, in simple terms, the connection of several databases in such a way that the union is transparent. Furthermore, applications can access a data repository index which contains information on data, data structures, data-use rules, data knowledge and other applications. The federation makes this possible from any application domain to any other using any data for any reason. A grand concept that is close to reality.

Tools for the analysis and management of large, spatially-oriented data sets can have a significant impact on the productivity of spatial analysis professionals. Although commercially available spatial information systems have offered assistance, it is recognized that improved analytic and data management methods are needed for significant advances to be made.

New analytic tools including artificial neural network simulation technology, fuzzy rule extraction technology, symbolic, and fuzzy rule-based inference systems are beginning to appear; their power, applicability and functionality will be enhanced. These offer an expanded scope for analysis of large spatial databases. Systems will enable results from these methods to be used in concert with results obtained by traditional statistical methods, and the experience of experts.

Effective planning and management requires the ready availability of large quantities of information covering the spatial domain of interest and facilities for manipulating these data. In recent years, spatial information systems have come to the fore to provide assistance in data handling and display, but for the most part these systems are limited to the relatively simple operations of buffering, overlay, and query. More in-depth, quantitative analyses, such as the standard statistical techniques of cluster analysis or regression analysis, are usually conducted as conceptually distinct 'batch' operations that stand apart from each other. Although these techniques can be useful, there is a need for an increased emphasis on the development of new tools for spatial analysis. This need was documented some time ago (Openshaw, 1991) but developments are still relatively simple and do need to be extended. In particular, Openshaw (1991) and Goodchild (1992) feel that current spatial information systems do not satisfactorily provide the integrated environment conducive to explanatory investigations that need to be carried out, particularly by land-use experts. Thus it is not only necessary to collect accurate spatial datasets, but also to devise methods for managing and analytically surveying the vast quantities of data that are being collected at an ever increasing rate. As Openshaw has remarked (Openshaw, 1991):

'. . . The real challenge is, therefore, to discover how to trawl through these geographical databases in search of interesting results with a high level of efficiency by devising new forms of spatial analytical methods.'

There are two key points that must be emphasized with regard to these new systems:

1. An environment for exploration is the central issue and needs to be researched and developed. This can be achieved not merely by offering a collection of disparate tools that can be accessed from a menu, but by building a set of tools that can be used in a methodical and iterative fashion for feedback and analysis. Essential to this approach will be:

- Neural network technology which will be employed to 'trawl' through some of the spatial data.
- Fuzzy (approximate reasoning) inference rules which will be extracted from the neural network models.
- A fuzzy rule-based production system that can carry out approximate reasoning in the application domain.

2. Once the rules have been derived it will be vital that they are 'interpreted' or 'explained' by an expert in the particular field of study, e.g. a soil scientist, a forester or a horticulturalist, etc. So:

- These rules should be examined by the expert, who may then modify them and potentially supply additional rules on the basis of what the system has already 'suggested'.
- The expert can then carry out additional, more focused numerical analysis explorations, using both neural network and statistical techniques.

These steps can be iterated in a feedback loop and can converge to a more refined analytical model.

16.2.3 Bill Macmillan writes about real intelligence and GeoComputation

What is GeoComputation? Like all conceptual terms, it means whatever its inventors want it to mean. That is not to say that it can mean entirely different things to different people because invention, in this context, is a collective business. This essay is intended to make a contribution to the process of invention, in the belief that there are certain fundamental issues at stake. It is loosely based on the closing address to the first Geo-Computation conference in Leeds in 1996 and makes reference to some of the papers presented at that meeting and at GeoComputation '97 in Otago.

One view of GC – one might say the authorized view since it springs from the work of the Leeds group who coined the term – is that it has to do with spatial data, high performance computing, and artificial intelligence. I do not wish to argue against the inclusion of any of these characteristics. But – and as Stan Openshaw might say, it's a big but – I do want to argue for the inclusion of something else. To complement the artificial intelligence of data-led GC, we need the real intelligence of theory-led GC.

The competing claims of induction and deduction are far from new. However, the development of high performance computing has created a new environment in which those claims have to be re-established. A strong case has been made elsewhere for inductive work but the argument runs over at times as a case against theory building. It may be appropriate, therefore, to begin defensively.

It relies, at least in part, on the notion that models and the generalizations that underpin them are necessarily simple, making them incapable of capturing the complexity of the real world. This is misconceived. The reasons for model simplicity are partly to do with the purposes of the modelling exercise (good models have an economy of design that makes them as simple as those purposes will allow) and partly computational (handling very large sets of conditions is a computationally-demanding task). Early models had to be simple because the implications of complex conditions could not be computed but this is not the case today.

One example that is reasonably well known should help to make the point. For largely pedagogic purposes, Rusty Dodson and others at UCSB produced a semi-realistic version of the von Thünen model in the early 1990s using the law-like mechanisms of the standard model but with heterogeneous landscapes in the form of IDRISI files. In such an environment, it is possible to compute the implications of the standard mechanisms under, to a first approximation, any conditions. The result, over a series of landscapes, is a series of unique images in which the operation of the standard mechanisms can be seen 'but dimly' to quote von Thünen.

Another set of examples that are much better known are the various geographical computer games ranging from SimEarth and its sister products to Civilization. Every game in Civilization operates under the same rules but on different initial conditions (this observation applies to all members of the genre). The configurations that emerge are always different, and there is endless variety and peculiarity. Evidently, such characteristics are not inconsistent with constancy in the set of rules.

The computational reasons for model simplicity go beyond the constraints on the richness of conditions. The rules that are taken to operate on the conditions are also computationally conditioned. Dodgson's reformulation of the von Thünen model was referred to above as semi-realistic because it made no attempt to relieve the constraints implicit in the rules. Von Thünen himself pushed the constraints of his day pretty hard, generating equation systems whose solution was famously difficult. Nevertheless, he operated within, and his thinking was constrained by, the computational environment of his day. Thus, he conceptualized processes in terms of continuous functions representing aggregate behaviour and formulated equations that could be solved by hand.

One consequence of increasing computational power, then, has been to allow us to expand our knowledge of the implications of established rule sets from simple to complex environments. Another consequence has to do with the rule sets themselves: they are computationally conditioned as well. A certain amount has been done in a GC environment to rethink rule systems. A number of new conceptualizations have been developed, including cellular processes, which have been applied to a wide variety of phenomena from the spread of forest fires and innovations to city growth (Batty, 1996;

Garriga and Ratick, 1996; Park and Wagner, 1996). For the most part, the pictures generated in these applications have been acceptably life-like. However, in terms of developing our understanding, cellular processes are better suited to some applications than others. For example, forest fires and slope processes (Coulthard *et al.*, 1996, 1997) seem to be good applications because they can be thought of as cellular phenomena. That is, it is reasonable to think of the state of any given spatial unit at a given time as being dependent upon those of its neighbours and itself at a previous time. Innovations appear to be less promising because they are adopted or not adopted by people rather than places and the movement and contact patterns between people tend to be more complex than basic cellular processes will allow. Building basic cellular models of city growth is problematic for similar reasons.

The key test of the success of such applications is whether the rule systems they produce give us a better understanding of the processes which generate the patterns we observe. In cellular slope process modelling, what is being captured by the rules are the processes themselves. In city growth modelling, the rule system is, at best, one which is capable of replicating observed patterns.

It might be argued that this is all that we require. If, say, a computer can be trained using neural networks to replicate the history of urban growth in a given region with a high degree of accuracy and can then generate a forecast of future growth, what more do we need? The answer is straightforward: for intelligent action, we need an understanding of the system. The computer's artificial intelligence may be helpful and stimulating but it is no substitute for the real thing. To believe otherwise is to think that the computer 'understands' the process so we don't have to.

The introduction of intelligent agents is a step in the right direction (Dibble, 1996; Portugali and Beneson, 1997). Having agents with various properties who can occupy cells with other properties is intrinsically more appealing in an urban context than having an unpopulated but autonomously evolving cellular landscape (Batty, 1996). It allows us to deploy the understanding we have of agent behaviour. Just as physical laws form the basis of the cellular rules in slope process models, so law-like propositions on economic and social behaviour form the basis of the rules governing intelligent agent behaviour. The great attraction of this strategy is that we are not obliged to work back from presumptions about outcomes to behavioural rules. The neoclassical tradition of treating markets as spaceless entities in which aggregate supply and aggregate demand are adjusted through price movements until they are equal is a product of 19th century intellectual technology, as noted above in connection with the work of von Thünen. Our formalizations of agent behaviour have been predicated, in the main, on the presumed need to generate market clearing outcomes. Thus, neoclassical economics has tended to skirt round inconvenient problems like indivisible commodities, set-up costs, internal economies of scale, agglomeration

This is a most significant jump in HPC performances. It will, unavoidably, create major new and probably very imperfectly identified opportunities for applying unbelievable amounts of computation power.

If GC is all about seeking computational solutions to various hard or interesting or significant problems, and if it is still restrained or limited in what it can do by machine speeds, then fairly soon these constraints will go. All manner of currently impossible things will suddenly become feasible. Those that will succeed best in this high performance computer powered world will be those who have or can develop new applications that need it. Of course, this is a computer-driven view but, as ever, what you want to do and the problems you study are critically determined (implicitly or explicitly) by those technologies (and philosophies) you know are available now as tools that you can use now. When in the near future all existing computing speed and memory sizes suddenly increase by a few orders of magnitude, it will be the entrepreneurial 'geocomputationalists' who will respond first.

It's quite simple really. The principal opportunity is as follows. Computation provides a generic paradigm for whatever it is that you do now or may wish to do in the future. No subject, no method, no model and no tool that is computable will escape. Of course by 2050, historians will look back in amazement at a time when people ever thought otherwise.

Computation is a key, core, generic, and widely applicable technology that is revolutionizing science. GeoComputation is merely the 'geo' version of it. One of the nicest features is that, at present, little has been done. This leaves virtually 99.99% of this potentially massive field of interest almost totally untouched. It leaves so much for 'geocomputationalists' of a great diversity of backgrounds to research, discuss, criticize, and develop in the years ahead. It was good to be there at the birth but its future development is a matter for subsequent generations to pursue. The hope is that this book will be a help in fostering this process.

References

Astroth, J. H. (1995) 'OGIS: planting the seeds for rapid market growth', *GeoInfo Systems*, 5, 1, pp. 58–59.
Batty, M. (1996) 'The evolution of cities', *GeoComputation '96: Proceedings 1st International Conference on GeoComputation*, Leeds: University of Leeds, 17–19 September 1996, 1, pp. 42–43.
Benwell, G. L. (1996) 'Spatial databases – creative future concepts and use', *GeoComputation '96: Proceedings 1st International Conference on GeoComputation*, Leeds: University of Leeds, 17–19 September 1996, 1, pp. 48–53.
Couclelis, H. (1998) 'Geocomputation in context', in Longley, P., Brooks, S., McDonnell, R. and Macmillan, B. (eds) *Geocomputation: A Primer*, Chichester: Wiley, pp. 17–30.
Coulthard, T. J., Kirkby, M. J. and Macklin, M. (1996) 'A cellular automaton fluvial and slope model of landscape evolution', *GeoComputation '96: Proceedings 1st*

International Conference on GeoComputation, Leeds: University of Leeds, 17–19 September 1996, 1, pp. 168–185.

Coulthard, T. J., Kirkby, M. J. and Macklin, M. (1997) 'Modelling hydraulic, sediment transport and slope processes, at a catchment scale, using a cellular automaton approach', *GeoComputation '97: Proceedings 2nd International Conference on GeoComputation*, Dunedin, New Zealand: University of Otago, 26–29 August 1997, pp. 309–318.

Dey, S. and Roberts, S. A. (1996) 'Combining spatial and relational databases for knowledge discovery', *GeoComputation '96: Proceedings 1st International Conference on GeoComputation*, Leeds: University of Leeds, 17–19 September 1996, 1, pp. 188–197.

Dibble, C. (1996) 'Theory in a complex world: agent-based simulations of geographic systems', *GeoComputation '96: Proceedings 1st International Conference on GeoComputation*, Leeds: University of Leeds, 17–19 September 1996, 1, pp. 210–212.

Diplock, G. (1996) 'Building new spatial interaction models using genetic programming and a supercomputer', *GeoComputation '96: Proceedings 1st International Conference on GeoComputation*, Leeds: University of Leeds, 17–19 September 1996, 1, pp. 213–226.

Garriga, H. M. and Ratick, S. J. (1996) 'Simulating land use allocation decisions with cellular automata in a geographic information system', *GeoComputation '96: Proceedings 1st International Conference on GeoComputation*, Leeds: University of Leeds, 17–19 September 1996, 1, pp. 371–384.

Goodchild, M. (1992) 'Integrating GIS and spatial data analysis: problems and possibilities', *International Journal of Geographic Information Systems*, 6, 5, pp. 407–423.

Kasabov, N. K., Purvis, M. K., Zhang, F. and Benwell, G. L. (1996) 'Neuro-fuzzy engineering for spatial information processing', *Australian Journal of Intelligent Information Processing Systems*, 3, 2, pp. 34–44.

Mahnic, V. (1991) 'Using grid files for a relational database management system', *Microprocessing and Microprogramming*, 31, pp. 13–18.

Marr, A. J., Pascoe, R. T. and Benwell, G. L. (1997) 'Interoperable GIS and spatial process modelling'. *GeoComputation '97: Proceedings 2nd International Conference on GeoComputation*, Dunedin, New Zealand: University of Otago, 26–29 August 1997, pp. 183–190.

Nievergelt, J., Hinterberger, H. and Sevcik, C. K. (1984) 'The grid file: an adaptable symmetric multikey file structure', *ACM Transactions on Database Systems*, 9, 1, pp. 38–71.

Openshaw, S. (1991) 'Developing appropriate spatial analysis methods for GIS', in Maguire, D., Goodchild, M. F. and Rhind, D. W. (eds) *Geographical Information Systems*, Vol. 1, Essex: Longman, pp. 389–402.

Openshaw S. (1996) 'Neural network and fuzzy logic models of spatial interaction', *GeoComputation '96: Proceedings 1st International Conference on GeoComputation*, Leeds: University of Leeds, 17–19 September 1996, 2, pp. 663–664.

Park, S. and Wagner, D. F. (1996) 'Cellular automata as analytical engines in geographical information systems', *GeoComputation '96: Proceedings 1st International Conference on GeoComputation*, Leeds: University of Leeds, 17–19 September 1996, 2, pp. 670–688.